Stochastic Processes and Functional Analysis
New Perspectives

CONTEMPORARY MATHEMATICS

774

Stochastic Processes and Functional Analysis

New Perspectives

AMS Special Session
Celebrating M.M. Rao's Many Mathematical Contributions as he Turns 90 Years Old
November 9–10, 2019
University of California
Riverside, California

Randall J. Swift
Alan Krinik
Jennifer M. Switkes
Jason H. Park
Editors

AMS AMERICAN MATHEMATICAL SOCIETY

2020 *Mathematics Subject Classification.* Primary 46-02, 46-06, 60-02, 60-06, 60C05, 60G07, 60J27, 62-02, 62M15.

For additional information and updates on this book, visit
www.ams.org/bookpages/conm-774

Library of Congress Cataloging-in-Publication Data

Names: Swift, Randall J., editor.
Title: Stochastic processes and functional analysis : new perspectives / Randall J. Swift [and three others], editors.
Description: Providence, Rhode Island : American Mathematical Society, [2021] | Series: Contemporary mathematics, 0271-4132 ; 774 | "AMS Special Session on Celebrating M.M. Rao's Many Mathematical Contributions as he Turns 90 Years Old, November 9–10, 2019 University of California Riverside, California | Includes bibliographical references."
Identifiers: LCCN 2021017497 | ISBN 9781470459826 (paperback) | 9781470467166 (ebook)
Subjects: LCSH: Rao, M. M. (Malempati Madhusudana), 1929- | Stochastic processes–Congresses. | Functional analysis–Congresses. | AMS: Functional analysis – Research exposition (monographs, survey articles). | Functional analysis – Proceedings, conferences, collections, etc. | Probability theory and stochastic processes – Research exposition (monographs, survey articles). | Probability theory and stochastic processes – Proceedings, conferences, collections, etc.. | Probability theory and stochastic processes – Combinatorial probability – Combinatorial probability. | Probability theory and stochastic processes – Stochastic processes – General theory of processes. | Probability theory and stochastic processes – Markov processes – Continuous-time Markov processes on discrete state | Statistics – Research exposition (monographs, survey articles). | Statistics – Inference from stochastic processes – Spectral analysis.
Classification: LCC QA274.A1 S76654 2021 | DDC 519.2/3–dc23
LC record available at https://lccn.loc.gov/2021017497

10 9 8 7 6 5 4 3 2 1 26 25 24 23 22 21

Contents

Preface

An AMS Special Session in honor of M.M. Rao was held at the 2019 Fall Western Sectional Meeting held at the University of California, Riverside, November 9–10. That Special Session, titled "Celebrating M.M. Rao's many mathematical contributions as he turns 90 years old" was organized by Professor Jerome Goldstein, University of Memphis and California State Polytechnic University, Pomona Professors Michael Green, Alan Krinik, Randall Swift and Jennifer Switkes.

Professor M.M. Rao has had a long and distinguished research career. His research spans the areas of probability, statistics, stochastic processes, Banach space theory, measure theory and differential equations - both deterministic and stochastic. The purpose of the Special Session was to celebrate a lifetime of mathematical achievement and highlight the key role played by abstract analysis in simplifying and solving fundamental problems in stochastic theory.

The Sessions were a wonderful success, bringing together a diverse group of research mathematicians whose work has been influenced by M.M.'s work and, in turn, have influenced his work. Several of his UC Riverside colleagues attended the talks and several spoke. Four of his former students also spoke. The Sessions were engaging with lively discussions and questions from the Session audience.

This volume contains a collection of these talks given at the Sessions and begins with Professor Rao's talk, "Stochastic Equations". We have included images of the slides that he used. Here, we hope to record the incredible passion and energy he has for mathematics. His unbounded enthusiasm is clear in the photos and the love he has for his students shown in his talk.

This volume also includes a biography of M.M. Rao, a complete bibliography of his published writings and a list of his Ph.D. students. This collection complements the two Festschrift volumes *Stochastic Processes and Functional Analysis* (1996) and *Stochastic Processes and Functional Analysis: Recent Advances* (2004), that were published to honor his 65th and 75th birthdays. We dedicate this collection, to honor his 90th birthday, in celebration of a mathematical life.

R. J. Swift
A.C. Krinik
J.M. Switkes
J. H. Park

Stochastic Equations

M.M. Rao

Professor M.M. Rao, University of California, Riverside, November 9, 2019. (Photo courtesy of R. J. Swift.)

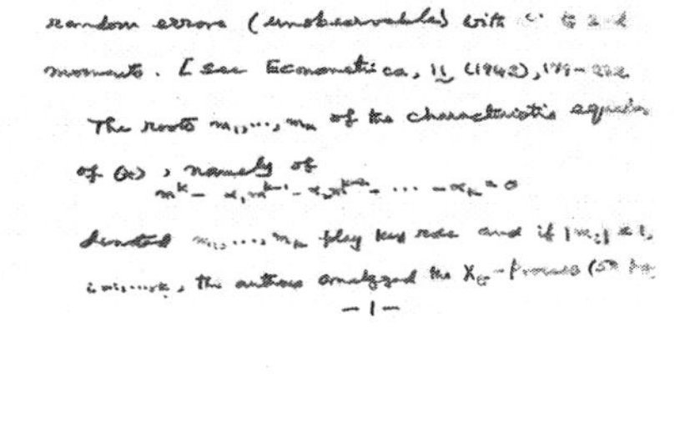

Stochastic Equations: (M.M. Rao)

1. In early 1940's, motivated by some important Economics problems, H. B. Mann and A. Wald considered a process $\{X_t, t \in I\}$ which satisfies a linear SDE (stochastic difference equation), given by the relation:

$$X_t = \alpha_1 X_{t-1} + \cdots + \alpha_k X_{t-k} + u_t,\ k \geq 1. \quad \cdots (*)$$

the $\alpha_1, \ldots, \alpha_k$ are real constants, the u_t's i.i.d. random errors (unobservable) with [illegible] moments. [See Econometrica, 11 (1943), 173–220.

The roots $m_1, \ldots, m_k$ of the characteristic equation of (*), namely of

$$m^k - \alpha_1 m^{k-1} - \alpha_2 m^{k-2} - \cdots - \alpha_k = 0$$

denoted $m_1, \ldots, m_k$ play key role and if $|m_i| < 1$, $i = 1, \ldots, k$, the authors analyzed the X_t-process (5th [illegible]

– 1 –

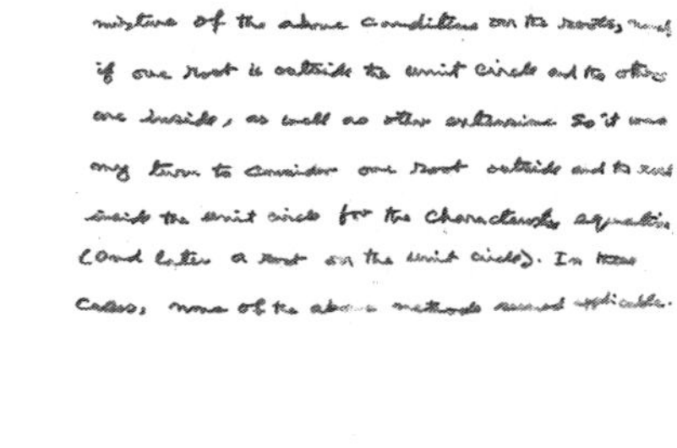

– 2 –

2. The analogous behavior of the solution process X_t of (*) is detailed by T. W. Anderson in 1952 (Ann. Math. Statist. 30 (1959)), who obtained the corresponding analysis of the process. The subject is of interest in Econometrics and Statistics. The location of roots in and out of the unit circle seemed crucial for these studies.

The next key step here is to study if a mixture of the above conditions on the roots, namely if one root is outside the unit circle and the others are inside, as well as other extensions. So it was my turn to consider one root outside and the rest inside the unit circle for the characteristic equation (and later a root on the unit circle). In these cases, none of the above methods seemed applicable.

(Photo courtesy of R. J. Swift.)

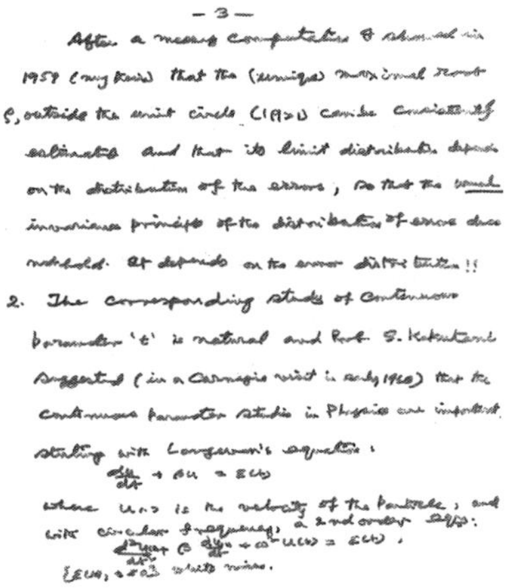

— 3 —

After a messy computation I showed in 1959 (my thesis) that the (unique) maximal root ρ, outside the unit circle (|ρ|>1) can be consistently estimated and that its limit distribution depends on the distribution of the errors, so that the usual invariance principle of the distribution of errors does not hold. It depends on the error distribution!!

2. The corresponding study of continuous parameter 't' is natural and Prof. S. Kakutani suggested (in a Carnegie visit in early 1960) that the continuous parameter studies in Physics are important, starting with Langevin's equation:

$\frac{du}{dt} + \beta u = \xi(t)$

where u(t) is the velocity of the particle; and with circular frequency, a 2nd order eqn:

$\frac{d^2u(t)}{dt^2} + \beta \frac{du}{dt} + \omega^2 u(t) = \xi(t)$,

$\{\xi(t), t \geq 0\}$ white noise.

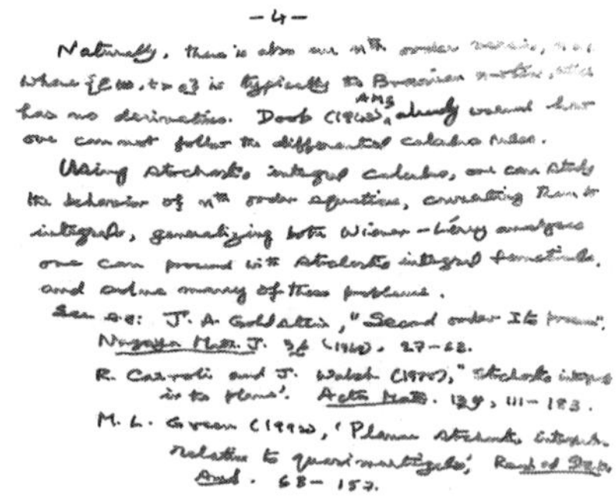

— 4 —

Naturally, there is also an nth order version, [illegible] where $\{\xi(t), t \geq 0\}$ is typically the Brownian motion, which has no derivatives. Doob (1942, AMS) already warned that one cannot follow the differential calculus rules.

Using stochastic integral calculus, one can study the behavior of nth order equations, converting them to integrals, generalizing both Wiener–Lévy analyses one can proceed with stochastic integral formulas, and solve many of these problems.

See e.g.: J. A. Goldstein, "Second order Itô process", Nagoya Math. J. 36 (1969), 27–63.

R. Cairoli and J. Walsh (1975), "Stochastic integrals in the plane". Acta Math. 134, 111–183.

M. L. Green (1997), 'Planar stochastic integration relative to quasi-martingales', Real and Stoch. Anal. 65–157.

(Photo courtesy of R. J. Swift.)

Annotated Typeset Slides

In the early 1940's, motivated by some important Economics problems, H.B. Mann and A. Wald **[9]** considered a process $\{X_t, t \geq 1\}$ which satisfies a linear stochastic difference equation (sde) given by the relation:

$$X_t = \alpha_1 X_{t-1} + \ldots + \alpha_k X_{t-k} + u_k \tag{1}$$

where the $\alpha_1, \ldots, \alpha_k$ are real constants, and the u_k are i.i.d. random errors (unobservables) with finite second moments.

The roots $m_1, \ldots, m_k$ of the characteristic equation of (1), namely of

$$m^k - \alpha_1 m^{k-1} - \alpha_2 m^{k-2} - \ldots - \alpha_k = 0$$

play a key role and if $|m_i| < 1$ for $i = 1, \ldots, k$, the author analyzed the X_t process.

The analogous behavior of the solution process X_t of (1) is detailed by T. W. Anderson in 1952 **[1]** who obtained the corresponding analysis of the process. The subject is of interest in Economics and Statistics. The location of the roots in and out of the unit circle seemed crucial for these studies.

The next key step here is to study if a mixture of the above conditions on the roots, namely if one root is outside the unit circle and the others are inside, as well as other extensions. So it was my turn to consider one root outside and the rest inside the unit circle for the characteristic equation (and later a root on the unit circle). In these cases none of the above methods seemed applicable.

(Photo courtesy of R. J. Swift.)

After a messy computation I showed in 1959 (my thesis, **[11]**) that the unique maximal root ρ, outside the unit circle ($|\rho| > 1$) can be consistently estimated and that its limit distribution depends on the distribution of the errors, so that the usual invariance principle of the distribution of errors does not hold. It depends on the error distribution!!

The corresponding study of continuous parameter "t" is natural and Professor Shizuo Kakutani suggested (in a Carnegie visit in early 1960) that the continuous parameter studies in Physics are important, starting with Langevin's equation

$$\frac{du}{dt} + \beta u = \varepsilon(t)$$

where $u(\cdot)$ is the velocity of the particle, and with circular frequency, a second order equation

$$\frac{d^2u(t)}{dt^2} + \beta\frac{du(t)}{dt} + \omega^2 u(t) = \varepsilon(t),$$

with $\{\varepsilon(t), t \geq 0\}$ white noise.

Naturally, there is also an nth order version, $n \geq 1$ where $\{\varepsilon(t), t \geq 0\}$ is typically Brownian motion, which has no derivatives. Doob **[5]** in 1944 already worked how one cannot follow the differential calculus rules. My first student R. Borchors (1964, **[2]**) studied the first order case and then J. Goldstein (1967, **[6]**) studied higher order (both nonlinear) cases. Their work led us to consider harmonizable processes and fields, my students Kelsh (1978, **[8]**), Chang (1983, **[4]**), Mehlman (1990, **[10]**), Swift (1992, **[14]**) and Soedjak (1996, **[13]**) developed the existing theory. This is summarized in my latest monograph *Stochastic Processes: Harmonizable Theory* **[12]**.

Using stochastic integral calculus, one can study the behavior of nth order equations, generalizing both Wiener and Lévy analyses. One can proceed with stochastic integral functionals and solve many of these problems. Some of these ideas where pursued by my students Brennan (1978, **[3]**) and Green (1995, **[7]**).

References

[1] T. W. Anderson & D. A. Darling, Asymptotic theory of certain "goodness of fit" criteria based on stochastic processes. *Ann. Math. Statistics* 23, 193-212, 1952.

[2] D. R. Borchers, "Second order stochastic differential equations and related Ito processes." Ph. D. Dissertation, Carnegie-Mellon University, 1964.

[3] M. D. Brennan, "Planar semi-martingales and stochastic integrals." Ph. D. Dissertation, University of California, Riverside, 1978.

[4] D. K. Chang, "Bimeasures, harmonizable process and filtering." Ph. D. Dissertation, University of California, Riverside, 1983.

[5] J. L. Doob, The elementary Gaussian processes. *Ann. Math. Statistics*, 15, 229–282, 1944.

[6] J. A. Goldstein , "Stochastic differential equations and nonlinear semi-groups." Ph. D. Dissertation, Carnegie-Mellon University, 1967.

[7] M. L. Green, "Multi-parameter semi-martingale integrals and boundedness principles." Ph. D. Dissertation, University of California, Riverside, 1995.

[8] J. P. Kelsh, "Linear analysis of harmonizable time series." Ph. D. Dissertation, University of California, Riverside, 1978.

[9] H. B. Mann & A. Wald, On the Statistical Treatment of Linear Stochastic Difference Equations, *Econometrica* Vol. 11, No. 3/4, 173-220, 1943.

[10] M. H. Mehlman, "Moving average representation and prediction for multidimensional strongly harmonizable process." Ph. D. Dissertation, University of California, Riverside, 1990.

[11] M.M. Rao, Properties of Maximum Likelihood Estimators in Nonstable Stochastic Difference Equations, Ph.D. Dissertation, University of Minnesota-Minneapolis 1959.

[12] M.M. Rao, *Stochastic Processes: Harmonizable Theory*, World Scientific, Singapore, 340 pages, 2020.

[13] H. Soedjak, "Estimation problems for harmonizable random processes and fields." Ph. D. Dissertation, University of California, Riverside, 1996.

[14] R. J. Swift, "Structural and sample path analysis of harmonizable random fields." Ph. D. Dissertation, University of California, Riverside, 1992.

Biography of M. M. Rao

M.M. Rao was born Malempati Madhusudana Rao in the village of Nimmagadda in the state of Andhra Pradesh in India on June 6, 1929. He came to the United States after completing his studies at the College of Andhra University and the Presidency College of Madras University. He obtained his Ph.D in 1959 at the University of Minnesota under the supervision of Monroe Donsker (as well as Bernard R. Gelbaum, Leonid Hurwicz, and I. Richard Savage).

His first academic appointment was at Carnegie Institute of Technology (now called Carnegie Mellon University) in 1959. In 1972, he joined the faculty at the University of California, Riverside where he remained until 2020. He has held visiting positions at the Institute for Advanced Study (Princeton), the Indian Statistical Institute, University of Vienna, University of Strassbourg, and the Mathematical Sciences Research Institute (Berkeley).

In 1966 he married Durgamba Kolluru in India. They have twin daughters Leela and Uma and one granddaughter.

M. M. and Durgamba
(Photo courtesy of R. J. Swift.)

M.M.'s research interests were initially in probability and mathematical statistics, but his intense mathematical interest and natural curiosity found him pursuing a wide range of mathematical analysis including stochastic processes, functional analysis, ergodic theory and related asymptotics, differential equations and difference equations. His breadth of interest is mirrored by his students, many of whom are recognized as experts in diverse fields such as measure theory, operator theory, partial differential equations and stochastic processes.

M.M. has always strived for complete understanding and generality in mathematics and rarely accepts less from others. This view of mathematics has played a central role in his teaching. M.M. Rao is truly a gifted lecturer and he has inspired many generations of students. He is a demanding Ph.D. advisor that expects the most from his students. The guidance and mentoring he provides them has led to many of his students becoming successful mathematicians.

M.M. is a prolific writer. His first published writings were not on mathematics, but rather Indian poetry. He wrote poetry in his late teenage years and had a collection of his poems published when he was 21. His mathematical research publications are many and span six decades. His most recent work, a research monograph on harmonizable processes, appeared in October of 2020.

(Photo courtesy of R. J. Swift.)

Published Writings of M. M. Rao

[1] Note on a remark of Wald, Amer. Math. Monthly 65 (1958), 277-278.
[2] Lower bounds for risk functions in estimation, Proc. Nat'l Acad. of Sciences 45 (1959), 1168-1171.
[3] Estimation by periodogram, Trabajos de Estadistica 11 (1960), 123-137.
[4] Two probability limit theorems and an application, Indagationes Mathematicae 23 (1961), 551-559.
[5] Theory of lower bounds for risk functions in estimation, Mathematische Annalon 143 (1961), 379-398.
[6] Consistency and limit distributions of estimators of parameters in explosive stochastic difference equations, Annals of Math. Stat. 32 (1961), 195-218.
[7] Some remarks on independence of statistics, Trabajos de Estadistica 12 (1961), 19-26.
[8] Remarks on a multivariate gamma distribution, Amer. Math. Monthly 68 (1961), 342-346 (with P. R. Krishnaiah).
[9] Theory of order statistics, Mathematische Annalen 147 (1962), 298-312.
[10] Nonsymmetric projections in Hilbert Space, Pacific J. Math. 12 (1962), 343-357, (with V. J. Mizel).
[11] Characterizing normal law and a nonlinear integral equation, J. Math. Mecli. 12 (1963), 869-880.
[12] Inference in stochastic processes-I, Teoria Veroyatnastei i ee Primeneniya 8 (1963), 282-298.
[13] Some inference theorems in stochastic processes, Bull. Amer. Math. Sec. 68 (1963), 72-77.
[14] Discriminant analysis, Annals of Inst. of Stat. Math. 15 (1963), 11-24.
[15] Bayes estimation with convex loss, Annals of Math. Stat. 34 (1963), 839-846, (with M.H. DeGroot).
[16] Stochastic give-and-take, J. Math. Anal. & Appl. 7 (1963), 489-498, (with M.H. DeGroot)
[17] Averagings and quadratic equations in operators, Carnegie-Mellon Univesity Technical Report # 9 (1963) 27 pages, (with V. J. Mizel)
[18] Projections, generalized inverses, and quadratic forms, J. Math. Anal. & Appl. 9 (1964), 1-11, (with J. S. Chipman).
[19] Decomposition of vector measures, Proceedings of Nat'l. Acad. of Sciences 51 (1964), 771-774.
[20] Decomposition of vector measures, Proceedings of Nat'l. Acad. of Sciences 51 (1964), 771-774, Erratum, 52 (1964), p. 864.
[21] Linear functionals on Orlicz spaces, Nieuw Archief voor Wiskunde 312 (1964), 77-98.

[22] The treatment of linear restrictions in regression analysis, Econometrica 32 (1964), 198-209, (with J.S. Chipman).

[23] Conditional expectations and closed projections, Indagationes Mathematicae 27 (1965), 100-112.

[24] Smoothness of Orlicz spaces-I and II, Indagationes Mathematicae 27 (1965), 671-680, 681-690.

[25] Existence and determination of optimal estimators relative to convex loss, Annals of Inst. of Stat. Math 17 (1965), 113-147.

[26] Interpolation, ergodicity, and martingales, J. of Math. & Mech. 16 (1965), 543-567.

[27] Inference in stochastic processes-II, Zeitschrift fur Wahrscheinlichkeitstheorie 5 (1966), 317-335.

[28] Approximations to some statistical tests, Trabajos de Estadistica 17 (1966), 85-100.

[29] Multidimensional information inequalities and prediction, Proceedings of Int'l. Symposium on Multivariate Anal., Academic Press, (1966) 287-313, (with M.H. DeGroot).

[30] Convolutions of vector fields and interpolation, Proceedings of Nat'l. Acad. Sciences 57 (1967), 222-226.

[31] Abstract Lebesgue-Radon-Nikodym theorems, Annali di Matematica Pura ed Applicata (4) 76 (1967), 107-132.

[32] Characterizing Hilbert space by smoothness, Indagationes Mathematicae 29 (1967), 132-135.

[33] Notes on pointwise convergence of closed martingales, Indagationes Mathematicae 29 (1967), 170-176.

[34] Inference in stochastic processes-III, Zeitschrift fur Wahrscheinlichkeitstheorie 8 (1967), 49-72.

[35] Characterization and extension of generalized harmonizable random fields, Proceedings Nat'l. Acad. Sciences 58 (1967), 1213-1219.

[36] Local functionals and generalized random fields, Bull. Amer. Math. Soc. 74 (1968), 288-293.

[37] Extensions of the Hausdorff-Young theorem, Israel J. of Math. 6 (1968), 133-149.

[38] Linear functionals on Orlicz spaces: General theory, Pacific J. Math. 25 (1968), 553-585.

[39] Almost every Orlicz space is isomorphic to a strictly convex Orlicz space, Proceedings Amer. Math. Soc. 19 (1968), 377-379.

[40] Predictions nonlineares et martingales d'operateurs, Comptes rendus (Academie des Sciences, Paris), Ser. A, 267 (1968) 122-124.

[41] Representation theory of multidimensional generalized random fields, Proceedings 2d Int'l. Sympt. Multivariate Anal., Academic Press (1969), 411-436.

[42] Operateurs de moyednes et moyennes conditionnelles, C.R. Acad. Sciences, Paris, Ser. A, 268 (1969), 795-797.

[43] Produits tensoriels et espaces de fontioiis, C.R. Acad. Sci., Paris 268 (1969), 1599-1601.

[44] Stone-Weierstrass theorems for function spaces, J. Math. Anal. 25 (1969), 362-371.

[45] Contractive projections and prediction operators, Bull. Amer. Math. Sec. 75 (1969), 1369-1373.
[46] Generalized martingales, Proceedings lst Midwestern Symp. on Ergodic Theory Prob., Lecture Notes in Math., Springer-Verlag, 160 (1970), 241-261.
[47] Linear operations, tensor products and contractive projections in function spaces, Studia Math. 38, 131-186, Addendum 48 (1970), 307-308.
[48] Approximately tame algebras of operators, Bull. Acad. Pol. Sci., Ser. Math. 19 (1971), 43-47.
[49] Abstract nonlinear prediction and operator martingales, J. Multivariate Anal. 1 (1971), 129-157, Erratum, 9. p. 646.
[50] Local functionals and generalized random fields with independent values, Teor. Vero- jatnost., Prem. 16 (1971), 466-483.
[51] Projective limits of probability spaces, J. Multivariate Anal. 1 (1971), 28-57.
[52] Contractive projections and conditional expectations, J. Multivariate Anal. 2 (1972), 262-381, (with N. Dinculeanu).
[53] Prediction sequences in smooth Banach spaces, Ann. Inst. Henri Poincare, Ser. B, 8 (1972), 319-332.
[54] Notes on characterizing Hilbert space by smoothness and smooth Orlicz spaces, J. Math. Anal. & Appl. 37 (1972), 228-234.
[55] Abstract martingales and ergodic theory, Proc. 3rd Symp. on Multivariate Anal., Academic Press (1973), 100-116.
[56] Remarks on a Radon-Nikodym theorem for vector measures, Proc. Symp. on Vector & Operator Valued Measures and Appi., Academic Press (1973), 303-317.
[57] Inference in stochastic processes-IV: Predictors and projections, Sankhya, Ser. A 36 (1974), 63-120.
[58] Inference in stochastic processes-V: Admissible means, Sankhya, Ser. A. 37 (1974), 538-549.
[59] Extensions of stochastic transformations, Trab. Estadistica 26 (1975), 473-485.
[60] Conditional measures and operators, J. Multivariate Anal. 5 (1975), 330-413.
[61] Compact operators and tensor products, Bull. Acad. Pol. Sci. Ser. Math. 23 (1975), 1175-1179.
[62] Two characterizations of conditional probability, Proc. Amer. Math. Sec. 59 (1976), 75-80.
[63] Conjugate series, convergence and martingales, Rev. Roum. Math. Pures et Appl. 22 (1977), 219-254.
[64] Inference in stochastic processes-VI: Translates and densities, Proc. 4th Symp. Multivariate Anal., North Holland, (1977), 311-324.
[65] Bistochastic operators, Commentationes Mathematicae, Vol. 21 March, (1978), 301-313.
[66] Asymptotic distribution of an estimator of the boundary parameter of an unstable process, Ann. Statistics 6 (1978), 185-190.
[67] Convariance analysis of nonstationary time series, Developments in Statistics I (1978), 171-225.
[68] Non L^1-bounded martingales, Stochastic Control Theory and Stochastic Differential Systems, Lecture Notes in Control and Information Sciences, 16 (1979), 527-538, Springer Verlag.
[69] Processus lineaires sur $C_{00}(G)$, C. R. Acad. Sci., Paris, 289 (1979), 139-141.
[70] Convolutions of vector fields-I, Math. Zeitschrift, 174 (1980), 63-79.

[71] Asymptotic distribution of an estimator of the boundary parameter of an unstable process, Ann. Statistics 6 (1978), 185-190, Correction, Ann. Statistics 8 (1980), 1403.

[72] Local Functionals on $C_{00}(G)$ and probability, J. Functional Analysis 39 (1980), 23-41.

[73] Local functionals, Proceedings of Oberwolfach Conference on Measure Theory, Lecture Notes in Math. 794, Springer-Verlag (1980), 484-496.

[74] Structure and convexity of Orlicz spaces of vector fields, Proceedings of the F.B. Jones Conference on General Topology and Modern Analysis, University of California, Riverside (1981), 457-473.

[75] Representation of weakly harmonizable processes, Proc. Nat. Acad. Sci., 79, No. 9 (1981), 5288-5289.

[76] Stochastic processes and cylindrical probabilities, Sankhya, Ser. A (1981), 149-169.

[77] Application and extension of Cramer's Theorem on distributions of ratios, In Contributions to Statistics and Probability, North Holland (1981), 617-633.

[78] Harmonizable processes: structure theory, L'Enseignement Mathematique, 28 (1982), 295-351.

[79] Domination problem for vector measures and applications to non-stationary processes, Oberwolfach Measure Theory Proceedings, Springer Lecture Notes in Math. 945 (1982), 296-313.

[80] Bimeasures and sampling theorems for weakly harmonizable processes, Stochastic Anal. Appl. 1 (1983), 21-55, (with D.K. Chang).

[81] Filtering and smoothing of nonstationary processes, Proceedings of the ONR workshop on "Signal Processing", Marcel-Dekker Publishing (1984), 59-65.

[82] The spectral domain of multivariate harmonizable processes, Proc. Nat. Acad. Sci. U.S.A. 81 (1984), 4611-4612.

[83] Harmonizable, Cramér, and Karhunen classes of processes, Handbook in Vol. 5 (1985), 279-310.

[84] Bimeasures and nonstationary processes, Real and Stochastic Analysis, Wiley & Sons (1986), 7-118, (with D.K. Chang).

[85] A commentary on "On equivalence of infinite product measures", in S. Kakutani's selected works, Birkhauser Boston Series (1986), 377-379.

[86] Probability, Academic Press, Inc., New York, Encyclopedia of Physical Science and Technology, Vol. 11 (1987), pp. 290-310.

[87] Special representations of weakly harmonizable processes, Stochastic Anal. (1988), 169-189, (with D.K. Chang).

[88] Paradoxes in conditional probability, J. Multivariate Anal., 27, (1988), pp. 434-446.

[89] Harmonizable signal extraction, filtering and sampling Springer-Verlag, Topics in Non-Guassian Signal Processing, Vol. II (1989), pp. 98-117.

[90] A view of harmonizable processes, North-Holland, New York, in Statistical Data Analysis and Inference (1989), pp. 597-615.

[91] Bimeasures and harmonizable processes; (analysis, classification, and representation), Springer-Verlag Lecture Notes in Math., 1379, (1989), pp. 254-298.

[92] Sampling and prediction for harmonizable isotropic random fields, J. Col Analysis, Information & System Sciences, Vol 16 (1991), pp. 207-220.

[93] $L^{2,2}$ - boundedness, harmonizability and filtering, Stochastic Anal. App., (1992), pp. 323-342.
[94] Probability (expanded for 2nd ed.), Encyclopedia of Physical Science and Technology Vol 13 (1992), pp. 491-512.
[95] Stochastic integration: a unified approach, C. R. Acad. Sci., Paris, Vol 314 (Series 1), (1992), pp. 629-633.
[96] A projective limit theorem for probability spaces and applications, Theor. Prob. and Appl., Vol 38 (1993), (with V. V. Sazonov, in Russian), pp. 345-355.
[97] Exact evaluation of conditional expectations in the Kolmogorov model, Indian J. Math., Vol 35 (1993) pp 57-70.
[98] An approach to stochastic integration (a generalized and unified treatment) in Multivariate Analysis: Future Directions, Elsivier Science Publishers, The Netherlands (1993), pp. 347-374.
[99] Harmonizable processes and inference: unbiased prediction for stochastic flows, J. Statistic. Planning and Inf., Vol 39 (1994), pp. 187-209.
[100] Some problems of real and stochastic analysis arising from applications, Stochastic Processes and Functional Analysis, J. A. Goldstein, N. E. Gretsky, J.J. Uhl, editors, Marcel Dekker Inc. (1997), 1-15.
[101] Packing in Orlicz sequence spaces, (with Z. D. Ren), Studia Math. 126 (1997), no. 3, 235-251.
[102] Second order nonlinear stochastic differential equations, Nonlinear Analysis, Vol. 30, no. 5 (1997) 3147-3151.
[103] Higher order stochastic differential equations. Real and Stochastic Analysis, CRC Press, Boca Raton, FL, (1997), 225-302.
[104] Nonlinear prediction with increasing loss. J. N. Srivastava: felicitation volume. J. Combin. Inform. System Sci. 23 (1998), no. 1-4, 187-192.
[105] Characterizing covariances and means of harmonizable processes. Infinite Dimensional Analysis and Quantum Probability, Kyoto (2000), 363-381.
[106] Multidimensional Orlicz space interpolation with changing measures. Peetre 65 Proceedings, Lund, Sweden, (2000).
[107] Representations of conditional means. Dedicated to Professor Nicholas Vakhania on the occasion of his 70th birthday. Georgian Math. J. 8 (2001), no. 2, 363-376.
[108] Convolutions of vector fields. II. Random walk models. Proceedings of the Third World Congress of Nonlinear Analysts, Part 6 (Catania, 2000). Nonlinear Anal. 47 (2001), no. 6, 3599-3615.
[109] Martingales and some applications. Shanbhag, D. N. (ed.) et al., Stochastic processes: Theory and methods. Amsterdam: North-Holland/Elsevier. Handbook Statistics 19,(2001) 765-816.
[110] Probability (revised and expanded for 3rd ed.), Encyclopedia of Physical Science and Technology (2002), pp. 87-109.
[111] Representation and estimation for harmonizable type processes. IEEE, (2002) 1559-1564.
[112] A commentary on "Une Théorie Unifiée des martingales et des moyennes ergodigues", C.R. Acad. Sci 252 (1961) p. 2064-2066, in Rota's Saleta, Birkhauser Boston (2002).

[113] Evolution operators in stochastic processes and inference. Evolution Equation, G. R. Goldstein, R. Nagel, S. Romanelli, editors, Marcel Dekker Inc. (2003), 357-372.
[114] Stochastic analysis and function spaces. Recent Advances in Stochastic Processes and Functional Analysis, A.C. Krinik, R.J. Swift, editors, Marcel Dekker Inc. (2004), 1-25.
[115] Convolutions of vector fields. III. Amenability and spectral properties. Real and stochastic analysis, (2004), 375–401.
[116] Characterizations of harmonizable fields. Nonlinear Anal. 63 (2005), no. 5-7, 935–947.
[117] Structure of Karhunen processes. J. Comb. Inf. Syst. Sci. 31 (2006), no. 1-4, 187–207.
[118] Exploring ramifications of the equation $E(Y \mid X) = X$. J. Stat. Theory Pract. 1 (2007), no. 1, 73–88.
[119] Integral representations of second order processes. Nonlinear Anal. 69 (2008), no. 3, 979–986.
[120] Random measures and applications. Stoch. Anal. Appl. 27 (2009), no. 5, 1014–1076.
[121] Quadratic equations in Hilbertian operators and applications, (with V.J. Mizel), Internat. J. Math. 20 (2009), no. 11, 1431–1454.
[122] Linear regression for random measures. Advances in multivariate statistical methods, (2009), 131–144.
[123] Applications and aspects of random measures. Nonlinear Anal. 71 (2009), 1513–1518.
[124] Characterization and duality of projective and direct limits of measures and applications. Internat. J. Math. 22 (2011), no. 8, 1089 - 1119.
[125] Infinite dimensional stationary random fields over a locally compact abelian group. Internat. J. Math. 23 (2012), no. 4, 23 pp
[126] Harmonic and probabilistic approaches to zeros of Riemann's zeta function. Stoch. Anal. Appl. 30 (2012), no. 5, 906 - 915.
[127] Integration with vector valued measures. Discrete Contin. Dyn. Syst. 33 (2013), no. 11-12, 5429 - 5440.
[128] Entropy, SDE-LDP and Fenchel-Legendre-Orlicz classes. Real and Stochastic analysis, (2014), 431 - 501.
[129] Stochastic Equations, Stochastic Processes and Functional Analysis, (2021).
[130] From Additive to Second order Processes, (with R. J Swift), Stochastic Processes and Functional Analysis, (2021).

Books Edited

[1] General Topology and Modern Analysis. Proceedings of the F.B. Jones Conference, Academic Press, Inc., New York (1981), 514 pages, (Edited jointly with L.F. McCauley).
[2] Handbook in Statistics, Volume 5, Time Series in the Time Domain, (Edited jointly with E.J. Hannan, P.R. Krishnaiah), North-Holland Publishing Co., Amsterdam (1985).
[3] Real and Stochastic Analysis, (Editor), Wiley & Sons, New York (1986), 347 pages.

[4] Multivariate Statistics and Probability, (Edited jointly with C.R. Rao), Academic Press Inc., Boston (1989), 565 pages.
[5] Real and stochastic analysis. Recent advances. (Editor) Boca Raton, FL, CRC Press. (1997), 393 pages.
[6] Real and stochastic analysis. New Perspectives. (Editor) Birkhauser Boston, MA, (2004), 405 pages.
[7] Real and Stochastic Analysis. Current Trends. (Editor) World Scientific, Singapore, (2014), 576 pages.

Books Written

[1] Stochastic Processes and Integration. Sijthoff & Noordhoff International Publishers, Alpehn aan den Rijn, The Netherlands, (1979), 460 pages.
[2] Foundations of Stochastic Analysis, Academic Press, Inc., New York, (1981), 295 pages. Reprinted by Dover Publications, (2011).
[3] Probability Theory with Applications, Academic Press, Inc. New York, (1984), 495 pages.
[4] Measure Theory and Integration, Wiley-Interscience, New York (1987), 540 pages.
[5] Theory of Orlicz Spaces (jointly with Z. D. Ren), Marcel Dekker Inc., New York (1991), 449 pages.
[6] Conditional Measures and Applications, Marcel Dekker Inc., New York (1993), 417 pages.
[7] Stochastic Processes: General Theory, Kluwer Academic Publishers, The Netherlands (1995), 620 pages.
[8] Stochastic Processes: Inference Theory, Kluwer Academic Publishers, The Netherlands (2000), 645 pages.
[9] Applications of Orlicz Spaces (jointly with Z. D. Ren), Marcel Dekker Inc., New York (2002), 464 pages.
[10] Measure Theory and Integration, (Revised and enlarged second edition), Marcel Dekker, Inc., New York, (2004), 761 pages.
[11] Conditional Measures and Applications, (Revised second edition), Chapman & Hall/CRC, Boca Raton, FL, (2005), 483 pages.
[12] Probability Theory with Applications, (jointly with R. J. Swift), (Revised and enlarged second edition), Springer, New York, (2006), 527 pages.
[13] Random and Vector Measures, World Scientific, Singapore, (2011), 550 pages.
[14] Stochastic Processes: Harmonizable Theory, Singapore, (2020), 340 pages.

Ph.D. Theses Completed Under the Direction of M.M. Rao

At Carnegie-Mellon University:

Dietmar R. Borchers (1964), "Second order stochastic differential equations and related Ito processes."
J. Jerry Uhl. Jr (1966), "Orlicz spaces of additive set functions and set martingales."
Jerome A. Goldstein (1967), "Stochastic differential equations and nonlinear semigroups."
Neil E. Gretsky (1967), "Representation theorems on Banach function spaces."
William T. Kraynek (1968), "Interpolation of sub-linear operators on generalized Orlicz and Hardy spaces."
Robert L. Rosenberg (1968), "Compactness in Orlicz spaces based on sets of probability measures."
George Y. H. Chi (1969), "Nonlinear prediction and multiplicity of generalized random processes."

At University of California, Riverside:

Vera Darlean Briggs (1973), "Densities for infinitely divisible processes."
Stephen V. Noltie (1975), "Integral representations of chains and vector measures."
Theodore R. Hillmann (1977), "Besicovitch - Orlicz spaces of almost periodic functions."
Michael D. Brennan (1978), "Planar semi-martingales and stochastic integrals."
James P. Kelsh (1978), "Linear analysis of harmonizable time series."
Alan C. Krinik (1978), "Stroock - Varadhan theory of diffusion in a Hilbert space and likelihood ratios."
Derek K. Chang (1983), "Bimeasures, harmonizable process and filtering."
Marc H. Mehlman (1990), "Moving average representation and prediction for multidimensional strongly harmonizable process."
Randall J. Swift (1992), "Structural and sample path analysis of harmonizable random fields."
Michael L. Green (1995), "Multi-parameter semi-martingale integrals and boundedness principles."
Heroe Soedjak (1996), "Estimation problems for harmonizable random processes and fields."
Jason H. Park (2015), "Random Measure Algebras Under Convolution."

M. M. with some of his students. From left to right: Randall Swift (1992), Alan Krinik (1978), Marc Mehlman (1990) and Jason Park (2015). (Photo courtesy of R. J. Swift.)

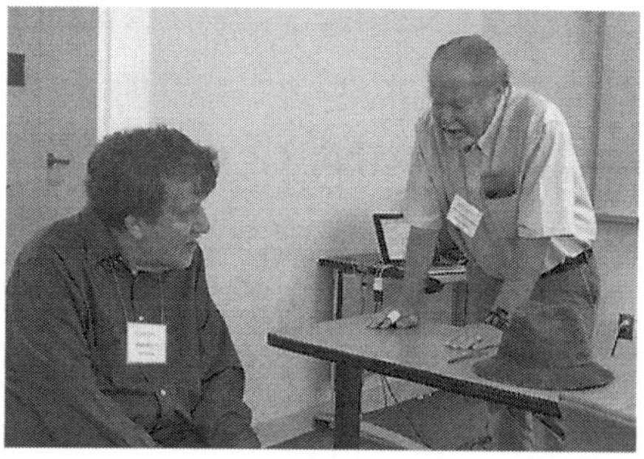

Alan Krinik and Jerry Goldstein
(Photo courtesy of R. J. Swift)

M. M. and Marc Mehlman
(Photo courtesy of R. J. Swift)

Randall Swift and Alan Krinik

Michel Lapidus and M. M.
(Photo courtesy of R. J. Swift.)

Durgamba, M. M. and Uma (Photo courtesy of R. J. Swift.)

After the banquet. From left to right: Gisèle Goldstein, Alan Krinik, Jerry Goldstein, M. M., Randall Swift, Jason Park, Marc Mehlman. (Photo courtesy of R. J. Swift.)

Celebrating M.M. Rao's Many Mathematical Contributions

American Mathematical Society
Fall Western Sectional Meeting
University of California, Riverside
November 9-10, 2019

Special Session on Celebrating M.M. Rao's Many Mathematical Contributions as he Turns 90 Years Old

Organizers:

Jerome Goldstein, University of Memphis
Michael Green, Alan Krinik, Randall Swift & Jennifer Switkes, Cal Poly Pomona

Speakers and Presentation Titles

Saturday, November 9, 2019

Search for Optimum Quadratic Forms as Estimators of Variance Components in Linear Mixed Effects Models
Subir Ghosh
University of California, Riverside

Banach space valued weak second order stochastic processes.
Yûichirô Kakihara
California State University, San Bernardino

Sharp Large Deviations for Random Projections of L^p Balls.
*Liao Yin-Ting** & *Kavita Ramanan*
Brown University

Diffusion limits for Shortest Remaining Processing Time Queues.
*Amber Puha**
California State University San Marcos
Sayan Banerjee & *Amarjit Budhiraja*
University of North Carolina

Convergence rates to stationarity for reflecting Brownian motions.
Sayan Banerjee & Amarjit Budhiraja*
University of North Carolina, Chapel Hill

From Additive Processes to Second-Order Processes.
Randall Swift
California State Polytechnic University Pomona

A Stochastic Predator-Prey Model through a Log-Normal Moment Closure Technique.
Jennifer M Switkes, Tanawat Trakoolthai, & Diana Curtis*
California State Polytechnic University Pomona

Stochastic Equations.
M. M. Rao
University of California, Riverside

How strong can the Parrondo effect be?
*Stewart N. Ethier**
University of Utah
Jiyeon Lee
Yeungnam University

Instantaneous blowup (IBU): Old and new results.
Jerome Goldstein
University of Memphis

New Results in Mathematical Finance.
Gisèle Ruiz Goldstein
University of Memphis

Sunday, November 10, 2019

Stick-breaking processes, clumping, and Markov chain occupation laws.
Zach Dietz
Cincinnati, OH
*William Lippitt & Sunder Sethuraman**
University of Arizona

Dueling bandit problems.
Erol Pekoz
Boston University
*Sheldon Ross & Zhengyu Zhang**
University of Southern California

The Boltzmann-Enskog process for hard and soft potentials.
*Padmanabhan Sundar**
Louisiana State University

Martin Friesen, & Barbara Rüdiger
Bergische Universität Wuppertal, Germany

Generating functions as tinker toys: Building connections from simple combinatorial structures to asymptotic behavior for a class of random processes with time-varying transition rates.
Barbara Margolius
Cleveland State University

Relating the Workload-barrier M/D/1 Queue, a Renewal Process, and an $< s, S >$ Inventory.
*Percy H. Brill**
University of Windsor, Windsor, Ontario, Canada
Mei Ling Huang
Brock University, St. Catharines Ontario, Canada

Efficient computation of transition probabilities and statistical estimation for general birth-death processes.
Forrest W Crawford
Yale University

Generalized ballot box problem and finite Markov chains with catastrophe-like transitions.
Alan Krinik, Saif A. Aljashamy, David Perez, Jeffrey Yeh, Aaron Kim, Jeremy Lin*, Thuy Vu Dieu Lu*, Mac Elroyd Fernandez & Mark Dela*
California State Polytechnic University Pomona

Analysis, Mathematical Physics and Randomness
Michel Lapidus
University of California, Riverside

Lorenz Order with Common Finite Support.
Barry C. Arnold
University of California, Riverside

Relations between irreducible and absorbing Markov chains.
Gerardo Rubino
INRIA, France

Numerically Solving a Rank-Based Forward Backward Stochastic Differential Equation by Applying the Least-Squares Monte Carlo Method
Mark Dela
California State Polytechnic University, Pomona

Algebra of Random Measures.
Jason Hong Jae Park,
University of Nevada, Las Vegas

Summation and Integration in Hyperspaces.
Mark Burgin
University of California, Los Angeles

* denotes Session speaker

Subir Ghosh
(Photo courtesy of R. J. Swift.)

Barry Arnold
(Photo courtesy of R. J. Swift.)

Yûichirô Kakihara
(Photo courtesy of R. J. Swift.)

Jennifer Switkes
(Photo courtesy of R. J. Swift.)

Randall Swift

Jason Park
(Photo courtesy of R. J. Swift.)

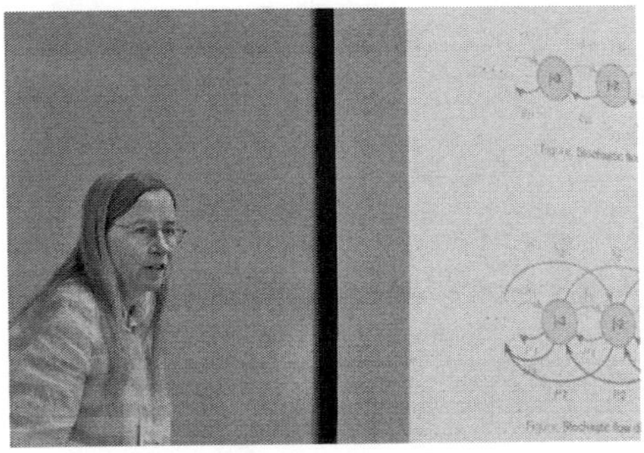

Barbara Margolius
(Photo courtesy of R. J. Swift.)

Stewart Ethier
(Photo courtesy of R. J. Swift.)

Alan Krinik
(Photo courtesy of R. J. Swift.)

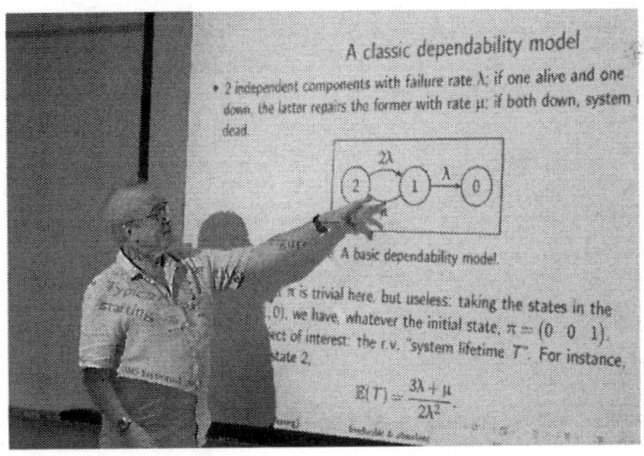

Gerardo Rubino
(Photo courtesy of R. J. Swift.)

Gisèle Goldstein
(Photo courtesy of R. J. Swift.)

Jerry Goldstein
(Photo courtesy of R. J. Swift.)

Sayan Banerjee
(Photo courtesy of R. J. Swift.)

Sunder Sethuraman
(Photo courtesy of R. J. Swift.)

Contemporary Mathematics
Volume **774**, 2021
https://doi.org/10.1090/conm/774/15564

Sufficient conditions for Lorenz ordering with common finite support

Barry C. Arnold

Abstract. Arnold and Gokhale (2017) provided a characterization of the Lorenz inequality order between distributions with common finite support. In the more general Lorenz order context, a variety of partial orders are often used to verify the existence of Lorenz ordering. In this paper we investigate whether parallel results can be identified within the common finite support context.

1. Introduction

The Lorenz order is classically defined on the class of all non-negative random variables with positive finite expectations. In this paper our focus is on the Lorenz order restricted to a class of random variables with common finite support. In particular we are concerned with the extent to which certain sufficient conditions for Lorenz order in the general context continue to be useful in the restricted finite support setting. While progressive or Robin Hood transfers can be used to characterize Lorenz order in the general setting, Arnold and Gokhale (2017) identified analogous operations that reduce inequality in common finite support cases. These so-called Robin Hood exchanges play a parallel role in the finite support setting to that played by Robin Hood transfers in the more general setting. We begin our investigation with a review of needed concepts dealing with majorization and the usual Lorenz order.

2. The usual Lorenz order and the role of Robin Hood

More detailed discussion of the topics in this Section may be found in Chapter 17 of Marshall, Olkin and Arnold (2011). Consider a population of n individuals, the ith member of which has income of x_i income units. A key desirable feature of suitable inequality measures of such populations is one associated with what may be called a Robin Hood transfer. Robin Hood was known for taking money from the rich and giving it to the poor. It is quite generally accepted that such an operation will reduce inequality, and any reasonable inequality measure should be reduced by such an operation. If one accepts this view then one is led to the use of the majorization partial order, defined by Hardy, Littlewood and Polya (1934),

2020 *Mathematics Subject Classification.* Primary 60E15; Secondary 91B82.

Key words and phrases. Robin Hood exchange, Robin Hood transfer, majorization, star order, sign change order, density crossing order.

to compare n-dimensional income vectors. In the following discussion, use will be made of notation used in the study of order statistics and the elements of the income vector $\underline{x}$ arranged in non-decreasing order will be denoted by $x_{1:n}, x_{2:n}, \ldots, x_{n:n}$.

DEFINITION 2.1. **Majorization.** An n-dimensional vector $\underline{x}$ is said to be majorized by an n-dimensional vector $\underline{y}$, written $\underline{x} \prec \underline{y}$, if

$$\begin{aligned} &\textstyle\sum_{i=1}^{k} x_{i:n} \geq \sum_{i=1}^{k} y_{i:n}, \quad k = 1, \ldots, n-1, \\ &\textstyle\sum_{i=1}^{n} x_{i:n} = \sum_{i=1}^{n} y_{i:n}. \end{aligned} \tag{2.1}$$

If $\underline{x} \prec \underline{y}$, then $\underline{x}$ exhibits less inequality than does $\underline{y}$. A Robin Hood operation consists of taking some money from an individual and giving it to a relatively poorer individual, without taking so much as to reverse the income order between the two individuals, thus reducing inequality in the population.

Hardy, Littlewood and Polya showed that if $\underline{x} \prec \underline{y}$, then $\underline{x}$ can be obtained from $\underline{y}$ by Robin Hood in a finite number (actually $n-1$) of operations.

On the basis of this result, one can say that, if one accepts that Robin Hood's operations reduce inequality, then one must accept the majorization partial order as an appropriate inequality ordering.

Another available characterization of majorization that is often useful is the following.

THEOREM 2.2. *$\underline{x} \prec \underline{y}$ if and only if $\sum_{i=1}^{n} g(x_i) \leq \sum_{i=1}^{n} g(y_i)$ for all continuous convex functions g.*

The majorization partial order only relates the inequality in two populations of the same size and with the same total. An extension is required if we are to make more general comparisons. To this end, we consider the Lorenz curve as defined by Lorenz (1905) as follows.

Consider a population of n individuals. Let x_i denote the wealth of individual i, $i = 1, \ldots, n$. Assume that all x_i's are non-negative and that not all of them are equal to 0. Order the individuals from poorest to richest to obtain $x_{1:n}, \ldots, x_{n:n}$. Now plot the points

$$(k/n, \sum_{i=1}^{k} x_{i:n} / \sum_{i=1}^{n} x_{i:n}), \quad k = 0, \ldots, n.$$

Join these $n+1$ points by line segments to obtain a curve connecting the origin with the point $(1,1)$. This is the Lorenz curve corresponding to the vector $\underline{x}$, to be denoted by $L_{\underline{x}}(u)$. Unless all the x_i's are equal, the Lorenz curve will be convex and will lie under the straight line joining $(0,0)$ to $(1,1)$. The associated Lorenz order relating vectors $\underline{x}$ and $\underline{y}$, of possibly different dimensions and possibly different totals, is denoted by $\underline{x} \leq_L \underline{y}$ and is defined as follows.

DEFINITION 2.3. **Lorenz Ordering.** $\underline{x} \leq_L \underline{y}$ if $L_{\underline{x}}(u) \geq L_{\underline{y}}(u)$ for all $u \in [0,1]$.

Lorenz curves can cross once or several times so that the Lorenz order is only a partial order.

Observe that if $\underline{x}$ and $\underline{y}$ are of the same dimension and if $\sum_{i=1}^{n} x_i = \sum_{i=1}^{n} y_i$ then $\underline{x} \leq_L \underline{y}$ if and only if $\underline{x} \prec \underline{y}$. Thus the Lorenz order appears as a natural extension of the majorization partial order, allowing us to compare populations with different sizes with different wealth totals.

Further extension is possible and desirable. It is possible to associate a non-negative random variable X with a given vector $\underline{x}$ with non-negative elements not all zero, by defining

$$P(X = x_i) = 1/n, \quad i = 1, 2, \ldots, n. \tag{2.2}$$

The distribution function of X is recognizable as the empirical distribution function of the set of numbers $x_1, x_2, \ldots, x_n$.

Gastwirth (1971) proposed the following definition of the Lorenz curve defined on the class $\mathcal{L}_+$ of non-negative random variables with finite positive expectations.

DEFINITION 2.4. The Lorenz curve L of a random variable $X \in \mathcal{L}_+$ is

$$L_X(u) = \frac{\int_0^u F_X^{-1}(y)dy}{\int_0^1 F_X^{-1}(y)dy} = \frac{\int_0^u F_X^{-1}(y)dy}{E(X)}, \quad 0 \le u \le 1, \tag{2.3}$$

where

$$\begin{aligned} F_X^{-1}(y) &= \sup\{x : F_X(x) \le y\}, \qquad 0 \le y < 1, \\ &= \sup\{x : F_X(x) < 1\}, \qquad y = 1, \end{aligned}$$

is the right continuous inverse distribution function (or quantile function) of the random variable X.

Observe that if X is the random variable associated with the vector $\underline{x}$ as in (2.2), then the corresponding Gastwirth Lorenz curve $L_X(u)$ and the curve suggested by Lorenz, $L_{\underline{x}}(u)$, are identical. The Lorenz order can then be extended to allow comparison of random variables as follows.

DEFINITION 2.5. For $X, Y \in \mathcal{L}_+$, with corresponding Lorenz curves L_X and L_Y, X is less than Y in the Lorenz order, written as $X \le_L Y$ if $L_X(u) \ge L_Y(u)$ for all $u \in [0, 1]$.

It is possible to prove a natural extensions of Theorems 1, applying to the Lorenz order rather than being restricted to majorization.

THEOREM 2.6. *For $X, Y \in \mathcal{L}_+$, $X \le_L Y$ if and only if $E(g(X/E(X))) \le E(g(Y/E(Y)))$ for every continuous convex function g such that the expectations exist.*

There is another available characterization of the Lorenz order in terms of "angle" functions (originally stated by Hardy, Littlewood and Polya in the context of majorization). Thus

THEOREM 2.7. *Suppose that $X, Y \in \mathcal{L}_+$ with $E(X) = E(Y)$. Then $X \le_L Y$ if and only if $E[(X - c)^+] \le E[(Y - c)^+]$ for every $c > 0$.*

This result will be particularly useful when we turn to consider the common finite support case.

3. Other partial orders defined on $\mathcal{L}_+$

Let $X, Y \in \mathcal{L}$ with corresponding distribution functions F_X and F_Y. Star-shaped ordering or, more briefly, star ordering is defined as follows.

DEFINITION 3.1. We say that X is star-shaped with respect to Y, and write $X \leq_* Y$ if $F_X^{-1}(u)/F_Y^{-1}(u)$ is a non-increasing function of u.

Since $F_{cX}^{-1}(u) = cF_X^{-1}(u)$ for any positive c and any $X \in \mathcal{L}_+$, it is obvious that $*$-ordering is scale invariant.

We can use $*$-ordering to verify that Lorenz ordering obtains as a consequence of the following result.

THEOREM 3.2. *Suppose $X, Y \in \mathcal{L}_+$. If $X \leq_* Y$, then $X \leq_L Y$.*

The proof of theorem 3.2 depends on the fact that $*$-ordering implies that $F_X^{-1}(v) - F_Y^{-1}(v)$ had only one sign change $(+, -)$ on the interval $[0, 1]$. This sign-change property is sufficient for Lorenz ordering.

THEOREM 3.3. *Suppose $X, Y \in \mathcal{L}_+$ and that $[F_X^{-1}(v)/E(X)] - [F_Y^{-1}(v)/E(Y)]$ has at most one sign change (from $+$ to $-$) as v ranges from 0 to 1. It follows that $X \leq_L Y$.*

We may then introduce the following definition of a partial order that occupies an intermediate position between $\leq_*$ and $\leq_L$.

DEFINITION 3.4. We will say that X is sign-change ordered with respect to Y and write $X \leq_{s.c.} Y$, if $[F_X^{-1}(v)/E(X)] - [F_Y^{-1}(v)/E(Y)]$ has at most one sign change (from $+$ to $-$) as v ranges from 0 to 1.

A simple sufficient condition for sign change ordering can be stated in terms of density crossings (assuming densities exist). Thus,

THEOREM 3.5. *Let $X, Y \in \mathcal{L}_+$ have corresponding densities $f_X(x)$ and $f_Y(y)$ (with respect to a convenient dominating measure on $\mathbb{R}^+$, in the most abstract setting). If the function*

$$E(X)f_X(E(X)x) - E(Y)f_Y(E(Y)x) \tag{3.1}$$

has two sign changes (from $-$ to $+$ to $-$) as x ranges from 0 to ∞, then $X \leq_{s.c.} Y$.

Verification that $X \leq_L Y$ is frequently most easily done by using the density crossing argument (Theorem 3.5) or by showing $*$-ordering obtains (if $F_X^{-1}(u)$ and $F_Y^{-1}(u)$ are available in convenient tractable forms).

4. When X and Y have common finite support

For a fixed positive integer n and a fixed set of n distinct numbers $0 < x_1 < x_2 < \cdots < x_{n-1} < x_n$, consider the class $\mathcal{L}_{\underline{x}}^{(n)}$ of all random variables with support $\{x_1, x_2, \ldots, x_n\}$. A random variable $X(\underline{p})$ in this class can be associated with a probability vector $\underline{p} = (p_1, p_2, \ldots, p_n)$ where $p_i = P(X(\underline{p}) = x_i)$, $i = 1, 2, \ldots, n$. How must two probability vectors $\underline{p}$ and $\underline{q}$ be related in order that $X(\underline{p}) \leq_L X(\underline{q})$. The first result in this direction is perhaps surprising.

THEOREM 4.1. *For $X(\underline{p}), X(\underline{q}) \in \mathcal{L}_{\underline{x}}^{(n)}$, if $X(\underline{p}) \leq_L X(\underline{q})$ then $E(X(\underline{p})) = E(X(\underline{q}))$.*

PROOF. Suppose that $E(X(p)) > E(X(q))$. Consider the local behavior of the corresponding Lorenz curves in a neighborhood of 0 and in a neighborhood of 1. First, in a neighborhood of 0, we have $L_{X(p)} < L_{X(q)}$, while in a neighborhood of 1, we have $L_{X(p)} > L_{X(q)}$. But this implies that the Lorenz curves must cross and cannot be nested. □

Now for two random variables with equal means, such as these, we can use the "angle" characterization of the Lorenz order (Theorem 2.7).

We have: $X(\underline{p}) \leq_L X(\underline{q})$ (with equal means) iff

$$E[(X(\underline{p}) - c)^+] \leq E[(X(\underline{q}) - c)^+] \text{ for every } c \in (0, \infty).$$

However, since $X(\underline{p})$ and $X(\underline{q})$ only take on the values $x_1, x_2, \ldots, x_n$ we have

$$X(\underline{p}) \leq_L X(\underline{q}) \text{ iff } E[(X(\underline{p}) - x_i)^+] \leq E[(X(\underline{q}) - x_i)^+] \text{ for } i = 1, 2, \ldots, n-1,$$

i.e., if $\sum_{j=i+1}^n (x_j - x_i)p_j \leq \sum_{j=i+1}^n (x_j - x_i)q_j$ for $i = 1, 2, \ldots, n-1$. Equivalently if $\sum_{j=i+1}^n (x_j - x_i)(p_j - q_j) \leq 0$, for $i = 1, 2, \ldots, n-1$ We can write this in the form $A(\underline{x})(\underline{p} - \underline{q}) \leq 0$ for a suitable matrix $A(\underline{x})$.

To summarize, we have

$$X(\underline{p}) \leq_L X(\underline{q}) \text{ iff } E(X(\underline{p})) = \sum_{j=1}^n x_i p_i = \sum_{j=1}^n x_i q_i = E(X(\underline{q}))$$

and

$$A(\underline{x})(\underline{p} - \underline{q}) \leq 0.$$

5. Robin Hood's role in the common finite support setting

We introduce the concept of an exchange to be applied to a probability vector $\underline{p}$. A vector $\underline{\delta}$ will be called an exchange if it satisfies $\sum_{i=1}^n \delta_i = 0$. and $\underline{p} + \underline{\delta} \geq \underline{0}$. The result of the application of an exchange $\underline{\delta}$ to probability vector $\underline{p}$ is a new probability vector $\underline{p}^* = \underline{p} + \underline{\delta}$. An exchange $\underline{\delta} \neq \underline{0}$ is inequality reducing if $X(\underline{p}^*) \leq_L X(\underline{p})$.

In order to be a mean-preserving exchange it is necessary that $\underline{\delta}$ has at least 3 non-zero coordinates. Simple exchanges with exactly 3 non-zero coordinates, will be called Robin Hood Exchanges, provided that they are mean-preserving and inequality attenuating. A Robin Hood exchange will thus have, for some indices $j < k < \ell$, $\delta_k = \psi > 0$, $\delta_j = -(1-\alpha)\psi$ and $\delta_\ell = -\alpha\psi$ where α is selected to preserve the mean and ψ is not too large. Thus we must have $p_j \geq (1-\alpha)\psi$ and $p_\ell \geq \alpha\psi$ so that the post-exchange vector is a probability vector.

Suppose that $\underline{p} \neq \underline{q}$ and $X(\underline{p}) \leq_L X(\underline{q})$. It is then possible to choose indices $r < k < m$ such that $p_r < q_r$, $p_m < q_m$, and $p_k > q_k$, and with no indices t with $r < t < m$ and $p_t = q_t$.Consider the Robin Hood exchange defined by

$$\alpha = \frac{p_k x_k - p_r x_r}{p_m x_m - p_r x_r}$$

to preserve the mean, and then choose $\psi = min\{(p_k - q_k), (1-\alpha)^{-1}(q_r - p_r), \alpha^{-1}(q_m - p_m)\}$. Application of this Robin Hood exchange to $\underline{q}$ will yield a new probability vector $\underline{q}*$ with

$$X(\underline{p}) \leq_L X(\underline{q}*) \leq_L X(\underline{q}),$$

and $N(\underline{q}*, \underline{p}) < N(\underline{q}, \underline{p})$, where we have used the notation

$$N(\underline{p}^{(1)}, \underline{p}^{(2)}) = \sum_{i=1}^n I(p_i^{(1)} \neq p_i^{(2)}).$$

A finite number of such exchanges will bring us to $\underline{p}$.

Arnold and Gokhale (2017) identified the key role of such exchanges in the common finite support setting, paralleling the role of Robin Hood transfers in majorization scenarios. Thus we have:

If $\underline{p} \neq \underline{q}$ and $X(\underline{p}) \leq_L X(\underline{q})$. t and $p_k > q_k$, then $\underline{p}$ can be obtained by applying a finite number of Robin Hood exchanges to $\underline{q}$.

6. Are the usual sufficient conditions for Lorenz ordering useful in the common finite support situation?

The Arnold-Gokhale contribution yields some challenging methods available to determine whether $X(\underline{p}) \leq_L X(\underline{q})$. First check that the means are equal. We could then try to identify the appropriate matrix $A(\underline{x})$ associated with the angle function condition for Lorenz ordering. Or, we could try to identify a particular sequence of Robin Hood exchanges that will transform $\underline{q}$ into $\underline{p}$. Or if all else fails, plot the two Lorenz curves and determine whether they are nested.

Attractive alternatives would involve consideration of *-ordering, sign change ordering and density crossing ordering, all of which are known to imply Lorenz ordering. We will first consider *-ordering. The quantile function (or inverse distribution function) of a random variable $X(\underline{p}) \in \mathcal{L}^{(n)}$ is given by

$$\begin{aligned} F^{-1}_{X(\underline{p})}(u) =& x_1, \quad 0 < u \leq p_1 \\ =& x_2, \quad p_1 < u \leq p_1 + p_2 \\ =& x_3, \quad p_1 + p_2 \leq u < p_1 + p_2 + p_3, \\ & etc. \end{aligned}$$

In order to determine whether $X(\underline{p}) \leq_* X(\underline{q})$ we must check to see that $F^{-1}_{X(\underline{p})}(u)/F^{-1}_{X(\underline{q})}(u)$ is a non-increasing function of u.

Suppose that $p_1 < q_1$. It follows that

$$\begin{aligned} \frac{F^{-1}_{X(\underline{p})}(u)}{F^{-1}_{X(\underline{q})}(u)} =& 1, \quad 0 < u < p_1, \\ =& \frac{x_2}{x_1} > 1, \quad p_1 < u < q_1, \\ & etc. \end{aligned}$$

So that, in this case, $F^{-1}_{X(\underline{p})}(u)/F^{-1}_{X(\underline{q})}(u)$ is not a non-increasing function of u.

Suppose that $q_1 < p_1$ there must be a first value of $j > 1$ such that $\sum_{i=1}^{j} q_i > p_1$.. In such a case we will have

$$\begin{aligned} \frac{F^{-1}_{X(\underline{p})}(u)}{F^{-1}_{X(\underline{q})}(u)} =& 1, \quad 0 < u < q_1, \\ =& \frac{x_1}{x_2}, \quad q_1 < u < q_1 + q_2, \\ =& \frac{x_1}{x_3}, \quad q_1 + q_2 < u < q_1 + q_2 + q_3, \\ & \vdots \\ =& \frac{x_1}{x_{j-1}}, \quad q_1 + q_2 + \cdots + q_{j-1} < u << p_1, \\ =& \frac{x_2}{x_{j-1}} > \frac{x_1}{x_{j-1}}, \quad p_1 < u < min\{q_1 + q_2 + \cdots + q_j, p_2\}, \\ & etc. \end{aligned}$$

It is then evident that $F^{-1}_{X(\underline{p})}(u)/F^{-1}_{X(\underline{q})}(u)$ is not a non-increasing function of u in this case also.

Finally, if $p_1 = q_1$, there must be first value of $j > 1$ for which $p_j \neq q_j$ and as in the case where $p_1 \neq q_1$ we will be able to verify that $F^{-1}_{X(\underline{p})}(u)/F^{-1}_{X(\underline{q})}(u)$ is not a non-increasing function of u on the interval $(p_1 + p_2 + \cdots + p_{j-1}, 1)$.

Our final conclusion is that for no pair of vectors $\underline{p} \neq \underline{q}$ will we have $X(\underline{p}) \leq_* X(\underline{q})$.

Next, we will consider sign-change ordering. Here the situation is more agreeable. Assuming, as we must, that $E(X(\underline{p})) = E(X(\underline{q}))$, we can identify a simple sufficient condition for $X(\underline{p}) \leq_L X(\underline{q})$. This ordering will obtain provided that there exists a value $u*$ such that $F^{-1}_{X(\underline{p})}(u) - F^{-1}_{X(\underline{q})}(u)$ is ≤ 0 for $u \leq u*$ and is ≥ 0 for $u > u*$. This condition is readily checked by considering the partial sums of the coordinates of the vectors $\underline{p}$ and $\underline{q}$.

It turns out that the density crossing ordering, sufficient for Lorenz ordering, is the most easy to check. Once more assuming equal means, i.e., $E(X(\underline{p})) = E(X(\underline{q}))$, to ensure that $X(\underline{p}) \leq_L X(\underline{q})$, it is sufficient to identify two integers $1 < j_1 < j_2 < n$ such that $p_j \leq q_j$ for $j < j_1$ and for $j > j_2$, while $p_j \geq q_j$ for $j_1 \leq j \leq j_2$.

7. Discussion

Either sign change ordering or density crossing ordering appear to be the tools of choice to verify Lorenz ordering in the common finite support case. The role of Robin Hood exchanges is an aid to understanding Lorenz ordering in that context, but identifying a suitable exchange sequence is not a convenient way to confirm Lorenz dominance.

It is noteworthy that the actual values $x_1 < x_2 < \ldots < x_n$ involved in the discussion are crucial to determine whether $E(X(\underline{p})) = E(X(\underline{q}))$, but beyond that, do not play a role in determining Lorenz dominance.

References

[1] Barry C. Arnold and D. V. Gokhale, *Lorenz order with common finite support*, Metron **75** (2017), no. 2, 215–226, DOI 10.1007/s40300-016-0101-z. MR3695006

[2] J. L. Gastwirth, *A general definition of the Lorenz curve*, Econometrica 39 (1971), 1037-1039.

[3] G. H. Hardy, J. E. Littlewood, and G. Pólya, *Inequalities*, Cambridge Mathematical Library, Cambridge University Press, Cambridge, 1988. Reprint of the 1952 edition. MR944909

[4] M.O. Lorenz, *Methods of measuring the concentration of wealth*, Publication of the American Statistical Association 9 (1905), 209-219.

[5] Albert W. Marshall, Ingram Olkin, and Barry C. Arnold, *Inequalities: theory of majorization and its applications*, 2nd ed., Springer Series in Statistics, Springer, New York, 2011, DOI 10.1007/978-0-387-68276-1. MR2759813

Department of Statistics

Current address: University of California, Riverside, California

Email address: barry.arnold@ucr.edu

Contemporary Mathematics
Volume **774**, 2021
https://doi.org/10.1090/conm/774/15565

Ergodicity and steady state analysis for interference queueing networks

Sayan Banerjee and Abishek Sankararaman

ABSTRACT. We analyze an interacting queueing network on $\mathbb{Z}^d$ that was introduced in Sankararaman, Baccelli and Foss (2019) as a model for wireless networks. We show that the marginals of the minimal stationary distribution have exponential tails. This is used to furnish asymptotics for the maximum steady state queue length in growing boxes around the origin. We also establish a decay of correlations which shows that the minimal stationary distribution is strongly mixing, and hence, ergodic with respect to translations on $\mathbb{Z}^d$.

1. Introduction and model

In this paper, we consider the Interference Queueing Network model introduced in [**9**]. The model consists of an infinite collection of queues, each placed at a grid point of a d dimensional grid $\mathbb{Z}^d$. Each queue has arrivals according to an independent Poisson process with intensity λ. The departures across queues are however coupled by the *interference* they cause to each other, parametrized by a sequence $\{a_i\}_{i\in\mathbb{Z}^d}$, where $a_i \geq 0$ and $a_i = a_{-i}$, for all $i \in \mathbb{Z}^d$ and $\sum_{i\in\mathbb{Z}^d} a_i < \infty$. For ease of exposition, and without loss of generality, we shall assume that $a_0 = 1$. The state of the network at time $t \in \mathbb{R}$ is encoded by the collection of processes $\{X_i(t)\}_{i\in\mathbb{Z}^d} \in \mathbb{N}_0^{\mathbb{Z}^d}$, where $X_i(t)$ denotes the queue length at site $i \in \mathbb{Z}^d$ at time t. Conditional on the queue lengths $\{X_i(t)\}_{i\in\mathbb{Z}^d}$, the departures across queues are independent with rate of departure from any queue $i \in \mathbb{Z}^d$ at time $t \in \mathbb{R}$ given by $\frac{X_i(t)}{\sum_{j\in\mathbb{Z}^d} a_j X_{i-j}(t)}$. Here, and in the rest of the paper, we adopt the convention that $0/0 = 0$. Under these conditions, Proposition 4.1 of [**9**] gives that the process is well-defined in a path-wise sense, even when the interference sequence has infinite support, namely $a_i > 0$ for infinitely many $i \in \mathbb{Z}^d$. Thus, the evolution of the queues are coupled, in a translation-invariant fashion, where the service rate at a queue is lower if the queue lengths of its neighbors, as measured by the interference sequence $(a_i)_{i\in\mathbb{Z}^d}$, are larger.

2020 *Mathematics Subject Classification.* Primary 60K25, 60K35; Secondary 60B10, 90B18, 28D05.

Key words and phrases. Wireless networks, interference, queues, Coupling-From-The-Past, stationary distribution, strongly mixing, ergodicity.

Most of this work was done when the second author was a PhD student at UT Austin and he thanks François Baccelli for supporting him through the Simons Foundation grant (# 197892) awarded to The University of Texas at Austin. The first author was partially supported by a Junior Faculty Development Award made by UNC, Chapel Hill.

Formally, we work on a probability space containing the collection of processes $(\mathcal{A}_i, \mathcal{D}_i)_{i\in\mathbb{Z}^d}$, where $\{\mathcal{A}_i\}_{i\in\mathbb{Z}^d}$ are independent Poisson Point Processes (PPP) on $\mathbb{R}$ with intensity λ; and $\{\mathcal{D}_i\}_{i\in\mathbb{Z}^d}$ are independent PPP of unit intensity on $\mathbb{R}\times[0,1]$. For each $i \in \mathbb{Z}^d$, the epochs of $\mathcal{A}_i$ denote the instants of arrivals to queue i. Similarly, any atom of the process $\mathcal{D}_i$ of the form $(t,u) \in \mathbb{R}\times[0,1]$ denotes an event of potential departure from queue i; precisely, a departure occurs at time t from queue i if and only if $u \leq \frac{X_i(t)}{\sum_{j\in\mathbb{Z}^d} a_j X_{i-j}(t)}$. Thus, the queue length process $(\{X_i(t)\}_{i\in\mathbb{Z}^d})_{t\in\mathbb{R}}$ is a factor of the driving sequences $(\mathcal{A}_i, \mathcal{D}_i)_{i\in\mathbb{Z}^d}$. A proof of existence of the process is given in [**9**].

This model was introduced in [**9**], as a means to study the dynamics in large scale wireless networks [**8**]. In two or three dimensions, this model has a physical interpretation of a wireless network. Each grid point (queue) represents a 'region of geographical space', and each customer represents a wireless link, i.e., a transmitter-receiver pair. For analytical simplicity, the link length (the distance between transmitter and receiver) is assumed to be 0, so that a single customer represents both a transmitter and receiver. The stochastic system models the spatial birth-death dynamics of the wireless network, where links arrive randomly in space, with the transmitter having an independent file of exponentially distributed size that it wants to communicate to its receiver. A link (customer) subsequently exits the network after the transmitter finishes transmitting the file to its receiver. The duration for which a transmitter transmits (i.e., a customer stays in the network) is governed by the rate at which a transmitter can transmit the file. As wireless is a shared medium, the rate of file transfer at a link depends on the geometry of nearby concurrently transmitting links —if there are a lot of links in the vicinity, i.e., large interference, the rate of file transfer is lowered. In our system, the instantaneous rate of file transfer at a link in queue $i \in \mathbb{Z}^d$ is equal to the Signal to Noise plus Interference Ratio $\frac{1}{\sum_{i\mathbb{Z}^d} a_i x_{i-j}(t)}$. Here, all transmitters transmit at unit power which is received at its corresponding receiver without attenuation (numerator is 1). However, the corresponding receiver also receives power from other neighboring transmitters that reduces the rate of transmission. The interfering power is attenuated through space, with the attenuation factor given by the interference sequence $\{a_i\}_{i\in\mathbb{Z}^d}$. As there are $x_i(t)$ links in queue $i \in \mathbb{Z}^d$, and they all have independent file sizes, the total rate of departure at a queue is then $\frac{x_i(t)}{\sum_{j\in\mathbb{Z}^d} a_j x_{i-j}(t)}$. We refer the reader to [**9**], [**8**], [**10**] for more information on the origin of this stochastic model and its applications to understanding wireless networks.

Mathematically, this model lies at the interface between queueing networks and interacting particle systems. Most well known queueing networks with interactions between servers, like the Join-the-shortest-queue policy and Power-of-d-choices policy [**7, 11**], incorporate global interactions between servers and the interaction between any two fixed servers approaches zero (in a suitable sense) as the system size increases. On the other hand, well known interacting particle systems like the exclusion process, zero range process, contact process, voter model, Ising model, etc., [**5**] have strong nearest neighbor interactions but they often have explicit stationary measures and/or locally compact state space (each site can take one of finitely many values/configurations). This model has nearest neighbor interactions as well as locally non-compact state space (queue lengths are unbounded), thus making many tools from either of the above two broad fields inapplicable. In particular, stationary measures, if they exist, are far from explicit and natural aspects of the stationary dynamics of the process, like uniqueness of stationary measure, decay of correlations, typical and extremal behavior of queue lengths, and convergence rates to stationarity from arbitrary initial configurations, are non-trivial to analyze and quantify. Moreover, the ratio-type functional dependence of the service rates on neighboring queues makes obtaining quantitative estimates challenging and most of the analysis necessarily has to rely on 'soft' arguments using qualitative traits of the model. Recently, motivated by this model, the first author revisited an interacting particle system called the Potlatch process [**4**], which shares many aspects in common with this model, but the simpler functional form of rates enables one (see [**3**]) to quantify rates of convergence (locally and globally) to equilibrium. Similar models have also appeared in the economics literature to analyze opinion dynamics on social networks [**1**].

The paper [**9**] established stability criteria, namely that if $\lambda < \frac{1}{\sum_{j\in\mathbb{Z}^d} a_j}$, then there exists a translation invariant (on $\mathbb{Z}^d$) stationary distribution for the queue lengths. The crucial property of the dynamics noted in [**9**] was the following form of monotonicity: if at time $t \in \mathbb{R}$, there are two initial configurations $\{X_i(t)\}_{i\in\mathbb{Z}^d}$ and $\{X_i^{'}(t)\}_{i\in\mathbb{Z}^d}$ such that for all $i \in \mathbb{Z}^d$, $X_i(t) \leq X_i^{'}(t)$ (assuming the system starts at time $t \in \mathbb{R}$), and if the processes $\{X_i(s) : i \in \mathbb{Z}^d, s \geq t\}$ and $\{X_i^{'}(s) : i \in \mathbb{Z}^d, s \geq t\}$ are constructed using the same arrival and departure PPP $(\mathcal{A}_i, \mathcal{D}_i)_{i\in\mathbb{Z}^d}$, then under this coupling, almost surely, for all $s \geq t$ and all $i \in \mathbb{Z}^d$, $X_i(s) \leq X_i^{'}(s)$. Monotonicity is then used to define the following notion of stability. For each $t \geq 0$ and $s \geq -t$, denote by $\{X_{i;t}(s)\}_{i\in\mathbb{Z}^d} \in \mathbb{N}_0^{\mathbb{Z}^d}$ the queue lengths at time s, when the system was started with all queues being empty at time $-t$, i.e., for all $i \in \mathbb{Z}^d$, $X_{i;t}(-t) = 0$. Monotonicity implies that under the above (synchronous) coupling, such that, almost surely, for all $s \in \mathbb{R}$, for all $i \in \mathbb{Z}^d$, the map $t \mapsto X_{i;t}(s)$ is non-decreasing. The stationary version of the process is then defined as $\{X_{i;\infty}(s)\}_{i\in\mathbb{Z}^d}$, where for any $s \in \mathbb{R}$ and $i \in \mathbb{Z}^d$, $X_{i;\infty}(s) := \lim_{t\to\infty} X_{i;t}(s)$ in the almost sure sense. It was shown in [**9**] (see Proposition 4.3 there) that $\{X_{i;\infty}(s)\}_{i\in\mathbb{Z}^d}$ is indeed a stationary solution to the dynamics which is *minimal* in the sense that any other stationary solution stochastically dominates this solution in a coordinate-wise sense for all time. We will refer to this coupled 'backward' construction of the process $\{X_{i;t}(s)\}_{i\in\mathbb{Z}^d} : s \geq -t\}$ (for $t \geq 0$), as well as $\{X_{i;\infty}(s)\}_{i\in\mathbb{Z}^d} : s \in \mathbb{R}\}$, as the "Coupling-From-The-Past" (CFTP) construction.

In the rest of the paper, we shall assume that $\lambda < \frac{1}{\sum_{j\in\mathbb{Z}^d} a_j}$ and that the process $\{X_i(t) : i \in \mathbb{Z}^d, t \in \mathbb{R}\}$ is stationary and distributed according to the unique minimal stationary solution to the dynamics. Proposition 4.3 in [**9**] gives that for any $i \in \mathbb{Z}^d$ and $t \in \mathbb{R}$, the steady state queue length satisfies $\mathbb{E}[X_i(t)] = \frac{\lambda}{1-\lambda\sum_{j\in\mathbb{Z}^d} a_j}$. Subsequently, [**10**] established that for all $\lambda < \frac{1}{\sum_{j\in\mathbb{Z}^d} a_j}$, for all $i \in \mathbb{Z}^d$, $t \in \mathbb{R}$, $\mathbb{E}[(X_i(t))^2] < \infty$.

In this paper, we show that the marginals of the minimal stationary distribution, in fact, has exponential tails (Theorem 2.1). This is used to obtain asymptotics for the maximum queue length in steady state in growing boxes around the origin (Corollary 2.2). We further show a decay of correlations between the queue lengths of two sites as the distance between the sites increases (Theorem 2.3). This, in turn, implies that the stationary distribution is strongly mixing, and thus, ergodic with respect to translations on $\mathbb{Z}^d$. An ergodic theorem is presented in Corollary 2.5.

2. Main results

2.1. Exponential moments and stationary distribution tail bounds. The first result concerns the existence of exponential moments for queue lengths which, in turn, yields two-sided exponential tail bounds on the marginals of the minimal stationary distribution.

THEOREM 2.1. *For all $\lambda < \frac{1}{\sum_{j\in\mathbb{Z}^d} a_j}$, there exists a constant $c_0 > 0$, such that for all $c \in [0, c_0)$, all $i \in \mathbb{Z}^d$ and $t \in \mathbb{R}$,*

$$\mathbb{E}[e^{cX_i(t)}] < \infty. \tag{2.1}$$

Moreover, for all $\lambda < \frac{1}{\sum_{j\in\mathbb{Z}^d} a_j}$, there exist constants $c_1, c_2, x_0 > 0$, such that, for all $x \geq x_0$, $i \in \mathbb{Z}^d$ and $t \in \mathbb{R}$,

$$e^{-c_1 x} \leq \mathbb{P}[X_i(t) \geq x] \leq e^{-c_2 x}. \tag{2.2}$$

The above theorem can be used to derive the following asymptotics for the maximum queue length in steady state in growing boxes around the origin.

COROLLARY 2.2. *For every $\lambda < \frac{1}{\sum_{j\in\mathbb{Z}^d} a_j}$, there exist positive constants C_1, C_2, such that for any $t \in \mathbb{R}$,*

$$\lim_{N\to\infty} \mathbb{P}\left(C_1 \log N \leq \max_{i\in\mathbb{Z}^d:\|i\|_\infty \leq N} X_i(t) \leq C_2 \log N\right) = 1.$$

Theorem 2.1 and Corollary 2.2 are proved in Section 3.

2.2. Correlation decay and mixing of the stationary queue length process. The main result of this section shows that the stationary queue lengths at distinct sites show a decay of correlations in space as the distance between the sites increases. This, in fact, shows that the minimal stationary distribution is strongly mixing, and thus, ergodic with respect to translations on $\mathbb{Z}^d$. Our subsequent goal, which we will address in a future article, is to understand the quantitative decay of correlations in the system and its sensitivity on the interference sequence and the underlying dimension. Thus, we take a constructive approach to showing the decay of correlations.

Before stating the results, we briefly recall some notions from ergodic theory.

Let $\{X_i\}_{i\in\mathbb{Z}^d} \in \mathbb{N}_0^{\mathbb{Z}^d}$ be a sample from the minimal stationary distribution of the dynamics. The law of $\mathbf{X} := \{X_i\}_{i\in\mathbb{Z}^d}$ induces a natural probability measure $\boldsymbol{\mu}$ on $\left(\mathbb{N}_0^{\mathbb{Z}^d}, \mathcal{B}\left(\mathbb{N}_0^{\mathbb{Z}^d}\right)\right)$ given by $\boldsymbol{\mu}(A) := \mathbb{P}(\mathbf{X} \in A), A \in \mathcal{B}\left(\mathbb{N}_0^{\mathbb{Z}^d}\right)$. For any $n \in \mathbb{N}$, and $h \in \{1, \cdots, d\}$, define $n\mathbf{e}_h := (\underbrace{0, \cdots, 0}_{h-1}, n, \underbrace{0, \cdots, 0}_{d-h})$, namely the vector in $\mathbb{Z}^d$ of all 0's except the hth coordinate that takes value n. For $h \in \{1, \cdots, d\}$ and $i \in \mathbb{Z}^d$, let $\theta_h(i) := i + \mathbf{e}_h$ denote the unit translation map on $\mathbb{Z}^d$ along the h-th coordinate. Denote the associated transformation on $\mathbb{N}_0^{\mathbb{Z}^d}$ by $T_h(\mathbf{x}) := \mathbf{x} \circ \theta_h, \mathbf{x} := (x_i)_{i\in\mathbb{Z}^d} \in \mathbb{N}_0^{\mathbb{Z}^d}$, where $(\mathbf{x} \circ \theta_h)_i := x_{\theta_h(i)}, i \in \mathbb{Z}^d$. By the translation invariance of the dynamics, $\boldsymbol{\mu} \circ T_h^{-1} = \boldsymbol{\mu}$ for any $h \in \{1, \cdots d\}$. For any $h \in \{1, \cdots, d\}$, the quadruple $\mathcal{Q}_h := \left(\mathbb{N}_0^{\mathbb{Z}^d}, \mathcal{B}\left(\mathbb{N}_0^{\mathbb{Z}^d}\right), \boldsymbol{\mu}, T_h\right)$ is referred to as a probability preserving transformation (ppt). Recall that $\mathcal{Q}_h$ is called *strongly mixing* if for any $A, B \in \mathcal{B}\left(\mathbb{N}_0^{\mathbb{Z}^d}\right)$,

$$\lim_{n\to\infty} \boldsymbol{\mu}\left(A \cap T_h^{-n}B\right) = \boldsymbol{\mu}(A)\boldsymbol{\mu}(B), \tag{2.3}$$

where, for $n \in \mathbb{N}$, $T_h^{-n}(\cdot)$ is the map on $\mathbb{N}_0^{\mathbb{Z}^d}$ obtained by composing $T_h^{-1}(\cdot)$ n times. A set $A \in \mathcal{B}\left(\mathbb{N}_0^{\mathbb{Z}^d}\right)$ is called *invariant* under the family of transformations $\{T_h\}_{h=1}^d$ if $T_h^{-1}A = A$ for all $h \in \{1, \cdots, d\}$. The family $\{T_h\}_{h=1}^d$ is called *ergodic* if all invariant sets are trivial, that is, for any A invariant, $\boldsymbol{\mu}(A) = 0$ or 1. One can show that (see for eg. [**2**]), if $\mathcal{Q}_h$ is strongly mixing for each $h \in \{1, \cdots, d\}$, then the family $\{T_h\}_{h=1}^d$ is ergodic.

For any $K \in \mathbb{N}_0$, define $\mathbf{X}_{0,K} := \{X_i : i \in \mathbb{Z}^d, \|i\|_\infty \le K\}$, thought of as a random variable in $\mathbb{N}_0^{(2K+1)^d}$. Similarly, for $n \in \mathbb{N}$, $K \in \mathbb{N}_0$ and $h \in \{1, \cdots, d\}$, define $\mathbf{X}_{n\mathbf{e}_h,K} := \{X_i : i \in \mathbb{Z}^d, \|i - n\mathbf{e}_h\|_\infty \le K\}$.

THEOREM 2.3. *Fix any $K \in \mathbb{N}_0$, $h \in \{1, \cdots, d\}$ and $0 \le \lambda < \frac{1}{\sum_{j\in\mathbb{Z}^d} a_j}$. Let f, g be functions from $\mathbb{N}_0^{(2K+1)^d}$ to $\mathbb{R}$ such that $\mathbb{E}[f^2(\mathbf{X}_{0,K})] < \infty$ and $\mathbb{E}[g^2(\mathbf{X}_{0,K})] < \infty$. The following limit exists:*

$$\lim_{n\to\infty} \mathbb{E}[f(\mathbf{X}_{0,K})g(\mathbf{X}_{n\mathbf{e}_h,K})] = \mathbb{E}[f(\mathbf{X}_{0,K})]\mathbb{E}[g(\mathbf{X}_{0,K})]. \tag{2.4}$$

In particular, $\mathcal{Q}_h$ is strongly mixing for any $h \in \{1, \cdots, d\}$. Hence, the family $\{T_h\}_{h=1}^d$ is ergodic.

An immediate corollary is the following explicit formula for the asymptotic covariances of the stationary queue length processes.

COROLLARY 2.4.

$$\lim_{n\to\infty} \mathbb{E}[X_0 X_{n\mathbf{e}_h}] = (\mathbb{E}[X_0])^2 = \left(\frac{\lambda}{1 - \lambda\sum_{j\in\mathbb{Z}^d} a_j}\right)^2.$$

PROOF. Applying Theorem 2.3 with $K = 0$ and $f(\ell) = g(\ell) = \ell, \ell \in \mathbb{N}$, and using Proposition 4.3 of [**9**], yields this result. □

The ergodicity established in Theorem 2.3 directly implies the following version of the ergodic theorem. A sequence of finite subsets $\{F_r : r \in \mathbb{N}\}$ of $\mathbb{Z}^d$ with $\cup_{r\in\mathbb{N}} F_r = \mathbb{Z}^d$ is said to be a *Følner sequence* if

$$\lim_{r\to\infty} \frac{|(\theta_h F_r)\Delta F_r|}{|F_r|} = 0,$$

where $\theta_h F_r := \{f + \mathbf{e}_h : f \in F_r\}$ and $|\cdot|$ denotes set cardinality. The Følner sequence $\{F_r : r \in \mathbb{N}\}$ of $\mathbb{Z}^d$ is called *tempered* if there exists $C > 0$ such that for all $r \in \mathbb{N}$,

$$\left|\bigcup_{u<r} F_u^{-1} F_r\right| \le C|F_r|.$$

where $F_u^{-1}F_r = \{i - j : i \in F_r, j \in F_u\}$. For example, consider the sequence of boxes defined by

$$B_r = \{k = (k_1, \cdots, k_d) \in \mathbb{Z}^d : i_l^{(r)} \leq k_l < j_l^{(r)} \text{ for all } 1 \leq l \leq d\}, \ i^{(r)}, j^{(r)} \in \mathbb{Z}^d.$$

If $B_r \subset B_{r+1}$ for all $r \in \mathbb{N}$ and $\cup_{r\in\mathbb{N}} B_r = \mathbb{Z}^d$, then $\{B_r : r \in \mathbb{N}\}$ forms a tempered Følner sequence with $C = 2^d$. The next corollary follows from ergodic theorems for amenable groups [**6**, Theorems 1.1,1.3] upon noting that $\mathbb{Z}^d$ is an amenable (additive) group. For $i \in \mathbb{Z}^d$ and $\mathbf{x} \in \mathbb{N}_0^{\mathbb{Z}^d}$, define $\mathbf{x}_{(i)} \in \mathbb{N}_0^{\mathbb{Z}^d}$ by $(\mathbf{x}_{(i)})_l := \mathbf{x}_{l+i}, l \in \mathbb{Z}^d$.

COROLLARY 2.5. *For any $f \in L^1(\mu)$ and any Følner sequence $\{F_r : r \in \mathbb{N}\}$,*

$$\lim_{r\to\infty} \frac{1}{|F_r|} \sum_{i\in F_r} f\left(\mathbf{X}_{(i)}\right) = \int f(\mathbf{x})\mu(d\mathbf{x})$$

in L^1 with respect to μ. In addition, the above convergence holds almost surely with respect to μ if $\{F_r : r \in \mathbb{N}\}$ is a tempered Følner sequence (eg. $\{B_r : r \in \mathbb{N}\}$ defined above).

Theorem 2.3 is proved in Section 4.

3. Proof of Theorem 2.1 and Corollary 2.2

3.1. Proof of the Lower Bound in (2.2)**.** The proof of the lower bound in (2.2) follows from a coupling argument, such that all queue lengths in the process are bounded from below by a collection of independent $M/M/1$ queues at each of the sites. We shall construct a coupling using the same driving sequence $(\mathcal{A}_i, \mathcal{D}_i)_{i\in\mathbb{Z}^d}$. For each $t \geq 0$, define the collection of independent $M/M/1$ queues $\{Q_{i;t}(s) : s \geq -t, i \in \mathbb{Z}^d\}$ such that for each $i \in \mathbb{Z}^d$, arrival happens at epochs of $\mathcal{A}_i$ and departure happens at epochs of $\mathcal{D}_i$ (that is, for any atom (s', u') of $\mathcal{D}_i$, $s' \geq -t$, a departure happens from the queue $Q_{i;t}$ at time s'). The process $\{Q_{i;t}(s) : s \geq -t, i \in \mathbb{Z}^d\}$ is started empty at time $-t$, namely, almost-surely, $Q_{i;t}(-t) = 0$, for all $i \in \mathbb{Z}^d$. Now, construct the process $\{X_{i;t}(s) : s \geq -t, i \in \mathbb{Z}^d\}$ using the same driving processes $(\mathcal{A}_i, \mathcal{D}_i)_{i\in\mathbb{Z}^d}$ using the CFTP recipe in Section 1. Then, upon noting that for each $s \geq -t$, $\frac{X_i(s)}{\sum_{j\in\mathbb{Z}^d} a_j X_{i-j}(s)} \leq 1$, it follows that under this coupling, almost surely, $Q_{i;t}(s) \leq X_{i;t}(s)$ for all $s \geq -t$. For each $s \in \mathbb{R}$ and $i \in \mathbb{Z}^d$, by monotonicity of $Q_{i;t}(s)$ in t, which follows similarly as for $X_{i;t}(s)$, $Q_{i;t}(s)$ converges almost surely to a random variable $Q_{i,\infty}(s)$ as $t \to \infty$, where $Q_{i,\infty}(s)$ is distributed according to the stationary distribution of an $M/M/1$ queue with arrival rate λ and departure rate 1. Hence, by the coupling, for each $s \in \mathbb{R}$ and $i \in \mathbb{Z}^d$, $X_{i,\infty}(s)$ stochastically dominates $Q_{i,\infty}(s)$. The result now follows upon noting that the stationary queue length for a $M/M/1$ queue (with arrival rate less than departure rate) has exponential tails [**2**].

3.2. Proof of (2.1)**and the upper bound in** (2.2)**.** We shall prove the exponential moment bound (2.1), which automatically implies the upper bound in (2.2).

In order to proceed with the proof, we will need to set some notations. For any $L \in \mathbb{N}$, denote by $\{X_i^L(t) : t \in \mathbb{R}, i \in \mathbb{Z}^d\}$ the stationary process (distributed according to the minimal stationary regime) of the dynamics described in Section 1 constructed using a *truncated* interference sequence $(a_i^L := a_i \mathbf{1}_{\|i\|_\infty \leq L})_{i\in\mathbb{Z}^d}$. One can consider a natural coupling of all the processes $\{(X_i^L(t), X_i(t)), t \in \mathbb{R}, i \in \mathbb{Z}^d, L \in \mathbb{N}\}$, where the lack of superscript denotes the dynamics without any truncation of the interference sequence, such that, *(i)* for each $L \in \mathbb{N}$, the process $\{X_i^L(t) : t \in \mathbb{R}, i \in \mathbb{Z}^d\}$ is stationary, distributed according to its minimal stationary solution, *(ii)* the process $\{X_i(t) : t \in \mathbb{R}, i \in \mathbb{Z}^d\}$ is stationary distributed according to its minimal stationary solution, and *(iii)* almost-surely, for all $t \in \mathbb{R}$, $i \in \mathbb{Z}^d$ and $L_1 \leq L_2$, $X_i^{L_1}(t) \leq X_i^{L_2}(t) \leq X_i(t)$. This coupling can be constructed by using the same driving sequence $(\mathcal{A}_i, \mathcal{D}_i)_{i \in \mathbb{Z}^d}$ to construct all the processes involved using the CFTP recipe given in Section 1. Moreover, from the proof of Proposition 4.3 of [**9**], almost surely, for all $t \in \mathbb{R}$, and $i \in \mathbb{Z}^d$, $X_i^L(t) \nearrow X_i(t)$ as $L \nearrow \infty$.

We will also need other modifications to the dynamics to restrict it on the torus, in order to carry out the proof. For any $n \in \mathbb{N}$, let $\mathbb{B}_n := [-n, n]^d \subset \mathbb{Z}^d$ denote the rectangle of side length $2n$ centered around the origin and let $\mathbb{T}_n$ denote the n-torus in $\mathbb{Z}^d$ seen as $\mathbb{B}_n$ with opposite faces identified. For any $L \in \mathbb{N}$, denote by $\{Y_i^{(n),L}(\cdot)\}_{i \in \mathbb{Z}_d}$ the process described in Section 1 whose dynamics are *restricted* to the torus $\mathbb{T}_n$ as follows. For each $i \in \mathbb{B}_n$, the arrival process into the queue $Y_i^{(n),L}$ follows the PPP $\mathcal{A}_i$ as before, but departure happens at rate $\frac{Y_i^{(n),L}(t)}{\sum_{j \in \mathbb{Z}^d} a_j^L Y_{(i-j) \bmod \mathbb{B}_n}^{(n),L}(t)}$ at time t. Set $Y_i^{(n),L}(\cdot) \equiv 0$ if $i \notin \mathbb{B}_n$. We define another process $(Z_i^{(n),L}(\cdot))_{i \in \mathbb{B}_n}$ to be the process described in Section 1, but with the dynamics restricted to the set $\mathbb{B}_n$ with the edge effects. More precisely, set $Z_i^{(n),L}(\cdot) \equiv 0$ if $i \notin \mathbb{B}_n$, and for each $i \in \mathbb{B}_n$, the arrival process into the queue $Z_i^{(n),L}$ follows the PPP $\mathcal{A}_i$ as before, but departure happens at rate $\frac{Z_i^{(n),L}(t)}{\sum_{j \in \mathbb{Z}^d} a_j^L Z_{i-j}^{(n),L}(t)}$ at time t.

From the monotonicity in the dynamics, there exists a coupling of all the processes (described as before using common PPP $\mathcal{A}_i$ and $\mathcal{D}_i$ respectively for arrival and departure) such that, they are all individually stationary and almost surely, for all n, L, all $i \in \mathbb{B}_n$ and all $t \in \mathbb{R}$, $X_i^L(t) \geq Z_i^{(n),L}(t)$ and $Y_i^{(n),L}(t) \geq Z_i^{(n),L}(t)$.

The main results established in [**9**] (see Theorem 1.1, Theorem 5.2, and Remark 5.5 there) show that if $\lambda < \frac{1}{\sum_{j \in \mathbb{Z}^d} a_j}$, then for all L and $n > L$, the two processes $\{Y_i^{(n),L}(\cdot)\}_{i \in \mathbb{B}_n}$ and $\{Z_i^{(n),L}(\cdot)\}_{i \in \mathbb{B}_n}$ admit a unique stationary solution and the process $\{X_i^L(\cdot)\}_{i \in \mathbb{Z}^d}$ has a non-trivial minimal stationary solution. From monotonicity, one can construct a coupling of the processes $\{(X_i^L(\cdot), Y_i^{(n),L}(\cdot), Z_i^{(n),L}(\cdot)) : i \in \mathbb{Z}^d$, and $n, L \in \mathbb{N}\}$ such that, they are all individually stationary (with $\{X_i^L(t)\}_{i \in \mathbb{Z}^d}$ having the minimal stationary distribution for every t, L) and, almost surely,

- For each fixed L and $n > L$, $Y_i^{(n),L}(t) \geq Z_i^{(n),L}(t)$, for all $i \in \mathbb{B}_n$ and all $t \in \mathbb{R}$,
- $n \to Z_0^{(n),L}(t)$ is non-decreasing, for all $t \in \mathbb{R}$,
- $\lim_{n \to \infty} Z_0^{(n),L}(t) = X_0^L(t)$, for all $t \in \mathbb{R}$,
- $L \to X_i^L(t)$ is non-decreasing and $X_i^L(t) \nearrow X_i(t)$ as $L \to \infty$, for all $i \in \mathbb{Z}^d$ and all $t \in \mathbb{R}$.

The third property above follows from Proposition 7.3 of [**9**]. The fourth property follows from monotonicity and the proof of Proposition 4.3 of [**9**]. In the rest of the proof, we shall assume that $\lambda < \frac{1}{\sum_{j \in \mathbb{Z}^d} a_j}$ and the processes $\{X_i^L(\cdot)\}_{i \in \mathbb{Z}^d}$,

$\{Y_i^{(n),L}(\cdot)\}_{i\in\mathbb{B}_n}$ and $\{Z_i^{(n),L}(\cdot)\}_{i\in\mathbb{B}_n}$ are all individually stationary and satisfy the above properties. The following is the key technical result needed for the proof of (2.1).

PROPOSITION 3.1. *Let* $\lambda < \frac{1}{\sum_{j\in\mathbb{Z}^d} a_j}$ *and* $c_0 > 0$ *such that* $c_0 e^{c_0} = \frac{\frac{1}{\sum_{j\in\mathbb{Z}^d} a_j} - \lambda}{\lambda+1} =: D$. *Then for all* $c \in [0, c_0)$, $L \in \mathbb{N}$ *and* $n > L$, $\mathbb{E}[e^{cY_0^{(n),L}(0)}] \le \frac{D}{D-ce^c} < \infty$.

Before giving a proof of the above proposition, we shall see how this concludes the proof of the exponential moment bound (2.1), and thus the upper bound in (2.2).

PROOF OF (2.1). By the first property above, almost surely, for all $i \in \mathbb{B}_n, L \in \mathbb{N}$ and $t \in \mathbb{R}$, we have, $Y_i^{(n),L}(t) \ge Z_i^{(n),L}(t)$. Thus, Proposition 3.1 implies that, for all $0 \le c < c_0$,

$$\sup_{n>L} \mathbb{E}[e^{cZ_0^{(n),L}(t)}] \le \sup_{n>L} \mathbb{E}[e^{cY_0^{(n),L}(t)}] \le \frac{D}{D-ce^c}.$$

As, almost surely, for any $t \in \mathbb{R}$, $L \in \mathbb{N}$, $n \to Z_0^{(n),L}(t)$ is non-decreasing and $\lim_{n\to\infty} Z_0^{(n),L}(t) = X_0^L(t)$, monotone convergence theorem establishes that, for all $c \in [0, c_0)$,

$$\mathbb{E}[e^{cX_0^L(t)}] = \lim_{n\to\infty} \mathbb{E}[e^{cZ_0^{(n),L}(t)}] \le \sup_{n>L} \mathbb{E}[e^{cZ_0^{(n),L}(t)}] \le \sup_{n>L} \mathbb{E}[e^{cY_0^{(n),L}(t)}] \le \frac{D}{D-ce^c}.$$

Since, for each $L \in \mathbb{N}$, the process $\{X_i^L(\cdot) : i \in \mathbb{Z}^d\}$ is stationary and, almost surely, for any $t \in \mathbb{R}$, $i \in \mathbb{Z}^d$, $X_i^L(t) \nearrow X_i(t)$ as $L \to \infty$, and $\sup_{L\in\mathbb{N}} \mathbb{E}[e^{cX_0^L(t)}] \le \frac{D}{D-ce^c} < \infty$, yet another application of the monotone convergence theorem yields that $\mathbb{E}[e^{cX_0(t)}] = \lim_{L\to\infty} \mathbb{E}[e^{cX_0^L(t)}] \le \frac{D}{D-ce^c} < \infty$. □

We set some notation and state two technical lemmas before proving Proposition 3.1. We will fix a $L \in \mathbb{N}$ and drop the superscript L notation to lighten the notational burden. For each $n \in \mathbb{N}$ and $k \ge 1$, denote by $\mu_k^{(n)} := \mathbb{E}[(Y_0^{(n)}(0))^k]$, recalling that $\{Y_i^{(n)}(0)\}_{i\in\mathbb{B}_n}$ is distributed according to the stationary distribution of the process $\{Y_i^{(n)}(\cdot)\}_{i\in\mathbb{B}_n}$. Observe that Theorem 5.2 from [**9**] immediately yields that for all $n \in \mathbb{N}$ and all $k \ge 1$, $\mu_k^{(n)} < \infty$ since $\lambda < \frac{1}{\sum_{j\in\mathbb{Z}^d} a_j}$. We state two useful lemmas.

LEMMA 3.2. *Let* $(y_i)_{i\in\mathbb{B}_n}$ *be any non-negative sequence of real numbers. For any* $i \in \mathbb{B}_n$, *define* $R_i := \frac{y_i}{\sum_{k\in\mathbb{Z}^d} a_k y_{(i-k) \bmod \mathbb{B}_n}}$. *Then, for all* $j \ge 1$,

$$\sum_{i\in\mathbb{B}_n} R_i y_i^j \ge \frac{1}{\sum_{k\in\mathbb{Z}^d} a_k} \sum_{i\in\mathbb{B}_n} y_i^j.$$

LEMMA 3.3. *For all* $n \in \mathbb{N}$ *and* $k \ge 1$,

$$D(k+1)\mu_k^{(n)} \le \sum_{j=0}^{k-1} \binom{k+1}{j} \mu_j^{(n)}, \tag{3.1}$$

where D *is given in Proposition* 3.1.

Before proving the above two lemmas, we use them to prove Proposition 3.1.

Proof of Proposition 3.1. Let $n \in \mathbb{N}$ be arbitrary and fixed. Let $c_0 > 0$ be such that $c_0 e^{c_0} = D$, where D is defined in Proposition 3.1, and fix any $0 \leq c < c_0$. Let $m \geq 1$ be arbitrary. For $k \geq 1$, by multiplying both sides of equation (3.1) by $\frac{c^k}{k!}$,

$$D(k+1)\mu_k^{(n)}\frac{c^k}{k!} \leq \sum_{j=0}^{k-1}\binom{k+1}{j}\mu_j^{(n)}\frac{c^k}{k!}.$$

Simplifying, we obtain

$$D\mu_k^{(n)}\frac{c^k}{k!} \leq \sum_{j=0}^{k-1}\frac{1}{j!(k+1-j)!}c^k\mu_j^{(n)}.$$

For $m \in \mathbb{N}$, summing both sides from $k = 1$ through to m,

$$\begin{aligned}
D\sum_{k=1}^{m}\mu_k^{(n)}\frac{c^k}{k!} &\leq \sum_{k=1}^{m}\sum_{j=0}^{k-1}\frac{1}{j!(k+1-j)!}c^k\mu_j^{(n)} \\
&\stackrel{(a)}{=} \sum_{j=0}^{m-1}\frac{\mu_j^{(n)}}{j!}\sum_{k=j+1}^{m}\frac{1}{(k+1-j)!}c^k \\
&\stackrel{(b)}{=} \sum_{j=0}^{m-1}\frac{\mu_j^{(n)}}{j!}\sum_{u=0}^{m-j-1}\frac{c^{u+j+1}}{(u+2)!} \\
&\leq \sum_{j=0}^{m-1}\frac{\mu_j^{(n)}}{j!}c^{j+1}\sum_{u=0}^{\infty}\frac{c^u}{(u+2)!} \\
&\leq c\sum_{j=0}^{m-1}\frac{c^j\mu_j^{(n)}}{j!}\sum_{u=0}^{\infty}\frac{c^u}{u!} \\
&\leq c\sum_{j=0}^{m}\frac{c^j\mu_j^{(n)}}{j!}e^c. \qquad (3.2)
\end{aligned}$$

Step (a) follows from swapping the order of summations. Step (b) follows from the substitution $u = k - j - 1$. Define $S_m^{(n)} := \sum_{j=0}^{m}\frac{c^j\mu_j^{(n)}}{j!}$. Observe that $\mu_k^{(n)} \geq 0$, for all $k \geq 0$ and $n \in \mathbb{N}$, and thus $S_m^{(n)}$ is non-decreasing in m and the (possibly infinite) limit $\lim_{m\to\infty} S_m^{(n)}$ exists. The calculation in (3.2) yields that $D(S_m^{(n)} - 1) \leq ce^c S_m^{(n)}$, which on re-arranging yields that $S_m^{(n)} \leq \frac{D}{D-ce^c} < \infty$. Taking a limit in m, we see that $\lim_{m\to\infty} S_m^{(n)} \leq \frac{D}{D-ce^c} < \infty$. Thus, from Taylor's expansion and monotone convergence theorem, we have that $\mathbb{E}[e^{cY_0^{(n)}}] = \lim_{m\to\infty} S_m^{(n)} \leq \frac{D}{D-ce^c} < \infty$. Since the bound does not depend on n, and $n \in \mathbb{N}$ and $c \in [0, c_0)$ are arbitrary, the proof is concluded. □

We now give the proof of Lemma 3.2.

PROOF OF LEMMA 3.2. By a direct application of Cauchy-Schwartz inequality, we have

$$\left(\sum_{i\in\mathbb{B}_n} y_i^j\right)^2 \leq \left(\sum_{i\in\mathbb{B}_n} R_i y_i^j\right)\left(\sum_{i\in\mathbb{B}_n} \frac{y_i^j}{R_i}\right),$$

where recall that $0/0$ in the RHS is interpreted as 0. It thus suffices from the above bound to establish that $\sum_{i\in\mathbb{B}_n} \frac{y_i^j}{R_i} \leq (\sum_{k\in\mathbb{Z}^d} a_k) \sum_{i\in\mathbb{B}_n} y_i^j$. We do this as follows.

$$\begin{aligned}
\sum_{i\in\mathbb{B}_n} \frac{y_i^j}{R_i} &= \sum_{i\in\mathbb{B}_n} y_i^{j-1}\left(\sum_{k\in\mathbb{Z}_d} a_k y_{(i-k) \mod \mathbb{B}_n}\right) \\
&\overset{(a)}{\leq} \sum_{i\in\mathbb{B}_n}\sum_{k\in\mathbb{Z}^d} a_k \left(\frac{j-1}{j} y_i^j + \frac{1}{j} y^j_{(i-k) \mod \mathbb{B}_n}\right) \\
&= \frac{j-1}{j}\sum_{k\in\mathbb{Z}^d} a_k \sum_{i\in\mathbb{B}_n} y_i^j + \frac{1}{j}\sum_{k\in\mathbb{Z}^d} a_k \sum_{i\in\mathbb{B}_n} y^j_{(i-k) \mod \mathbb{B}_n} \\
&\overset{(b)}{=} \frac{j-1}{j}\sum_{k\in\mathbb{Z}^d} a_k \sum_{i\in\mathbb{B}_n} y_i^j + \frac{1}{j}\sum_{k\in\mathbb{Z}^d} a_k \sum_{i\in\mathbb{B}_n} y_i^j \\
&= \sum_{k\in\mathbb{Z}^d} a_k \sum_{i\in\mathbb{B}_n} y_i^j.
\end{aligned}$$

Step (a) follows from Young's inequality that for any $a, b \geq 0$, we have $ab \leq \frac{a^p}{p} + \frac{b^q}{q}$, for any $p, q > 0$ such that $p^{-1} + q^{-1} = 1$. Thus, $a^{j-1}b \leq (j-1)\frac{a^j}{j} + \frac{b^j}{j}$, where we set $p = \frac{j}{j-1}$ and $q = j$. Inequality (b) follows from the observation that, by translational symmetry of the torus, $\sum_{i\in\mathbb{B}_n} y^j_{(i-k) \mod \mathbb{B}_n} = \sum_{i\in\mathbb{B}_n} y_i^j$, for all $k \in \mathbb{Z}^d$. □

We are now ready to prove Lemma 3.3.

PROOF OF LEMMA 3.3. Let $\{Y_i^{(n)}\}_{i\in\mathbb{Z}^d}$ be a collection of random variables sampled from the (unique) stationary distribution of $\{Y_i^{(n)}(\cdot)\}_{i\in\mathbb{B}_n}$. For brevity, we shall drop the n superscript and write $Y_i := Y_i^{(n)}$ for all $i \in \mathbb{B}_n$, and $\mu_k^{(n)} = \mu_k$, for all $k \geq 0$, as n is fixed throughout the proof. We apply the rate-conservation equation to the Lyapunov function $V(\mathbf{y}) = \sum_{i\in\mathbb{B}_n} (y_i)^{k+1}$, writing $\mathbf{y} := \{y_i\}_{i\in\mathbb{B}_n}$. Since $\{Y_i\}_{i\in\mathbb{B}_n}$ is stationary, we have that $\mathbb{E}\left(\mathcal{L}V(\mathbf{Y})\right) = 0$, where $\mathcal{L}$ is the generator of the continuous time Markov process corresponding to our dynamics. This in particular yields that

$$\begin{aligned}
0 &= \lambda \sum_{i\in\mathbb{B}_n} \mathbb{E}[((Y_i+1)^{k+1} - Y_i^{k+1})] + \sum_{i\in\mathbb{B}_n} \mathbb{E}[R_i((Y_i-1)^{k+1} - Y_i^{k+1})] \\
&= \lambda \sum_{i\in\mathbb{B}_n}\sum_{j=0}^{k} \binom{k+1}{j} \mathbb{E}[Y_i^j] + \sum_{i\in\mathbb{B}_n}\sum_{j=0}^{k} \binom{k+1}{j} \mathbb{E}[R_i Y_i^j (-1)^{k+1-j}] \\
&= \sum_{i\in\mathbb{B}_n} (k+1)(\lambda\mathbb{E}[Y_i^k] - \mathbb{E}[R_i Y_i^k]) + \sum_{i\in\mathbb{B}_n}\sum_{j=0}^{k-1} \binom{k+1}{j} \mathbb{E}[(\lambda + R_i(-1)^{k+1-j})Y_i^j].
\end{aligned}$$

Now, rearranging the above equality, we obtain

$$\sum_{i\in\mathbb{B}_n}(k+1)(-\lambda\mathbb{E}[Y_i^k]+\mathbb{E}[R_iY_i^k]) = \sum_{i\in\mathbb{B}_n}\sum_{j=0}^{k-1}\binom{k+1}{j}\mathbb{E}[(\lambda+R_i(-1)^{k+1-j})Y_i^j]$$
$$\overset{(a)}{\leq}\sum_{i\in\mathbb{B}_n}(\lambda+1)\sum_{j=0}^{k-1}\binom{k+1}{j}\mathbb{E}[Y_i^j].$$

where step (a) follows from the fact that $0\leq R_i\leq 1$ for all $i\in\mathbb{B}_n$. Now, applying Lemma 3.2 to the LHS above,

$$\sum_{i\in\mathbb{B}_n}(k+1)\left(-\lambda+\frac{1}{\sum_{k\in\mathbb{Z}^d}a_k}\right)\mathbb{E}[Y_i^k]\leq\sum_{i\in\mathbb{B}_n}(\lambda+1)\sum_{j=0}^{k-1}\binom{k+1}{j}\mathbb{E}[Y_i^j].$$

Rearranging the last display concludes the proof as, by translation invariance, for all $1\leq j\leq k+1$ and all $i\in\mathbb{B}_n$, we have $\mathbb{E}[Y_i^j]=\mathbb{E}[Y_0^j]$. □

3.3. Proof of Corollary 2.2.

PROOF. Recall from Subsection 3.1 the coupling of $\{X_{i;\infty}(\cdot)\}_{i\in\mathbb{Z}^d}$ with a collection of stationary independent $M/M/1$ queues $\{Q_{i;\infty}(\cdot)\}_{i\in\mathbb{Z}^d}$, each queue having Poisson arrivals with rate λ and departures with rate 1, such that, almost surely, $X_{i;\infty}(t)\geq Q_{i;\infty}(t)$ for all $i\in\mathbb{Z}^d$ and $t\in\mathbb{R}$. For each $i\in\mathbb{Z}^d$ and $t\in\mathbb{R}$, the distribution of $1+Q_{i;\infty}(t)$ is Geometric$(1-\lambda)$ [**2**]. Thus, for any $C<d/\log(1/\lambda)$,

$$\mathbb{P}\left(\max_{i\in\mathbb{Z}^d:\|i\|_\infty\leq N}X_i(t)\geq\lfloor C\log N\rfloor\right)\geq\mathbb{P}\left(\max_{i\in\mathbb{Z}^d:\|i\|_\infty\leq N}Q_{i;\infty}(t)\geq\lfloor C\log N\rfloor\right)$$
$$\geq 1-\left(1-\lambda^{C\log N}\right)^{(2N+1)^d}$$
$$\geq 1-\exp\{-2^dN^{C\log\lambda+d}\}\to 1,\quad\text{as } N\to\infty. \tag{3.3}$$

Recall the constant c_2 appearing in the the upper bound of (2.2). For any $C'>\frac{d}{c_2}$, using the union bound and the upper bound in (2.2),

$$\mathbb{P}\left(\max_{i\in\mathbb{Z}^d:\|i\|_\infty\leq N}X_i(t)>C'\log N\right)\leq(2N+1)^de^{-c_2C'\log N} \tag{3.4}$$
$$=(2N+1)^dN^{-c_2C'}\to 0,\quad\text{as } N\to\infty.$$

The corollary now follows from (3.3) and (3.4). □

4. Proof of Theorem 2.3

PROOF. For this proof, let $\{X_i\}_{i\in\mathbb{Z}^d}\in\mathbb{N}_0^{\mathbb{Z}^d}$ be a sample from the stationary solution of the dynamics. Fix $K\in\mathbb{N}_0$. From the symmetry in the dynamics it suffices to prove (2.4) for $h=1$. Moreover, it suffices to consider f,g non-negative. The general case follows upon separately considering the positive and negative parts of f,g.

We first consider bounded $f(\cdot)$ and $g(\cdot)$. As before, we will proceed via a version of the dynamics with a truncated interference sequence. Consider a sequence L_n, such that for every $n\in\mathbb{N}$, $L_n\in\mathbb{N}$ and $\lim_{n\to\infty}L_n=\infty$ and $\lim_{n\to\infty}\frac{L_n}{n}=0$. Moreover, assume $n\mapsto L_n$ and $n\mapsto\lfloor\frac{n}{2}\rfloor-L_n$ are non-decreasing in n for $n\geq 2$. One valid choice of $\{L_n\}_{n\in\mathbb{N}}$ is $L_n:=\sqrt{\lfloor\frac{n}{2}\rfloor},n\geq 1$. As before, for each $n\in\mathbb{N}$, denote the truncated interference sequence by $\{a_i^{L_n}\}_{i\in\mathbb{Z}^d}$, where $a_i^{L_n}:=a_i\mathbf{1}_{\|i\|_\infty\leq L_n}$.

Let $n_0 \in \mathbb{N}$ be such that $n \geq 2L_n + 2K + 2$ for all $n \geq n_0$. For $n \geq n_0$, define $\mathcal{X}^{(n)} := \{(z_1, \cdots, z_d) \in \mathbb{Z}^d : \lfloor \frac{n}{2} \rfloor - L_n \leq z_1 \leq \lceil \frac{n}{2} \rceil + L_n\}$. Consider the CFTP construction (with the truncated interference sequence $\{a_i^{L_n}\}_{i \in \mathbb{Z}^d}$) of the dynamics, where the infinite system was started with all queues being empty at a time $-t \leq 0$ (i.e., t is positive). From this all empty state at time $-t$ in the past, the dynamics is run in forward time with no arrivals at sites in $\mathcal{X}^{(n)}$ and independent PP(λ) arrivals at other sites, and with departure rates governed by the truncated interference sequence $\{a_i^{L_n}\}_{i \in \mathbb{Z}^d}$. For any $i \in \mathbb{Z}^d$, $n \geq n_0$ and $t \geq 0$, denote by $X_i^{(n;t)}$ the queue length at site i at time 0 for this system. Monotonicity in the dynamics implies that for each $i \in \mathbb{Z}^d$, the map $t \mapsto X_i^{(n:t)}$ is non-decreasing and hence an almost sure limit $X_i^{(n)} := \lim_{t \to \infty} X_i^{(n;t)}$ exists. In other words, the random variable $X_i^{(n)}$ is defined to be the queue length at site i, at time 0, in the stationary regime of the infinite dynamics, constructed with the truncated interference sequence $\{a_i^{L_n}\}_{i \in \mathbb{Z}^d}$, and with the queues at sites in the set $\mathcal{X}^{(n)}$ "frozen" without activity with 0 customers at all time. For $n \geq n_0$, write $\mathbf{X}_{0,K}^{(n)} := \{X_i^{(n)} : i \in \mathbb{Z}^d, \|i\|_\infty \leq K\}$ and $\mathbf{X}_{n\mathbf{e}_1,K}^{(n)} := \{X_i^{(n)} : i \in \mathbb{Z}^d, \|i - n\mathbf{e}_1\|_\infty \leq K\}$. Also, recall $\mathbf{X}_{0,K} := \{X_i : i \in \mathbb{Z}^d, \|i\|_\infty \leq K\}$ and $\mathbf{X}_{n\mathbf{e}_1,K} := \{X_i : i \in \mathbb{Z}^d, \|i - n\mathbf{e}_1\|_\infty \leq K\}$.

We now collect several useful properties of the random variables $\mathbf{X}_{0,K}^{(n)}$ and $\mathbf{X}_{n\mathbf{e}_1,K}^{(n)}$. Under the synchronous coupling (same arrival and departure PPP), almost surely,

(1) For each $n \geq n_0$, $\mathbf{X}_{0,K}^{(n)} \leq \mathbf{X}_{0,K}$ and $\mathbf{X}_{n\mathbf{e}_1,K}^{(n)} \leq \mathbf{X}_{n\mathbf{e}_1,K}$ (here '$\leq$' denotes co-ordinate-wise ordering).
(2) The map $n \mapsto \mathbf{X}_{0,K}^{(n)}$, $n \geq n_0$, is non-decreasing and $\lim_{n \to \infty} \mathbf{X}_{0,K}^{(n)} = \mathbf{X}_{0,K}$.
(3) For all $n \geq n_0$, $\mathbf{X}_{0,K}^{(n)}$ and $\mathbf{X}_{n\mathbf{e}_1,K}^{(n)}$ are independent and identically distributed.

The first property and the first part of the second property follow from monotonicity of the dynamics. To verify the limit in the second property, first note by monotonicity that $\mathbf{X}_{0,K}^{(\infty)} := \lim_{n \to \infty} \mathbf{X}_{0,K}^{(n)}$ exists and, by property 1,

$$\mathbf{X}_{0,K}^{(\infty)} \leq \mathbf{X}_{0,K}. \tag{4.1}$$

We will now argue the reverse inequality. For $L \in \mathbb{N}$, let $\{X_i^{(n),L} : i \in \mathbb{Z}^d\}$ denote the queue lengths at time 0 under the CFTP construction for the stationary dynamics with the same arrival process $\mathcal{A}_i$ at sites $i \notin \mathcal{X}^{(n)}$, zero arrivals at sites in $\mathcal{X}^{(n)}$, and departure rate governed by the truncated interference sequence $(a_i^L := a_i \mathbf{1}_{\|i\|_\infty \leq L})_{i \in \mathbb{Z}^d}$. Denote by $\{Z_i^{(\lfloor \frac{n}{2} \rfloor - L_n),L} : i \in \mathbb{Z}^d\}$ the queue lengths at time 0 under the CFTP construction for the stationary dynamics with the same arrival process $\mathcal{A}_i$ at sites $i \in \mathbb{Z}^d$ with $\|i\|_\infty < \lfloor \frac{n}{2} \rfloor - L_n$, zero arrivals outside this set of sites, and departure governed by the interference sequence $(a_i^L)_{i \in \mathbb{Z}^d}$. Finally, denote by $\{X_i^L : i \in \mathbb{Z}^d\}$ the stationary queue lengths at time zero under the CFTP construction of the dynamics with arrival process $\mathcal{A}_i$ for all $i \in \mathbb{Z}^d$ but departure governed by the interference sequence $(a_i^L)_{i \in \mathbb{Z}^d}$. As before, let $\mathbf{X}_{0,K}^{(n),L} := \{X_i^{(n),L} : i \in \mathbb{Z}^d, \|i\|_\infty \leq K\}$, $\mathbf{Z}_{0,K}^{(\lfloor \frac{n}{2} \rfloor - L_n),L} := \{Z_i^{(\lfloor \frac{n}{2} \rfloor - L_n),L} : i \in \mathbb{Z}^d, \|i\|_\infty \leq K\}$ and $\mathbf{X}_{0,K}^L := \{X_i^L : i \in \mathbb{Z}^d, \|i\|_\infty \leq K\}$.

By monotonicity, $\mathbf{Z}_{0,K}^{(\lfloor\frac{n}{2}\rfloor-L_n),L} \leq \mathbf{X}_{0,K}^{(n),L} \leq \mathbf{X}_{0,K}^{L}$ for all $n \geq n_0$, and hence,

$$\mathbf{Z}_{0,K}^{(\infty),L} := \lim_{n\to\infty} \mathbf{Z}_{0,K}^{(\lfloor\frac{n}{2}\rfloor-L_n),L} \leq \mathbf{X}_{0,K}^{(\infty),L} := \lim_{n\to\infty} \mathbf{X}_{0,K}^{(n),L} \leq \mathbf{X}_{0,K}^{L}. \tag{4.2}$$

Moreover, as $\lfloor\frac{n}{2}\rfloor - L_n \to \infty$ as $n \to \infty$, by Proposition 7.3 of [**9**], $\mathbf{Z}_{0,K}^{(\infty),L} = \mathbf{X}_{0,K}^{L}$ and hence, by (4.2), for any $L \in \mathbb{N}$,

$$\mathbf{X}_{0,K}^{(\infty),L} = \mathbf{X}_{0,K}^{L}. \tag{4.3}$$

Again, by monotonicity, $\mathbf{X}_{0,K}^{(n),L} \leq \mathbf{X}_{0,K}^{(n)}$ for all n such that $L_n \geq L$, and hence, $\mathbf{X}_{0,K}^{(\infty),L} \leq \mathbf{X}_{0,K}^{(\infty)}$. Hence, from (4.3), for any $L \in \mathbb{N}$,

$$\mathbf{X}_{0,K}^{L} \leq \mathbf{X}_{0,K}^{(\infty)}. \tag{4.4}$$

Finally, from the proof of Proposition 4.3 in [**9**], almost surely, for any $i \in \mathbb{Z}^d$, $\lim_{L\to\infty} X_i^L = X_i$ and hence, by (4.4),

$$\mathbf{X}_{0,K} \leq \mathbf{X}_{0,K}^{(\infty)}. \tag{4.5}$$

The limit in property 2 above now follows from (4.1) and (4.5).

To obtain the third property, note that as $a_i^{L_n} = 0$ for all $\|i\|_\infty > L_n$, there are no interactions between queues on either side of the frozen queue(s). Thus, $\mathbf{X}_{0,K}^{(n)}$ and $\mathbf{X}_{n\mathbf{e}_1,K}^{(n)}$ are independent. The identical distribution follows from the symmetry of the sites in $\{i \in \mathbb{Z}^d, \|i\|_\infty \leq K\}$ and $\{i \in \mathbb{Z}^d, \|i - n\mathbf{e}_1\|_\infty \leq K\}$ with respect to the set $\mathcal{X}^{(n)}$ and the fact that $a_i = a_{-i}$, for all $i \in \mathbb{Z}^d$.

We now proceed as follows:

$$\begin{aligned}
&\mathbb{E}[f(\mathbf{X}_{0,K})g(\mathbf{X}_{n\mathbf{e}_1,K})] - \mathbb{E}[f(\mathbf{X}_{0,K}^{(n)})]\mathbb{E}[g(\mathbf{X}_{n\mathbf{e}_1,K}^{(n)})] \\
&\quad = \mathbb{E}[f(\mathbf{X}_{0,K})g(\mathbf{X}_{n\mathbf{e}_1,K})] - \mathbb{E}[f(\mathbf{X}_{0,K}^{(n)})g(\mathbf{X}_{n\mathbf{e}_1,K}^{(n)})] \\
&\quad = \mathbb{E}[f(\mathbf{X}_{0,K})(g(\mathbf{X}_{n\mathbf{e}_1,K}) - g(\mathbf{X}_{n\mathbf{e}_1,K}^{(n)}))] \\
&\qquad + \mathbb{E}[g(\mathbf{X}_{n\mathbf{e}_1,K}^{(n)})(f(\mathbf{X}_{0,K}) - f(\mathbf{X}_{0,K}^{(n)}))], \\
&\quad = \mathbb{E}[f(\mathbf{X}_{n\mathbf{e}_1,K})(g(\mathbf{X}_{0,K}) - g(\mathbf{X}_{0,K}^{(n)}))] \\
&\qquad + \mathbb{E}[g(\mathbf{X}_{n\mathbf{e}_1,K}^{(n)})(f(\mathbf{X}_{0,K}) - f(\mathbf{X}_{0,K}^{(n)}))].
\end{aligned} \tag{4.6}$$

The first equality follows since $\mathbf{X}_{0,K}^{(n)}$ and $\mathbf{X}_{n\mathbf{e}_1,K}^{(n)}$ are independent random variables. The second equality follows from adding and subtracting $\mathbb{E}\left(f(\mathbf{X}_{0,K})g(\mathbf{X}_{n\mathbf{e}_1,K}^{(n)})\right)$. The third equality follows as, by the symmetry of the sites in $\{i \in \mathbb{Z}^d, \|i\|_\infty \leq K\}$ and $\{i \in \mathbb{Z}^d, \|i - n\mathbf{e}_1\|_\infty \leq K\}$ with respect to the set $\mathcal{X}^{(n)}$ and the fact that $a_i = a_{-i}$, for all $i \in \mathbb{Z}^d$, the law of $(\mathbf{X}_{0,K}, \mathbf{X}_{n\mathbf{e}_1,K}, \mathbf{X}_{n\mathbf{e}_1,K}^{(n)})$ is the same as that of $(\mathbf{X}_{n\mathbf{e}_1,K}, \mathbf{X}_{0,K}, \mathbf{X}_{0,K}^{(n)})$.

As both f, g are bounded functions on $\mathbb{N}_0^{(2K+1)^d}$ and X_i and $X_i^{(n)}$ are integer valued random variables for all $i \in \mathbb{Z}^d$, using properties 2 and 3 above, dominated convergence theorem yields $\lim_{n\to\infty} \mathbb{E}[f(\mathbf{X}_{0,K}^{(n)})] = \mathbb{E}[f(\mathbf{X}_{0,K})]$,

$$\lim_{n\to\infty} \mathbb{E}[g(\mathbf{X}_{n\mathbf{e}_1,K}^{(n)})] = \lim_{n\to\infty} \mathbb{E}[g(\mathbf{X}_{0,K}^{(n)})] = \mathbb{E}[g(\mathbf{X}_{0,K})].$$

and

$$\lim_{n\to\infty}\mathbb{E}[f(\mathbf{X}_{n\mathbf{e}_1,K})(g(\mathbf{X}_{0,K})-g(\mathbf{X}^{(n)}_{0,K}))]=0=\lim_{n\to\infty}\mathbb{E}[g(\mathbf{X}^{(n)}_{n\mathbf{e}_1,K})(f(\mathbf{X}_{0,K})-f(\mathbf{X}^{(n)}_{0,K}))].$$

Using these limits in (4.6), we obtain (2.4) for all bounded functions f and g.

Now, we consider general f and g. For any $\epsilon > 0$, there exist simple functions $f^{(\epsilon)}$ and $g^{(\epsilon)}$ such that $\mathbb{E}\left(f(\mathbf{X}_{0,K})-f^{(\epsilon)}(\mathbf{X}_{0,K})\right)^2<\epsilon^2$ and $\mathbb{E}\left(g(\mathbf{X}_{0,K})-g^{(\epsilon)}(\mathbf{X}_{0,K})\right)^2 < \epsilon^2$. Now,

$$\begin{aligned}
&|\mathbb{E}\left(f(\mathbf{X}_{0,K})g(\mathbf{X}_{n\mathbf{e}_1,K})\right)-\mathbb{E}\left(f(\mathbf{X}_{0,K})\right)\mathbb{E}\left(g(\mathbf{X}_{n\mathbf{e}_1,K})\right)|\\
&\le\left|\mathbb{E}\left(f(\mathbf{X}_{0,K})g(\mathbf{X}_{n\mathbf{e}_1,K})\right)-\mathbb{E}\left(f^{(\epsilon)}(\mathbf{X}_{0,K})g^{(\epsilon)}(\mathbf{X}_{n\mathbf{e}_1,K})\right)\right|\\
&\quad+\left|\mathbb{E}\left(f^{(\epsilon)}(\mathbf{X}_{0,K})g^{(\epsilon)}(\mathbf{X}_{n\mathbf{e}_1,K})\right)-\mathbb{E}\left(f^{(\epsilon)}(\mathbf{X}_{0,K})\right)\mathbb{E}\left(g^{(\epsilon)}(\mathbf{X}_{n\mathbf{e}_1,K})\right)\right|\\
&\quad+\left|\mathbb{E}\left(f^{(\epsilon)}(\mathbf{X}_{0,K})\right)\mathbb{E}\left(g^{(\epsilon)}(\mathbf{X}_{n\mathbf{e}_1,K})\right)-\mathbb{E}\left(f(\mathbf{X}_{0,K})\right)\mathbb{E}\left(g(\mathbf{X}_{n\mathbf{e}_1,K})\right)\right|.
\end{aligned}\tag{4.7}$$

By triangle inequality, Cauchy-Schwartz inequality and translation invariance of the dynamics,

$$\begin{aligned}
&\left|\mathbb{E}\left(f(\mathbf{X}_{0,K})g(\mathbf{X}_{n\mathbf{e}_1,K})\right)-\mathbb{E}\left(f^{(\epsilon)}(\mathbf{X}_{0,K})g^{(\epsilon)}(\mathbf{X}_{n\mathbf{e}_1,K})\right)\right|\\
&\qquad+\left|\mathbb{E}\left(f^{(\epsilon)}(\mathbf{X}_{0,K})\right)\mathbb{E}\left(g^{(\epsilon)}(\mathbf{X}_{n\mathbf{e}_1,K})\right)-\mathbb{E}\left(f(\mathbf{X}_{0,K})\right)\mathbb{E}\left(g(\mathbf{X}_{n\mathbf{e}_1,K})\right)\right|\\
&\le 2\sqrt{\mathbb{E}\left(f(\mathbf{X}_{0,K})\right)^2}\sqrt{\mathbb{E}\left(g(\mathbf{X}_{0,K})-g^{(\epsilon)}(\mathbf{X}_{0,K})\right)^2}\\
&\qquad+2\sqrt{\mathbb{E}\left(g^{(\epsilon)}(\mathbf{X}_{0,K})\right)^2}\sqrt{\mathbb{E}\left(f(\mathbf{X}_{0,K})-f^{(\epsilon)}(\mathbf{X}_{0,K})\right)^2}\\
&\le 2\epsilon\left(\sqrt{\mathbb{E}\left(f(\mathbf{X}_{0,K})\right)^2}+\sqrt{\mathbb{E}\left(g(\mathbf{X}_{0,K})\right)^2}+\epsilon\right).
\end{aligned}\tag{4.8}$$

Moreover, as $f^{(\epsilon)}$ and $g^{(\epsilon)}$ are bounded,

$$\lim_{n\to\infty}\left|\mathbb{E}\left(f^{(\epsilon)}(\mathbf{X}_{0,K})g^{(\epsilon)}(\mathbf{X}_{n\mathbf{e}_1,K})\right)-\mathbb{E}\left(f^{(\epsilon)}(\mathbf{X}_{0,K})\right)\mathbb{E}\left(g^{(\epsilon)}(\mathbf{X}_{n\mathbf{e}_1,K})\right)\right|=0.\tag{4.9}$$

Using (4.8) and (4.9) in (4.7), we obtain

$$\begin{aligned}
&\limsup_{n\to\infty}|\mathbb{E}\left(f(\mathbf{X}_{0,K})g(\mathbf{X}_{n\mathbf{e}_1,K})\right)-\mathbb{E}\left(f(\mathbf{X}_{0,K})\right)\mathbb{E}\left(g(\mathbf{X}_{n\mathbf{e}_1,K})\right)|\\
&\qquad\le 2\epsilon\left(\sqrt{\mathbb{E}\left(f(\mathbf{X}_{0,K})\right)^2}+\sqrt{\mathbb{E}\left(g(\mathbf{X}_{0,K})\right)^2}+\epsilon\right).
\end{aligned}$$

As $\epsilon > 0$ is arbitrary, this completes the proof of (2.4).

Take any $h \in \{1,\cdots,d\}$. Upon taking f and g to be indicator functions of cylinder sets $\mathcal{F}_0 \subset \mathcal{B}\left(\mathbb{N}_0^{\mathbb{Z}^d}\right)$, (2.4) shows that (2.3) holds for all $A, B \in \mathcal{F}_0$. A standard argument using the 'good sets principle' can now be used to conclude that (2.3) holds for all $A, B \in \mathcal{B}\left(\mathbb{N}_0^{\mathbb{Z}^d}\right)$. This shows that $\mathcal{Q}_h$ is strongly mixing for all $h \in \{1,\cdots,d\}$. Hence, $\{T_h\}_{h=1}^d$ is ergodic. □

Acknowledgments

The second author thanks the first author for hosting him at UNC Chapel Hill, where a large part of this work was done.

References

[1] Daron Acemoğlu, Giacomo Como, Fabio Fagnani, and Asuman Ozdaglar, *Opinion fluctuations and disagreement in social networks*, Math. Oper. Res. **38** (2013), no. 1, 1–27, DOI 10.1287/moor.1120.0570. MR3029476

[2] François Baccelli and Pierre Brémaud, *Elements of queueing theory*, Applications of Mathematics (New York), vol. 26, Springer-Verlag, Berlin, 1994. Palm-martingale calculus and stochastic recurrences. MR1288301

[3] Sayan Banerjee and Krzysztof Burdzy, *Rates of convergence to equilibrium for potlatch and smoothing processes*, arXiv preprint arXiv:2001.09524, 2020.

[4] Thomas M. Liggett and Frank Spitzer, *Ergodic theorems for coupled random walks and other systems with locally interacting components*, Z. Wahrsch. Verw. Gebiete **56** (1981), no. 4, 443–468, DOI 10.1007/BF00531427. MR621659

[5] Thomas M. Liggett, *Interacting particle systems*, Grundlehren der Mathematischen Wissenschaften [Fundamental Principles of Mathematical Sciences], vol. 276, Springer-Verlag, New York, 1985, DOI 10.1007/978-1-4613-8542-4. MR776231

[6] Elon Lindenstrauss, *Pointwise theorems for amenable groups*, Electron. Res. Announc. Amer. Math. Soc. **5** (1999), 82–90, DOI 10.1090/S1079-6762-99-00065-7. MR1696824

[7] Michael Mitzenmacher, *The power of two choices in randomized load balancing*, *IEEE Transactions on Parallel and Distributed Systems*, 12(10):1094–1104, 2001.

[8] Abishek Sankararaman and François Baccelli, *Spatial birth-death wireless networks*, IEEE Trans. Inform. Theory **63** (2017), no. 6, 3964–3982, DOI 10.1109/TIT.2017.2669298. MR3677758

[9] Abishek Sankararaman, François Baccelli, and Sergey Foss, *Interference queueing networks on grids*, Ann. Appl. Probab. **29** (2019), no. 5, 2929–2987, DOI 10.1214/19-AAP1470. MR4019879

[10] Seva Shneer and Alexander Stolyar, *Stability and moment bounds under utility-maximising service allocations: Finite and infinite networks*, Adv. in Appl. Probab. **52** (2020), no. 2, 463–490, DOI 10.1017/apr.2020.8. MR4123643

[11] Mark van der Boor, Sem C. Borst, Johan S. H. van Leeuwaarden, and Debankur Mukherjee, *Scalable load balancing in networked systems: universality properties and stochastic coupling methods*, Proceedings of the International Congress of Mathematicians—Rio de Janeiro 2018. Vol. IV. Invited lectures, World Sci. Publ., Hackensack, NJ, 2018, pp. 3893–3923. MR3966556

Department of Statistics and Operations Research, University of North Carolina, Chapel Hill, North Carolina

Email address: `sayan@email.unc.edu`

Electrical Engineering and Computer Sciences Department, University of California, Berkeley, California

Email address: `abishek@berkeley.edu`

Contemporary Mathematics
Volume **774**, 2021
https://doi.org/10.1090/conm/774/15566

How strong can the Parrondo effect be? II

S. N. Ethier and Jiyeon Lee

Abstract. Parrondo's coin-tossing games comprise two games, A and B. The result of game A is determined by the toss of a fair coin. The result of game B is determined by the toss of a p_0-coin if capital is a multiple of r, and by the toss of a p_1-coin otherwise. In either game, the player wins one unit with heads and loses one unit with tails. Game B is fair if $(1-p_0)(1-p_1)^{r-1} = p_0\, p_1^{r-1}$. In a previous paper we showed that, if the parameters of game B, namely r, p_0, and p_1, are allowed to be arbitrary, subject to the fairness constraint, and if the two (fair) games A and B are played in an arbitrary periodic sequence, then the rate of profit can not only be positive (the so-called Parrondo effect), but also be arbitrarily close to 1 (i.e., 100%). Here we prove the same conclusion for a random sequence of the two games instead of a periodic one, that is, at each turn game A is played with probability γ and game B is played otherwise, where $\gamma \in (0,1)$ is arbitrary.

1. Introduction

The flashing Brownian ratchet of Ajdari and Prost (1992) is a stochastic model in statistical physics that is also of interest to biologists in connection with so-called molecular motors. In 1996 J. M. R. Parrondo proposed a toy model of the flashing Brownian ratchet involving two coin-tossing games. Both of the games, A and B, are individually fair or losing, whereas the random mixture (toss a fair coin to determine whether game A or game B is played) is winning, as are periodic sequences of the games, such as $AABB\,AABB\,AABB\,\cdots$.

Harmer and Abbott (1999) described the games explicitly. For simplicity, we omit the bias parameter, so that both games are fair. Let us define a p-coin to be a coin with probability p of heads. In Parrondo's original games, game A uses a fair coin, while game B uses two biased coins, a p_0-coin if capital is a multiple of 3 and a p_1-coin otherwise, where

$$p_0 = \frac{1}{10} \quad \text{and} \quad p_1 = \frac{3}{4}. \tag{1.1}$$

2020 *Mathematics Subject Classification.* Primary 60J10; Secondary 60F15.

Key words and phrases. Parrondo games, rate of profit, strong law of large numbers, stationary distribution, random walk on the n-cycle.

The first author was partially supported by a grant from the Simons Foundation (429675).

The second author was supported by the Basic Science Research Program through the National Research Foundation of Korea (NRF) funded by the Ministry of Education (NRF-2018R1D1A1B07042307).

The player wins one unit with heads and loses one unit with tails. Both games are fair, but the random mixture, denoted by $\frac{1}{2}A+\frac{1}{2}B$, has long-term cumulative profit per game played (hereafter, rate of profit)

$$\mu\left(\frac{1}{2}A+\frac{1}{2}B\right)=\frac{18}{709}\approx 0.0253879, \tag{1.2}$$

and the pattern $AABB$, repeated ad infinitum, has rate of profit

$$\mu(AABB)=\frac{4}{163}\approx 0.0245399. \tag{1.3}$$

Dinis (2008) found that the pattern $ABABB$ (or any cyclic permutation of it) has the highest rate of profit, namely

$$\mu(ABABB)=\frac{3613392}{47747645}\approx 0.0756769. \tag{1.4}$$

How large can these rates of profit be if we vary the parameters of the games, subject to a fairness constraint?

Game A is always the same fair-coin-tossing game. With $r\ge 3$ an integer, game B is a mod r capital-dependent game that uses two biased coins, a p_0-coin ($p_0<\frac{1}{2}$) if capital is a multiple of r, and a p_1-coin ($p_1>\frac{1}{2}$) otherwise. The probabilities p_0 and p_1 must be such that game B is fair, requiring the constraint

$$(1-p_0)(1-p_1)^{r-1}=p_0\,p_1^{r-1},$$

or equivalently,

$$p_0=\frac{\rho^{r-1}}{1+\rho^{r-1}}\quad\text{and}\quad p_1=\frac{1}{1+\rho} \tag{1.5}$$

for some $\rho\in(0,1)$. The special case of $r=3$ and $\rho=\frac{1}{3}$ gives (1.1).

The games are played randomly or periodically. Specifically, we consider the random mixture $\gamma A+(1-\gamma)B$ (game A is played with probability γ and game B is played otherwise) as well as the pattern $\Gamma(A,B)$, repeated ad infinitum. We denote the rate of profit by

$$\mu(r,\rho,\gamma A+(1-\gamma)B)\quad\text{or}\quad\mu(r,\rho,\Gamma(A,B)),$$

so that the rates of profit in (1.2)–(1.4) in this notation become $\mu(3,\frac{1}{3},\frac{1}{2}A+\frac{1}{2}B)$, $\mu(3,\frac{1}{3},AABB)$, and $\mu(3,\frac{1}{3},ABABB)$.

How large can $\mu(r,\rho,\gamma A+(1-\gamma)B)$ and $\mu(r,\rho,\Gamma(A,B))$ be? The answer, at least in the second case, is that it can be arbitrarily close to 1 (i.e., 100%):

THEOREM 1.1 (Ethier and Lee (2019)).

$$\sup_{r\ge 3,\ \rho\in(0,1),\ \Gamma(A,B)\ \textit{arbitrary}}\mu(r,\rho,\Gamma(A,B))=1.$$

In the first case the question was left open, and it is the aim of this paper to resolve that issue. It turns out that the conclusion is the same:

THEOREM 1.2.

$$\sup_{r\ge 3,\ \rho\in(0,1),\ \gamma\in(0,1)}\mu(r,\rho,\gamma A+(1-\gamma)B)=1.$$

This will be seen to be a consequence of Corollary 1.5 below.

We can compute $\mu(r,\rho,\gamma A+(1-\gamma)B)$ and $\mu(r,\rho,\Gamma(A,B))$ for $r\geq 3$, $\rho\in(0,1)$, $\gamma\in(0,1)$, and patterns $\Gamma(A,B)$. Indeed, the method of Ethier and Lee (2009) applies if r is odd, and generalizations of it apply if r is even; see Section 2 for details in the random mixture case and Ethier and Lee (2019) in the periodic pattern case. For example,

$$\mu\left(3,\rho,\frac{1}{2}A+\frac{1}{2}B\right)=\frac{9(1-\rho)^3(1+\rho)}{2(35+70\rho+78\rho^2+70\rho^3+35\rho^4)} \tag{1.6}$$

and

$$\mu(3,\rho,AABB)=\frac{3(1-\rho)^3(1+\rho)}{8(3+6\rho+7\rho^2+6\rho^3+3\rho^4)}. \tag{1.7}$$

These and other examples suggest that typically $\mu(r,\rho,\gamma A+(1-\gamma)B)$ and $\mu(r,\rho,\Gamma(A,B))$ are decreasing in ρ (for fixed r, γ, and $\Gamma(A,B)$), hence maximized at $\rho=0$. We excluded the case $\rho=0$ in (1.5), but now we want to include it. We find from (1.6) and (1.7) that

$$\mu\left(3,0,\frac{1}{2}A+\frac{1}{2}B\right)=\frac{9}{70}\approx 0.128571\quad\text{and}\quad \mu(3,0,AABB)=\frac{1}{8}=0.125,$$

which are substantial increases over $\mu(3,\frac{1}{3},\frac{1}{2}A+\frac{1}{2}B)$ and $\mu(3,\frac{1}{3},AABB)$ (see (1.2) and (1.3)). We can do slightly better by choosing γ optimally:

$$\mu(3,0,\gamma A+(1-\gamma)B)=\frac{3\gamma(1-\gamma)(2-\gamma)}{(2+\gamma)(4-\gamma)}, \tag{1.8}$$

so $\max_\gamma \mu(3,0,\gamma A+(1-\gamma)B)\approx 0.133369$, achieved at $\gamma\approx 0.407641$. Similarly, we can do considerably better by choosing the optimal pattern $ABABB$:

$$\mu(3,0,ABABB)=\frac{9}{25}=0.36. \tag{1.9}$$

Thus, we take $\rho=0$ in what follows.

Theorem 1.1 was shown to follow from the next theorem.

THEOREM 1.3 (Ethier and Lee (2019)). *Let $r\geq 3$ be an odd integer and s be a positive integer. Then*

$$\mu(r,0,(AB)^sB^{r-2})=\frac{r}{2s+r-2}\,\frac{2^s-1}{2^s+1},$$

regardless of initial capital.

Let $r\geq 4$ be an even integer and s be a positive integer. Then

$$\mu(r,0,(AB)^sB^{r-2})=\begin{cases}\dfrac{r}{2s+r-2}\sum_{k=0}^{s}\left\lceil\frac{2k}{r}\right\rceil\binom{s}{k}\frac{1}{2^s} & \text{if initial capital is even,}\\ 0 & \text{if initial capital is odd.}\end{cases}$$

The special case $(r,s)=(3,2)$ of this theorem is equivalent to (1.9).

Theorem 1.2 will be seen to follow from the next two results, the proofs of which are deferred to Section 4.

THEOREM 1.4. *Let $r\geq 3$ be an integer and $0<\gamma<1$. Then*

$$\mu(r,0,\gamma A+(1-\gamma)B)=\frac{r\gamma(1-\gamma)(2-\gamma)[(2-\gamma)^{r-2}-\gamma^{r-2}]}{2[(2-\gamma)^r-\gamma^r]+r\gamma(2-\gamma)[(2-\gamma)^{r-2}-\gamma^{r-2}]},$$

regardless of initial capital.

The special case $r = 3$ of this theorem is equivalent to (1.8).

COROLLARY 1.5. *For each integer* $r \geq 3$, *define* $\gamma_r := 2/\sqrt{r}$. *Then*

$$1 - \mu(r, 0, \gamma_r A + (1 - \gamma_r)B) \sim 2\gamma_r \text{ as } r \to \infty,$$

regardless of initial capital.

Table 1 illustrates these results.

TABLE 1. The rate of profit $\mu(r, 0, \gamma A + (1 - \gamma)B)$.

r	$\arg\max_\gamma \mu$	$1 - \max_\gamma \mu$	$\gamma_r := 2/\sqrt{r}$	$1 - \mu$ at $\gamma = \gamma_r$
10	0.366017	0.665064	0.632456	0.743544
100	0.165296	0.316931	0.200000	0.322034
1000	0.0594276	0.117089	0.0632456	0.117307
10000	0.0196059	0.0390196	0.0200000	0.0390273
100000	0.00628474	0.0125497	0.00632456	0.0125500
1000000	0.00199601	0.00399002	0.00200000	0.00399003

For the purpose of comparison, let us state a corollary to Theorem 1.3 that is analogous to Corollary 1.5.

COROLLARY 1.6. *For each integer* $r \geq 3$, *define* $s_r := \lfloor \log_2 r \rfloor - 1$. *Then*

$$1 - \mu(r, 0, (AB)^{s_r} B^{r-2}) \sim \frac{2s_r}{r} \text{ as } r \to \infty,$$

assuming initial capital is even if r *is even,*

Table 2 illustrates Theorem 1.3 and Corollary 1.6.

TABLE 2. The rate of profit $\mu(r, 0, (AB)^s B^{r-2})$, assuming initial capital is even.

r	$\arg\max_s \mu$	$1 - \max_s \mu$	$s_r := \lfloor \log_2 r \rfloor - 1$
10	$2, 3$	0.375000	2
100	5	0.103009	5
1000	8	0.0176590	8
10000	12	0.00243878	12
100000	15	0.000310431	15
1000000	18	0.0000378134	18

Ethier and Lee (2019) remarked that the rates of profit of periodic sequences tend to be larger than those of random sequences. Corollaries 1.5 and 1.6 yield a precise formulation of this conclusion.

2. SLLN for random sequences of games

Ethier and Lee (2009) proved a strong law of large numbers (SLLN) and a central limit theorem for random sequences of Parrondo games. It is only the SLLN that is needed here.

THEOREM 2.1 (Ethier and Lee (2009)). *Let $\boldsymbol{P}$ be the transition matrix for a Markov chain in a finite state space Σ. Assume that $\boldsymbol{P}$ is irreducible and aperiodic, and let the row vector $\boldsymbol{\pi}$ be the unique stationary distribution of $\boldsymbol{P}$. Given a real-valued function w on $\Sigma \times \Sigma$, define the payoff matrix $\boldsymbol{W} := (w(i,j))_{i,j\in\Sigma}$, and put*

$$\mu := \boldsymbol{\pi}\dot{\boldsymbol{P}}\mathbf{1},$$

where $\dot{\boldsymbol{P}} := \boldsymbol{P}\circ\boldsymbol{W}$ (the Hadamard, or entrywise, product), and $\mathbf{1}$ denotes a column vector of 1s with entries indexed by Σ. Let $\{X_n\}_{n\geq 0}$ be a Markov chain in Σ with transition matrix $\boldsymbol{P}$, and let the initial distribution be arbitrary. For each $n \geq 1$, define $\xi_n := w(X_{n-1}, X_n)$ and $S_n := \xi_1 + \cdots + \xi_n$. Then $\lim_{n\to\infty} n^{-1}S_n = \mu$ a.s.

We wish to apply Theorem 2.1 with

$$\Sigma = \{0, 1, \ldots, r-1\}$$

(r is the modulo number in game B), $\boldsymbol{P} := \gamma\boldsymbol{P}_A + (1-\gamma)\boldsymbol{P}_B$, where the $r \times r$ transition matrices $\boldsymbol{P}_A$ and $\boldsymbol{P}_B$ are given by

$$\boldsymbol{P}_A = \begin{pmatrix} 0 & \frac12 & 0 & 0 & \cdots & 0 & 0 & 0 & \frac12 \\ \frac12 & 0 & \frac12 & 0 & \cdots & 0 & 0 & 0 & 0 \\ 0 & \frac12 & 0 & \frac12 & \cdots & 0 & 0 & 0 & 0 \\ \vdots & \vdots & \vdots & \vdots & & \vdots & \vdots & \vdots & \vdots \\ 0 & 0 & 0 & 0 & \cdots & \frac12 & 0 & \frac12 & 0 \\ 0 & 0 & 0 & 0 & \cdots & 0 & \frac12 & 0 & \frac12 \\ \frac12 & 0 & 0 & 0 & \cdots & 0 & 0 & \frac12 & 0 \end{pmatrix}$$

and

$$\boldsymbol{P}_B = \begin{pmatrix} 0 & p_0 & 0 & 0 & \cdots & 0 & 0 & 0 & q_0 \\ q_1 & 0 & p_1 & 0 & \cdots & 0 & 0 & 0 & 0 \\ 0 & q_1 & 0 & p_1 & \cdots & 0 & 0 & 0 & 0 \\ \vdots & \vdots & \vdots & \vdots & & \vdots & \vdots & \vdots & \vdots \\ 0 & 0 & 0 & 0 & \cdots & q_1 & 0 & p_1 & 0 \\ 0 & 0 & 0 & 0 & \cdots & 0 & q_1 & 0 & p_1 \\ p_1 & 0 & 0 & 0 & \cdots & 0 & 0 & q_1 & 0 \end{pmatrix}$$

with p_0 and p_1 as in (1.5) and $q_0 := 1 - p_0$ and $q_1 := 1 - p_1$, and the $r \times r$ payoff matrix $\boldsymbol{W}$ is given by

$$\boldsymbol{W} = \begin{pmatrix} 0 & 1 & 0 & 0 & \cdots & 0 & 0 & 0 & -1 \\ -1 & 0 & 1 & 0 & \cdots & 0 & 0 & 0 & 0 \\ 0 & -1 & 0 & 1 & \cdots & 0 & 0 & 0 & 0 \\ \vdots & \vdots & \vdots & \vdots & & \vdots & \vdots & \vdots & \vdots \\ 0 & 0 & 0 & 0 & \cdots & -1 & 0 & 1 & 0 \\ 0 & 0 & 0 & 0 & \cdots & 0 & -1 & 0 & 1 \\ 1 & 0 & 0 & 0 & \cdots & 0 & 0 & -1 & 0 \end{pmatrix}.$$

The transition matrix $\boldsymbol{P}$ is irreducible and aperiodic if r is odd, in which case the theorem applies directly. But if r is even, then $\boldsymbol{P}$ is irreducible and periodic with period 2. In that case we need the following extension of Theorem 2.1.

THEOREM 2.2. *Theorem* 2.1 *holds with "is irreducible and aperiodic" replaced by "is irreducible and periodic with period 2".*

REMARK 2.3. An alternative proof of a strong law of large numbers for Parrondo games could be based on the renewal theorem; see Pyke (2003).

PROOF. The irreducibility and aperiodicity in Theorem 2.1 ensures that the Markov chain, with initial distribution equal to the unique stationary distribution, is a stationary strong mixing sequence (Bradley (2005), Theorem 3.1). Here we must deduce this property in a different way.

The assumption that $\boldsymbol{P} = (P_{ij})_{i,j\in\Sigma}$ is irreducible with period 2 implies that Σ is the disjoint union of Σ_1 and Σ_2, and transitions under $\boldsymbol{P}$ take Σ_1 to Σ_2 and Σ_2 to Σ_1. This tells us that $\boldsymbol{P}^2$ is reducible with two recurrent classes, Σ_1 and Σ_2, and no transient states. Let the row vectors $\boldsymbol{\pi}_1 = (\pi_1(i))_{i\in\Sigma}$ and $\boldsymbol{\pi}_2 = (\pi_2(j))_{j\in\Sigma}$ be the unique stationary distributions of $\boldsymbol{P}^2$ concentrated on Σ_1 and Σ_2, respectively. Then $\boldsymbol{\pi}_1\boldsymbol{P} = \boldsymbol{\pi}_2$ and $\boldsymbol{\pi}_2\boldsymbol{P} = \boldsymbol{\pi}_1$, and $\boldsymbol{\pi} := \frac{1}{2}(\boldsymbol{\pi}_1 + \boldsymbol{\pi}_2)$ is the unique stationary distribution of $\boldsymbol{P}$.

We consider two Markov chains, one in $\Sigma_1 \times \Sigma_2$ and the other in $\Sigma_2 \times \Sigma_1$, both denoted by $\{(X_0, X_1), (X_2, X_3), (X_4, X_5), \ldots\}$. The transition probabilities are of the form $P^*((i,j),(k,l)) := P_{jk}P_{kl}$ in both cases. To ensure that the Markov chains are irreducible, we change the state spaces to $S_1 := \{(i,j) \in \Sigma_1 \times \Sigma_2 : P_{ij} > 0\}$ and $S_2 := \{(j,k) \in \Sigma_2 \times \Sigma_1 : P_{jk} > 0\}$. The unique stationary distributions are $\boldsymbol{\pi}_1^*$ and $\boldsymbol{\pi}_2^*$ given by

$$\pi_1^*(i,j) = \pi_1(i)P_{ij} \quad \text{and} \quad \pi_2^*(j,k) = \pi_2(j)P_{jk}.$$

To check stationarity, we confirm that for each $(k,l) \in S_1$,

$$\begin{aligned}\sum_{(i,j)\in S_1} \pi_1^*(i,j)P^*((i,j),(k,l)) &= \sum_{j\in\Sigma_2}\sum_{i\in\Sigma_1} \pi_1(i)P_{ij}P_{jk}P_{kl} \\ &= \sum_{j\in\Sigma_2} \pi_2(j)P_{jk}P_{kl} = \pi_1(k)P_{kl} = \pi_1^*(k,l).\end{aligned}$$

An analogous calculation applies to $\boldsymbol{\pi}_2^*$.

We claim that $\boldsymbol{P}^*$ is irreducible and aperiodic on S_1 as well as on S_2. It suffices to show that all entries of $(\boldsymbol{P}^*)^n$ are positive on $S_1 \times S_1$ and on $S_2 \times S_2$ for sufficiently large n. Indeed, given $(i_0,j_0),(i_n,j_n) \in S_1$,

$$\begin{aligned}&(\boldsymbol{P}^*)^n_{(i_0,j_0)(i_n,j_n)} \\ &= \sum_{(i_1,j_1),(i_2,j_2),\ldots,(i_{n-1},j_{n-1})\in S_1} P^*((i_0,j_0),(i_1,j_1))P^*((i_1,j_1),(i_2,j_2))\cdots \\ &\quad \cdots P^*((i_{n-1},j_{n-1}),(i_n,j_n)) \\ &= \sum_{(i_1,j_1),(i_2,j_2),\ldots,(i_{n-1},j_{n-1})\in S_1} P_{j_0i_1}P_{i_1j_1}P_{j_1i_2}P_{i_2j_2}\cdots P_{j_{n-1}i_n}P_{i_nj_n} \\ &= \sum_{i_1\in\Sigma_1} P_{j_0i_1}(\boldsymbol{P})^{2(n-1)}_{i_1i_n}P_{i_nj_n} > 0\end{aligned}$$

since all entries of $\boldsymbol{P}^{2(n-1)}$ are positive on $\Sigma_1 \times \Sigma_1$ for sufficiently large n. A similar argument applies to S_2.

Now we compute mean profit at stationarity. Starting from $\boldsymbol{\pi}_1^*$ we have

$$\begin{aligned}
&E_{\boldsymbol{\pi}_1^*}[w(X_0, X_1) + w(X_1, X_2)]\\
&\quad = \sum_{(i,j)\in S_1} \sum_{(k,l)\in S_1} \pi_1^*(i,j) P^*((i,j),(k,l))[w(i,j) + w(j,k)]\\
&\quad = \sum_{i\in\Sigma_1} \sum_{j\in\Sigma_2} \sum_{k\in\Sigma_1} \pi_1(i) P_{ij} P_{jk}[w(i,j) + w(j,k)]\\
&\quad = \sum_{i,j\in\Sigma} \pi_1(i) P_{ij} w(i,j) + \sum_{j,k\in\Sigma} \pi_2(j) P_{jk} w(j,k)\\
&\quad = \boldsymbol{\pi}_1 \dot{\boldsymbol{P}} \mathbf{1} + \boldsymbol{\pi}_2 \dot{\boldsymbol{P}} \mathbf{1}\\
&\quad = 2\boldsymbol{\pi} \dot{\boldsymbol{P}} \mathbf{1},
\end{aligned}$$

and the same result holds starting from $\boldsymbol{\pi}_2^*$.

We conclude that, starting with initial distribution $\boldsymbol{\pi}_1^*$, $(X_0, X_1), (X_2, X_3), (X_4, X_5), \ldots$ is a stationary strong mixing sequence with a geometric rate, hence the same is true of $w(X_0, X_1) + w(X_1, X_2), w(X_2, X_3) + w(X_3, X_4), \ldots$.

As in Ethier and Lee (2009), the SLLN applies and

$$\lim_{n\to\infty} (2n)^{-1} S_{2n} = \frac{1}{2} 2\boldsymbol{\pi} \dot{\boldsymbol{P}} \mathbf{1} = \boldsymbol{\pi} \dot{\boldsymbol{P}} \mathbf{1} \text{ a.s.}$$

The same is true starting with initial distribution $\boldsymbol{\pi}_2^*$, and the coupling argument used by Ethier and Lee (2009) to permit an arbitrary initial state extends to this setting as well. □

3. Stationary distribution of the random walk on the n-cycle

We will need to find the stationary distribution of the general random walk on the n-cycle (n points arranged in a circle and labeled $0, 1, 2, \ldots, n-1$) with transition matrix

$$\boldsymbol{P} := \begin{pmatrix}
0 & p_0 & 0 & 0 & \cdots & 0 & 0 & 0 & q_0\\
q_1 & 0 & p_1 & 0 & \cdots & 0 & 0 & 0 & 0\\
0 & q_2 & 0 & p_2 & \cdots & 0 & 0 & 0 & 0\\
\vdots & \vdots & \vdots & \vdots & & \vdots & \vdots & \vdots & \vdots\\
0 & 0 & 0 & 0 & \cdots & q_{n-3} & 0 & p_{n-3} & 0\\
0 & 0 & 0 & 0 & \cdots & 0 & q_{n-2} & 0 & p_{n-2}\\
p_{n-1} & 0 & 0 & 0 & \cdots & 0 & 0 & q_{n-1} & 0
\end{pmatrix}, \tag{3.1}$$

where $p_i \in (0,1)$ and $q_i := 1 - p_i$. It is possible that a formula has appeared in the literature, but we were unable to find it. (We did find an erroneous formula.) We could derive a more general result with little additional effort by replacing the diagonal of $\boldsymbol{P}$ by $(r_0, r_1, \ldots, r_{n-1})$, where $p_i > 0$, $q_i > 0$, $r_i \geq 0$, and $p_i + q_i + r_i = 1$ ($i = 0, 1, \ldots, n-1$). But to minimize complications, we treat only the case of (3.1).

The transition matrix $\boldsymbol{P}$ is irreducible and its unique stationary distribution $\boldsymbol{\pi} = (\pi_0, \pi_1, \ldots, \pi_{n-1})$ satisfies $\boldsymbol{\pi} = \boldsymbol{\pi}\boldsymbol{P}$ or

$$\pi_i = \pi_{i-1} p_{i-1} + \pi_{i+1} q_{i+1}, \quad i = 1, 2, \ldots, n-1,$$

where $\pi_n := \pi_0$ and $q_n := q_0$, or

$$\pi_{i-1} p_{i-1} - \pi_i q_i = \pi_i p_i - \pi_{i+1} q_{i+1}, \quad i = 1, 2, \ldots, n-1.$$

Thus, $\pi_{i-1}p_{i-1} - \pi_i q_i = C$, a constant, for $i = 1, 2, \ldots, n$, where $\pi_n := \pi_0$ and $q_n := q_0$; alternatively,

$$\pi_i = -\frac{C}{q_i} + \frac{p_{i-1}}{q_i}\pi_{i-1}. \tag{3.2}$$

This is of the form $x_i = a_i + b_i x_{i-1}$, $i = 1, 2, \ldots$, the solution of which is

$$x_i = \sum_{j=1}^{i} a_j \prod_{k=j+1}^{i} b_k + \bigg(\prod_{j=1}^{i} b_j\bigg) x_0, \quad i = 1, 2, \ldots,$$

where empty products are 1. Applying this to (3.2), we find that

$$\begin{aligned}\pi_i &= -C\sum_{j=1}^{i}\frac{1}{q_j}\prod_{k=j+1}^{i}\frac{p_{k-1}}{q_k} + \bigg(\prod_{j=1}^{i}\frac{p_{j-1}}{q_j}\bigg)\pi_0\\ &= -C\frac{1}{q_i}\bigg(1+\sum_{j=1}^{i-1}\prod_{k=j}^{i-1}\frac{p_k}{q_k}\bigg) + \frac{q_0}{q_i}\bigg(\prod_{j=0}^{i-1}\frac{p_j}{q_j}\bigg)\pi_0, \quad i = 1, 2, 3, \ldots, n.\end{aligned}$$

In particular, C can be determined in terms of π_0 from the $i = n$ case (since $\pi_n := \pi_0$ and $q_n := q_0$). It is given by

$$C = q_0\bigg(\prod_{j=0}^{n-1}\frac{p_j}{q_j} - 1\bigg)\bigg(1+\sum_{j=1}^{n-1}\prod_{k=j}^{n-1}\frac{p_k}{q_k}\bigg)^{-1}\pi_0.$$

Defining $\Pi_0 := 1$ and

$$\Pi_i := -\frac{q_0}{q_i}\bigg(\prod_{j=0}^{n-1}\frac{p_j}{q_j} - 1\bigg)\bigg(1+\sum_{j=1}^{n-1}\prod_{k=j}^{n-1}\frac{p_k}{q_k}\bigg)^{-1}\bigg(1+\sum_{j=1}^{i-1}\prod_{k=j}^{i-1}\frac{p_k}{q_k}\bigg) + \frac{q_0}{q_i}\prod_{j=0}^{i-1}\frac{p_j}{q_j} \tag{3.3}$$

for $i = 1, 2, \ldots, n-1$, we find that $\pi_i = \Pi_i \pi_0$ for $i = 0, 1, \ldots, n-1$, and the following lemma is immediate.

LEMMA 3.1. *The unique stationary distribution* $\boldsymbol{\pi} = (\pi_0, \pi_1, \ldots, \pi_{n-1})$ *of the transition matrix* $\boldsymbol{P}$ *of* (3.1) *is given by*

$$\pi_i = \frac{\Pi_i}{\Pi_0 + \Pi_1 + \cdots + \Pi_{n-1}}, \quad i = 0, 1, \ldots, n-1,$$

where $\Pi_0 := 1$ *and* Π_i *is defined by* (3.3) *for* $i = 1, 2, \ldots, n-1$.

REMARK 3.2. Under the assumptions of the lemma, $\boldsymbol{\pi}$ is reversible if and only if $\prod_{j=0}^{n-1}(p_j/q_j) = 1$, in which case (3.3) simplifies considerably.

EXAMPLE 3.3. As a check of the formula, consider the case in which $p_0 = p_1 = \cdots = p_{n-1} = p \in (0,1)$ and $q_0 = q_1 = \cdots = q_{n-1} = q := 1-p$. Here the transition matrix is doubly stochastic, so the unique stationary distribution is discrete uniform on $\{0, 1, \ldots, n-1\}$. Indeed, algebraic simplification shows that $\Pi_0 = \Pi_1 = \cdots = \Pi_{n-1} = 1$.

EXAMPLE 3.4. Consider next the case in which $p_1 = p_2 = \cdots = p_{n-1} = p \in (0,1)$ and $q_1 = q_2 = \cdots = q_{n-1} = q := 1-p$. Of course p_0 and $q_0 := 1 - p_0$ may differ from p and q. Then $\Pi_0 := 1$ and

$$\Pi_i = -\bigg[\frac{(p_0/q)(p/q)^{n-1} - q_0/q}{(p/q)^n - 1}\bigg]((p/q)^i - 1) + (p_0/q)(p/q)^{i-1} \tag{3.4}$$

for $i = 1, 2, \dots, n-1$. It follows that

$$\begin{aligned}(3.5)\quad \sum_{i=0}^{n-1} \Pi_i &= 1 - \left[\frac{(p_0/q)(p/q)^{n-1} - q_0/q}{(p/q)^n - 1}\right]\left[\frac{(p/q)((p/q)^{n-1} - 1)}{p/q - 1} - (n-1)\right] \\ &\quad + \frac{(p_0/q)((p/q)^{n-1} - 1)}{p/q - 1} \\ &= 1 - \frac{p_0 - q_0}{p - q} + n\,\frac{p_0 p^{n-1} - q_0 q^{n-1}}{p^n - q^n},\end{aligned}$$

where the last step involves some algebra and we have implicitly assumed that $p \neq \frac{1}{2}$. In particular, π_0 is the reciprocal of (3.5). This result is useful in evaluating $\mu(r, \rho, \gamma A + (1-\gamma)B)$; see Section 4.

EXAMPLE 3.5. Consider finally the special case of Example 3.4 in which $p_0 = q$ and $q_0 = p$. Then (3.4) becomes

$$\Pi_i = -\left[\frac{(p/q)((p/q)^{n-2} - 1)}{(p/q)^n - 1}\right]((p/q)^i - 1) + (p/q)^{i-1}$$

for $i = 1, 2, \dots, n-1$, and (3.5) becomes

$$(3.6)\qquad \sum_{i=0}^{n-1} \Pi_i = 2 + npq\,\frac{p^{n-2} - q^{n-2}}{p^n - q^n}.$$

We have again implicitly assumed that $p \neq \frac{1}{2}$, and again π_0 is the reciprocal of (3.6). This result is useful in evaluating $\mu(r, 0, \gamma A + (1-\gamma)B)$; see Section 4.

4. Evaluation of rate of profit

Recall that mean profit has the form $\mu = \boldsymbol{\pi}\dot{\boldsymbol{P}}\boldsymbol{1}$, which we apply to $\boldsymbol{P} := \gamma \boldsymbol{P}_A + (1-\gamma)\boldsymbol{P}_B$.

To find $\mu(r, \rho, \gamma A + (1-\gamma)B)$, it suffices to note that $\boldsymbol{P}$ has the form (3.1) under the assumptions of Example 3.4 with $n := r$,

$$(4.1)\qquad p := \frac{\gamma}{2} + (1-\gamma)\frac{1}{1+\rho}, \quad \text{and} \quad p_0 := \frac{\gamma}{2} + (1-\gamma)\frac{\rho^{r-1}}{1+\rho^{r-1}},$$

where $0 < \rho < 1$. Thus,

$$(4.2)\qquad \mu(r, \rho, \gamma A + (1-\gamma)B) = \pi_0(p_0 - q_0) + (1 - \pi_0)(p - q),$$

with π_0 being the reciprocal of (3.5).

To find $\mu(r, 0, \gamma A + (1-\gamma)B)$, it suffices to note that $\boldsymbol{P}$ has the form (3.1) under the assumptions of Example 3.4 with $n := r$,

$$p := \frac{\gamma}{2} + (1-\gamma)1 = 1 - \frac{\gamma}{2}, \quad \text{and} \quad p_0 := \frac{\gamma}{2} + (1-\gamma)0 = \frac{\gamma}{2} = 1 - p = q.$$

We are therefore in the setting of Example 3.5, and

$$(4.3)\quad \mu(r, 0, \gamma A + (1-\gamma)B) = \pi_0(q - p) + (1 - \pi_0)(p - q) = (p - q)(1 - 2\pi_0),$$

with π_0 being the reciprocal of (3.6).

PROOF OF THEOREM 1.4. From (4.3) and (3.6) with $n = r$, we have

$$\begin{aligned}\mu(r, 0, \gamma A + (1-\gamma)B) &= (p-q)(1-2\pi_0)\\ &= (p-q)\left(1 - \frac{2(p^r - q^r)}{2(p^r - q^r) + rpq(p^{r-2} - q^{r-2})}\right)\\ &= \frac{rpq(p-q)(p^{r-2} - q^{r-2})}{2(p^r - q^r) + rpq(p^{r-2} - q^{r-2})},\end{aligned}$$

and the theorem follows by substituting $1 - \gamma/2$ and $\gamma/2$ for p and q. □

PROOF OF COROLLARY 1.5. We want to show that $\mu(r, 0, \gamma A + (1-\gamma)B)$ can be close to 1 by choosing $p := 1 - \gamma/2$ close to 1 and π_0 close to 0, which requires r large. So we consider a sequence $p \to 1$ as $r \to \infty$. In this case,

$$\begin{aligned}\pi_0 &= \frac{p^r - q^r}{2(p^r - q^r) + rpq(p^{r-2} - q^{r-2})}\\ &\sim \frac{p^r}{2p^r + rqp^{r-1}}\\ &= \frac{p}{2p + rq}.\end{aligned}$$

Now let us specify that $p = 1 - 1/\sqrt{r}$ (equivalently, $\gamma = 2/\sqrt{r}$). Then, by (4.3),

$$\begin{aligned}1 - \mu(r, 0, \gamma A + (1-\gamma)B) &= 1 - (p-q)(1-2\pi_0)\\ &\sim 1 - (1 - 2/\sqrt{r})\left(1 - \frac{2(1 - 1/\sqrt{r})}{2(1 - 1/\sqrt{r}) + \sqrt{r}}\right)\\ &\sim \frac{4}{\sqrt{r}},\end{aligned}$$

as required. □

PROOF OF COROLLARY 1.6. For even $r \geq 4$ and positive integers $s \leq r/2$, Theorem 1.3 implies that

$$\begin{aligned}&1 - \mu(r, 0, (AB)^s B^{r-2})\\ &\quad = 1 - \left(1 - \frac{2(s-1)}{r + 2(s-1)}\right)\left(1 - \frac{1}{2^s}\right)\\ &\quad = \frac{2s}{r + 2(s-1)} - \frac{2}{r + 2(s-1)} + \frac{1}{2^s} - \frac{2(s-1)}{r + 2(s-1)} \cdot \frac{1}{2^s},\end{aligned}$$

if initial capital is even. With s replaced by $s_r := \lfloor \log_2 r \rfloor - 1$, the first term is asymptotic to $2s_r/r$ as $r \to \infty$ and the remaining terms are $O(1/r)$.

For odd $r \geq 3$, the argument is essentially the same. □

PROOF OF THEOREM 1.2. It is enough to show that $\mu(r, \rho, \gamma A + (1-\gamma)B)$ is continuous at $\rho = 0$ for fixed r and γ. In fact, there is a complicated but explicit formula, given by (4.2), using (3.5) and (4.1), showing that it is a rational function of ρ. Therefore, we need only show that it does not have a pole at $\rho = 0$. In fact, Theorem 1.4 shows that $\mu(r, 0, \gamma A + (1-\gamma)B)$ is the ratio of two positive numbers, and this is sufficient. □

References

[1] A. Ajdari and J. Prost. *Drift induced by a spatially periodic potential of low symmetry: Pulsed dielectrophoresis.* C. R. Acad. Sci., Série 2 **315** (1992), 1635–1639.

[2] Richard C. Bradley, *Basic properties of strong mixing conditions. A survey and some open questions*, Probab. Surv. **2** (2005), 107–144, DOI 10.1214/154957805100000104. Update of, and a supplement to, the 1986 original. MR2178042

[3] Luis Dinis, *Optimal sequence for Parrondo games*, Phys. Rev. E (3) **77** (2008), no. 2, 021124, 6, DOI 10.1103/PhysRevE.77.021124. MR2453277

[4] S. N. Ethier and Jiyeon Lee, *Limit theorems for Parrondo's paradox*, Electron. J. Probab. **14** (2009), no. 62, 1827–1862, DOI 10.1214/EJP.v14-684. MR2540850

[5] S. N. Ethier and Jiyeon Lee, *How strong can the Parrondo effect be?*, J. Appl. Probab. **56** (2019), no. 4, 1198–1216, DOI 10.1017/jpr.2019.68. MR4041456

[6] G. P. Harmer and D. Abbott, *Parrondo's paradox*, Statist. Sci. **14** (1999), no. 2, 206–213, DOI 10.1214/ss/1009212247. MR1722065

[7] Ronald Pyke, *On random walks and diffusions related to Parrondo's games*, Mathematical statistics and applications: Festschrift for Constance van Eeden, IMS Lecture Notes Monogr. Ser., vol. 42, Inst. Math. Statist., Beachwood, OH, 2003, pp. 185–216. MR2138293

Department of Mathematics, University of Utah, 155 S. 1400 E., Salt Lake City, UT 84112

Email address: ethier@math.utah.edu

Department of Statistics, Yeungnam University, 280 Daehak-Ro, Gyeongsan, Gyeongbuk 38541, South Korea

Email address: leejy@yu.ac.kr

Contemporary Mathematics
Volume **774**, 2021
https://doi.org/10.1090/conm/774/15567

Binary response models comparison using the α-Chernoff divergence measure and exponential integral functions

Subir Ghosh and Hans Nyquist

In celebration of Professor M. M. Rao's influential contributions...

ABSTRACT. In this paper, the families of binary response models are describing the data on a response variable having two possible outcomes and p explanatory variables when the possible responses and their probabilities are functions of the explanatory variables. The α-Chernoff divergence measure and the Bhattacharyya divergence measure when $\alpha = 1/2$ are the criterion functions used for quantifying the dissimilarity between probability distributions by expressing the divergence measures in terms of the exponential integral functions. The dependences of odds ratio and hazard function on the explanatory variables are also a part of the modeling.

1. Introduction

Consider a Bernoulli random variable Y with $P(Y = 1) = \pi$ and $P(Y = 0) = 1-\pi$. Let $\mathbf{x}(p \times 1)$ be a known vector whose elements are values of the explanatory or predictor variables. At each $\mathbf{x}$, consider the Bernoulli random variable $Y(\mathbf{x})$ with $P(Y(\mathbf{x}) = 1) = \pi(\eta(\mathbf{x}))$ and $P(Y(\mathbf{x}) = 0) = 1 - \pi(\eta(\mathbf{x}))$. The mean $E(Y(\mathbf{x})) = P(Y(\mathbf{x}) = 1) = \pi(\eta(\mathbf{x}))$ and the variance $var(Y(\mathbf{x})) = \pi(\eta(\mathbf{x}))(1-\pi(\eta(\mathbf{x})))$, where $\pi(\eta(\mathbf{x}))$ is assumed to be continuous and differentiable. For generalized linear models (GLMs),

$$\eta(\mathbf{x}) = \psi(\mathbf{x}^\top \boldsymbol{\beta}) = \frac{\pi(\eta(\mathbf{x}))}{1 - \pi(\eta(\mathbf{x}))},$$

where ψ is a smooth monotonic invertible function and $\boldsymbol{\beta}$ is a $(p \times 1)$ vector of unknown parameters.

Consider a random variable $\mathbf{X}(p \times 1)$ and its realized value $\mathbf{x}$. The cumulative distribution function of $\eta(\mathbf{X})$ is $F(\eta(\mathbf{x}))$, $0 \leq F(\eta(\mathbf{x})) \leq 1$, which is assumed to be continuous and differentiable. By setting $\pi(\eta(\mathbf{x})) = F(\eta(\mathbf{x}))$ and writing for simplicity $\eta(\mathbf{x}) = \eta$ and $F(\eta(\mathbf{x})) = F(\eta)$, it follows that

$$\frac{d}{d\eta}\pi(\eta) = \frac{d}{d\eta}F(\eta) = f(\eta),$$

2020 *Mathematics Subject Classification.* Primary 62J12, 62N05, 65C20, 62B10; Secondary 62P30, 62P10, 62P20.

Key words and phrases. Binary response, divergence, exponential family, exponential integral functions, hazard function, odds ratio, probability distributions.

where $f(\eta)$ is the probability density function of η. The right-sided cumulative distribution function $1 - F(\eta)$ is called the survivor function $S(\eta)$ (Cox [**C5**], Cox and Oakes [**17**]). The hazard rate (or failure rate) $h(\eta)$ (Barlow, Marshall and Proschan [**6**], Barlow and Proschan [**7**], Cox and Oaks [**17**], Kalbfleisch and Prentice [**26**]) is defined as

$$h(\eta) = \frac{f(\eta)}{1 - F(\eta)} = -\frac{d}{d\eta} log(1 - F(\eta)) = -\frac{d}{d\eta} log(S(\eta)).$$

The Mills' ratio $m(\eta)$ (Mills [**M**], Small [**S1**]) is known as

$$m(\eta) = \frac{1 - F(\eta)}{f(\eta)} = \frac{S(\eta)}{f(\eta)} = \frac{1}{h(\eta)}.$$

The hazard rate $h(\eta)$ uniquely determines $F(\eta)$ or $(1 - F(\eta))$ as

$$1 - F(\eta) = exp\left\{ -\int_0^{\eta} h(u)du \right\}.$$

The $\Phi(\cdot)$ is the standard normal cumulative distribution function. Some popular choices of $F(\eta)$ in practice are given in Table 1.

TABLE 1. Some popular choices of $F(\eta)$

Name	$F(\eta) = \pi(\eta)$	Support
Logistic	$\frac{e^{\eta}}{1+e^{\eta}}$	$-\infty < \eta < \infty$
Probit	$\Phi(\eta)$	$-\infty < \eta < \infty$
Extreme Value	$1 - e^{-e^{-\eta}}$	$-\infty < \eta < \infty$
Lomax	$\frac{\eta}{1+\eta}$	$0 \leq \eta < \infty$
Exponential	$1 - e^{-\theta\eta}$	$0 \leq \eta < \infty, \theta > 0$
Weibull	$1 - e^{-\eta^{\theta}}$	$0 \leq \eta < \infty, \theta > 0$
Pareto	$1 - (a/\eta)^{\theta}$	$a \leq \eta < \infty, a > 0, \theta > 0$

For comparing two models $\pi(\eta) = F(\eta)$ in Table 1 to describe the data collected from an experiment or observational study, a problem of practical importance is the estimation of the mean $E(Y(\mathbf{x})) = \pi(\eta(\mathbf{x}))$ or the estimation of parameters in $\boldsymbol{\beta}$. Fisher[**F1**] introduced the concepts of *consistency, efficiency* and *sufficiency* of estimating functions and advocated the use of the *maximum likelihood* method. Darmois[**18**], Koopman[**27**], and Pitman[**P1**] characterized distributions admitting sufficient statistic of fixed dimensionality regardless of the sample size. In the research advancement made by Neyman[**N**], Rao[**33**], Halmos and Savage[**24**], Lehmann and Scheffe[**29**], Kullback and Leibler[**28**], Dynken[**19**], Bahadur[**1**], Basu[**8**], Barankin and Katz[**2**], Barankin[**3**], Fraser[**21**], Barankin and Maitra[**4**], Barndorff-Nielsen[**5**], Efron[**20**], Lehman[**30**] and many others, sufficiency was at the core of development. The distributions in the exponential family have sufficient statistics for their vector of parameters. Consequently, the exponential family of distributions is a desirable class of distributions to begin the search for finding the distribution describing the data best (Sundberg [**35**]). Section 2 presents the details about the exponential family of distributions relevant to this paper. Example 1 in Section 3 compares two exponential densities. Example 2 in Section 3 compares two Pareto densities with respect to the α-Chernoff divergence and the Bhattacharyya divergence measures. Example 3 in Section 5 compares a Lomax density with an exponential density. The α-Chernoff divergence measures are expressed in terms of the exponential integral functions for different values of α. This paper proposes two alternative classes of distributions characterized by the differential equations. Section 4 describes the first class, defined by the condition of $\pi(\eta)$ as an increasing function of η in terms of the differential equation

$$\frac{d}{d\eta}\pi(\eta) = c(1-\pi(\eta))^{\gamma}, \tag{1}$$

where c and γ are positive constants. Section 6 describes the second class, defined by the condition of $\pi(\eta)$ in terms of the differential equation

$$\begin{aligned} \frac{d}{d\eta}\pi(\eta) &= f(\eta) \\ &= \frac{f(\eta)}{(1-F(\eta))^{\gamma}}(1-F(\eta))^{\gamma} \\ &= c(\eta)(1-F(\eta))^{\gamma} \\ &= c(\eta)(1-\pi(\eta))^{\gamma}, \end{aligned} \tag{2}$$

where $c(\eta)$ is

$$c(\eta) = \frac{f(\eta)}{(1-F(\eta))^{\gamma}}. \tag{3}$$

2. Exponential family of models

DEFINITION 1. The family of probability density functions for the generalized linear models (GLMs), Nelder and Wedderburn[**32**], McCulloch and Nelder[**31**], and the exponential dispersion models, Jørgensen[**25**], is given by

$$f(\eta;\theta,\phi) = exp\left\{\frac{[\eta\theta - b(\theta)]}{a(\phi)} + c(\eta,\phi)\right\}$$

for some functions $a(\cdot)$, $b(\cdot)$ and $c(\cdot)$ and parameters θ and ϕ. The θ is called the *canonical parameter* and ϕ the *dispersion parameter*.

DEFINITION 2. When ϕ is known, the family of probability density functions in Definition 1 simplifies into the natural exponential family which is

$$\begin{aligned} f(\eta;\theta) &= f(\eta;\theta,\phi) \\ &= R(\theta,\phi)T(\eta,\phi)exp\{\eta Q(\theta,\phi)\} \\ &= R(\theta)T(\eta)exp\{\eta Q(\theta)\}, \quad \eta \in H, \end{aligned} \tag{4}$$

where $Q(\theta) = Q(\theta, \phi) = (\theta/a(\phi))$, $R(\theta) = R(\theta, \phi) = exp\{-b(\theta)/a(\phi)\}$ and $T(\eta) = T(\eta, \phi) = exp\{c(\eta, \phi)\}$. Moreover,

$$\begin{aligned} R(\theta) &= R(\theta, \phi) \\ &= \frac{1}{\int_{\eta \in H} T(\eta, \phi) exp\left\{\eta Q(\theta, \phi)\right\} d\eta} \\ &= \frac{1}{\int_{\eta \in H} T(\eta) exp\left\{\eta Q(\theta)\right\} d\eta}. \end{aligned}$$

The representation of the natural exponential family in Definition 2 may or may not be possible for the exponential family when ϕ is unknown.

Denote the cumulative distribution function by $F(\eta; \theta, \phi)$ and the hazard function by $h(\eta; \theta, \phi)$. The general closed form expression of $F(\eta; \theta, \phi)$ is hard to find. On the other hand, for some specific functional forms of $h(\eta; \theta, \phi)$, it is easy to find the closed form expression of $F(\eta; \theta, \phi)$. Following the Cox's proportional hazards model of survival analysis (Cox[**16**], Cox and Oakes[**17**]), it is natural to set

$$h(\eta; \theta, \phi) = h_0(\eta) g(\theta, \phi), \tag{5}$$

where $h_0(\eta)$ is an arbitrary baseline hazard rate. Two members of the natural exponential family with their $f(\eta; \theta)$, $F(\eta; \theta)$, and $h(\eta; \theta)$ are given below.

Assuming $a(\phi) = 1$ and $R(\theta) = exp\{-b(\theta)\} = -Q(\theta) = \theta$, the expression of $f(\eta; \theta, \phi)$ becomes the exponential probability density function as

$$f(\eta; \theta) = \theta exp\left\{-\theta\eta\right\} \quad I(\eta \geq 0), \tag{6}$$

where $I(\eta \geq 0)$ is the indicator function. It is easy to find the formula of the cumulative probability distribution function $F(\eta; \theta) = 1 - exp\left\{-\theta\eta\right\}$. The hazard function $h(\eta; \theta) = \theta$ which is not dependent on η and hence a constant for the exponential distribution.

For the Pareto density function

$$f(\eta; \theta) = \frac{\theta a^{\theta}}{\eta^{\theta+1}} I(\eta \geq a), \tag{7}$$

where $a \leq \eta < \infty, a > 0, \theta > 0$ and a is known. The $f(\eta; \theta)$ belongs to the natural exponential family when $logR(\theta) = log\theta + \theta loga$ and $\eta Q(\theta) + logT(\eta) = -(\theta+1)log\eta$. For the Pareto distribution in (7)

$$F(\eta; \theta) = 1 - \left(\frac{a}{\eta}\right)^{\theta}, \quad h(\eta; \theta) = \frac{\theta}{\eta} = \frac{\theta}{a}\left(1 - F(\eta; \theta)\right)^{\frac{1}{\theta}}.$$

When $T(\eta) = f_0(\eta), \eta \in H$, is a known probability density function, $f_0(\eta)$ can be treated as *embedded* in an exponential family of densities

$$f(\eta, \theta) = \frac{f_0(\eta) exp\left\{\eta Q(\theta)\right\}}{\int_{\eta \in H} f_0(\eta) exp\left\{\eta Q(\theta)\right\} d\eta} = R(\theta) T(\eta) exp\left\{\eta Q(\theta)\right\}, \qquad \eta \in H, \tag{8}$$

by *exponential tilting* (Sundberg[**35**]). Clearly, $Q(\theta) = Q(0) = 0$ and $f(\eta, \theta) = f(\eta, 0) = f_0(\eta)$ when $\theta = 0$.

For two members of exponential family of densities

$$f^{(i)}(\eta; \theta^{(i)}) = R^{(i)}(\theta^{(i)}) T^{(i)}(\eta) exp\left\{\eta Q^{(i)}(\theta^{(i)})\right\}, \qquad \eta \in H, \qquad i = 1, 2, \tag{9}$$

a *fusion* of densities (Goodman, Mahler, and Nguyen[**23**]) is

$$(10)\quad f(\eta;\theta^{(1)},\theta^{(2)}) = \frac{\left(f^{(1)}(\eta;\theta^{(1)})\right)^{\alpha}\left(f^{(2)}(\eta;\theta^{(2)})\right)^{1-\alpha}}{\int_{\eta\in H}\left(f^{(1)}(\eta;\theta^{(1)})\right)^{\alpha}\left(f^{(2)}(\eta;\theta^{(2)})\right)^{1-\alpha}d\eta}, \qquad 0\le\alpha\le 1.$$

3. The α-Chernoff divergence

DEFINITION 3. The *α-Chernoff divergence measure* (Chernoff[**13**],[**14**]) between two probability distributions with their probability density functions $f^{(i)}(\eta;\theta^{(i)})$, $i=1,2$, is defined as

$$(11)\quad \begin{aligned} &C_\alpha(f^{(1)},f^{(2)}) = -log\left(\int_{\eta\in H}\left(f^{(1)}(\eta;\theta^{(1)})\right)^{\alpha}\left(f^{(2)}(\eta;\theta^{(2)})\right)^{1-\alpha}d\eta\right),\\ &0\le\alpha\le 1.\end{aligned}$$

DEFINITION 4. When $\alpha=1/2$, the Chernoff divergence measure in Definition 3 becomes the *Bhattacharyya divergence measure* (Bhattacharyya[**10**],[**11**]), Kailath[**K1**])

$$(12)\quad B(f^{(1)},f^{(2)}) = -log\left(\int_{\eta\in H}\sqrt{(f^{(1)}(\eta;\theta^{(1)})f^{(2)}(\eta;\theta^{(2)}))}\;d\eta\right).$$

EXAMPLE 1. For two exponential probability densities in (6)

$$(13)\quad f^{(1)}(\eta)=0.2\quad e^{-0.2\eta}\quad I(\eta\ge 0), f^{(2)}(\eta)=0.5\quad e^{-0.5\eta}\quad I(\eta\ge 0),$$

having $\theta^{(1)}=0.2$ and $\theta^{(2)}=0.5$, it can be seen that

$$(14)\quad \begin{aligned} C_\alpha(f^{(1)},f^{(2)}) &= -log\left(\frac{(0.2)^{\alpha}(0.5)^{1-\alpha}}{[0.2\alpha+0.5(1-\alpha)]}\right),\\ B(f^{(1)},f^{(2)}) &= 0.1014704.\end{aligned}$$

It can be checked that the maximum value of $C_\alpha(f^{(1)},f^{(2)})$ is 0.1037472 at $\alpha=0.575317$ which is in between the values 0.55 and 0.6 of α. The value of $C_\alpha(f^{(1)},f^{(2)})$ is 0.1034873 at $\alpha=0.6$ and it is 0.1034823 at $\alpha=0.55$. Figure 2 presents the graphs of $C_\alpha(f^{(1)},f^{(2)})$ against α for $0\le\alpha\le 1$ and $0.55\le\alpha\le 0.60$.

EXAMPLE 2. For two Pareto probability densities in (7)

$$(15)\quad f^{(1)}(\eta)=\frac{192}{\eta^4}\quad I(\eta\ge 4), f^{(2)}(\eta)=\frac{1024}{\eta^5}\quad I(\eta\ge 4),$$

having $(\theta^{(1)}=3, a^{(1)}=4)$ and $(\theta^{(2)}=4, a^{(2)}=4)$, it can be seen that

$$(16)\quad \begin{aligned} C_\alpha(f^{(1)},f^{(2)}) &= -log\left(\frac{(192)^{\alpha}(1024)^{1-\alpha}}{(4-\alpha)4^{(4-\alpha)}}\right),\\ B(f^{(1)},f^{(2)}) &= 0.01030964.\end{aligned}$$

It can be checked that the maximum value of $C_\alpha(f^{(1)},f^{(2)})$ is 0.01033325 at $\alpha=0.5239485$ which is very close to the Bhattacharyya divergence $B(f^{(1)},f^{(2)})$ value in (16). The value of $C_\alpha(f^{(1)},f^{(2)})$ is 0.01009031 at $\alpha=0.6$. Figure 4 presents the graph of $C_\alpha(f^{(1)},f^{(2)})$ against $\alpha, 0\le\alpha\le 1$.

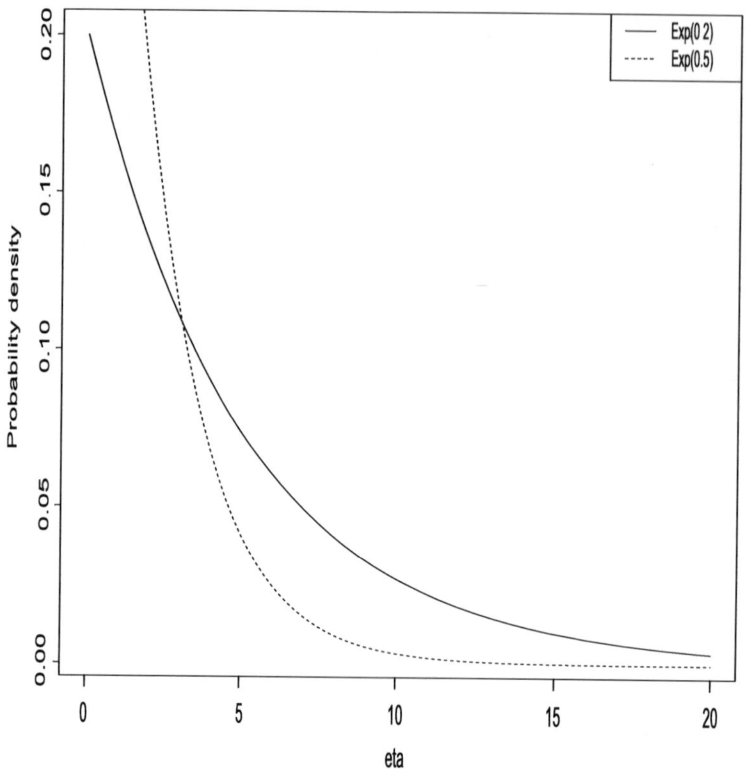

FIGURE 1. Plot of $f^{(1)}(\eta) = Exp(0.2)$ and $f^{(2)}(\eta) = Exp(0.5)$ against $\eta \in [0, 20]$

4. First family of models

Suppose that $\pi(\eta)$ satisfies the differential equation (1).

THEOREM 1. *A general solution of $\pi(\eta)$ satisfying the differential equation* (1) *for $\gamma = 1$ and $\pi(a) = 0$ is*

$$\pi(\eta) = 1 - e^{-c(\eta - a)}. \tag{17}$$

PROOF. When $\gamma = 1$, (1) becomes

$$\frac{d}{d\eta}\pi(\eta) = c(1 - \pi(\eta)),$$

or, equivalently,

$$\frac{d}{d\eta}(1 - \pi(\eta)) = (-c)(1 - \pi(\eta)). \tag{18}$$

A general solution of (18) is

$$1 - \pi(\eta) = de^{-c\eta}, \tag{19}$$

where d is a constant. The condition $\pi(a) = 0$ and the equation (19) imply that

$$d = e^{-ca}, \pi(\eta) = 1 - de^{-c\eta} = 1 - e^{-c(\eta - a)}.$$

□

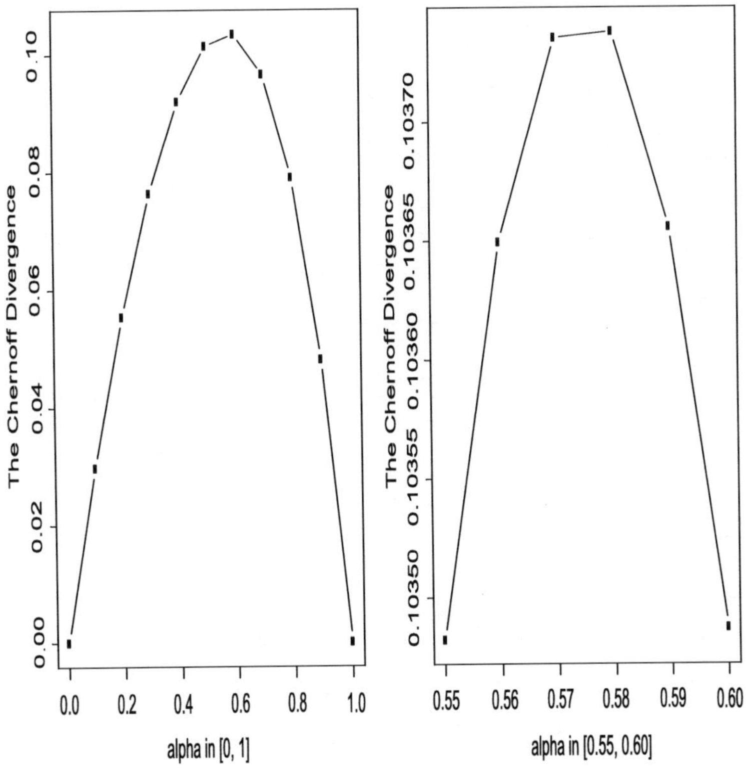

FIGURE 2. Plots of $C_\alpha(f^{(1)}, f^{(2)})$ against α in Example 1

THEOREM 2. *A general solution of* $\pi(\eta)$ *satisfying the differential equation* (1) *for* $\gamma \neq 1$ *and* $\pi(a) = 0$ *is*

$$\pi(\eta) = 1 - \left[1 - \frac{c(\eta - a)}{\lambda}\right]^{\lambda}, \qquad \text{for} \quad \gamma \neq 1 \quad \text{and} \quad \lambda = \frac{1}{(1-\gamma)}. \tag{20}$$

PROOF. When $\gamma \neq 1$, define

$$u(\eta) = (1 - \pi(\eta))^{1-\gamma}, \qquad \gamma \neq 1. \tag{21}$$

By using the chain rule and the equation (1), it follows from (21)

$$\begin{aligned} &\frac{d}{d\eta} u(\eta) \\ &= \left(\frac{d}{d\pi(\eta)} (1 - \pi(\eta))^{1-\gamma}\right) \frac{d}{d\eta} \pi(\eta) \\ &= -(1-\gamma)(1 - \pi(\eta))^{-\gamma} \frac{d}{d\eta} \pi(\eta) \\ &= -(1-\gamma)c. \end{aligned} \tag{22}$$

A general solution of (22) is

$$u(\eta) = -(1-\gamma)c\eta + d, \tag{23}$$

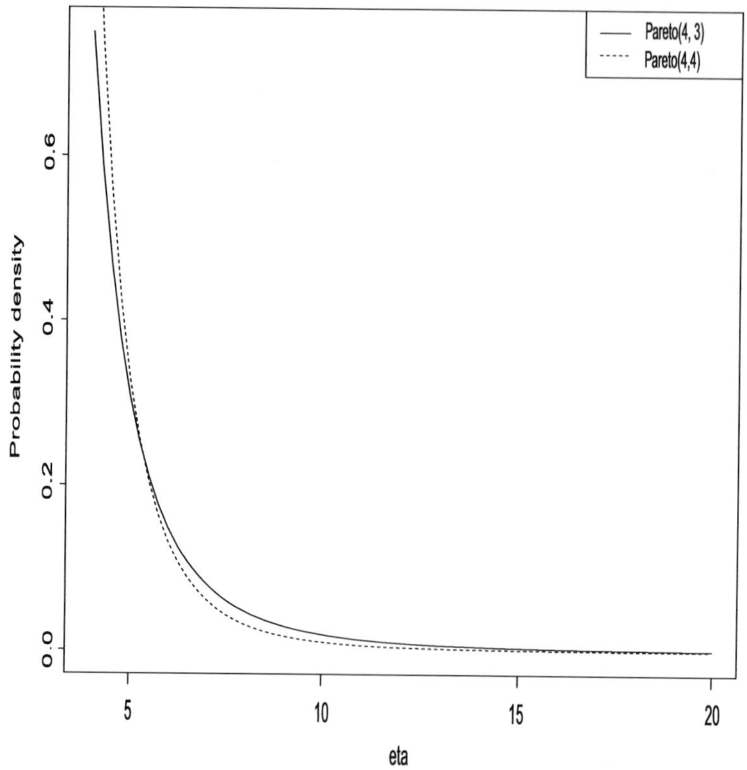

FIGURE 3. Plot of $f^{(1)}(\eta) = Pareto(3,4)$ and $f^{(2)}(\eta) = Pareto(4,4)$ against $\eta \in [4,20]$

where d is a constant. The condition $\pi(a) = 0$ and the equation (20) imply that $u(a) = 1$ and therefore, from (23), $d = 1 + (1-\gamma)c\ a$. It follows from (21) and (23)

$$u(\eta) = (1 - \pi(\eta))^{1-\gamma} = 1 - (1-\gamma)c(\eta - a). \tag{24}$$

It can be seen from (24)

$$\begin{aligned} \pi(\eta) &= 1 - (u(\eta))^{\frac{1}{\gamma-1}} \\ &= 1 - [1 - (1-\gamma)c(\eta - a)]^{\frac{1}{1-\gamma}} \\ &= 1 - \left[1 - \frac{c(\eta - a)}{\lambda}\right]^{\lambda}. \end{aligned}$$

□

When $\lambda \to \infty$, it follows from (20)

$$\lim_{\lambda\to\infty} \pi(\eta) = 1 - \lim_{\lambda\to\infty} \left[1 - \frac{c(\eta - a)}{\lambda}\right]^{\lambda} = 1 - e^{-c(\eta - a)}, \tag{25}$$

which is a general solution of (1) for $\gamma = 1$, given in (17).

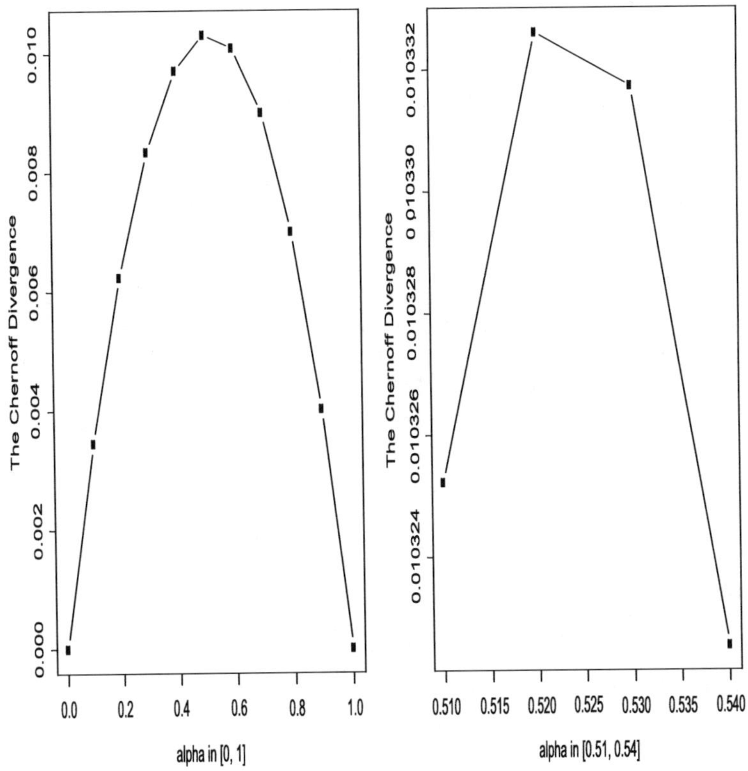

FIGURE 4. Plots of $C_\alpha(f^{(1)}, f^{(2)})$ against α in Example 2.

Assuming the constant c to be equal to 1 and a to be zero, it follows from Theorem 2 that a family of models emerges from the differential equation in (1)

$$\pi(\eta) = \begin{cases} 1 - e^{-\eta}, \text{for} & \gamma = 1, \\ 1 - \left[1 - \frac{\eta}{\lambda}\right]^{\lambda}, & \text{for} \quad \gamma \neq 1. \end{cases} \tag{26}$$

For $\gamma = 2$ or equivalently $\lambda = -1$, the Lomax distribution (Lomax[**L2**]) in Table 1 obtained from (26) is expressed as

$$\pi(\eta) = 1 - \frac{1}{1+\eta} = \frac{\eta}{1+\eta}, \qquad \frac{\pi(\eta)}{1-\pi(\eta)} = \eta. \tag{27}$$

The property $0 \leq \pi(\eta) \leq 1$ implies that $\eta \geq 0$. For all $\eta \geq 0$, $\pi(\eta) \leq 1$ and the "=" holds approximately for all practical considerations when η becomes very large. The $\pi(\eta) = 1/2$ when $\eta = 1$ and $\pi(\eta) = 0$ when $\eta = 0$. The model in (20) for $\gamma = 2$ or equivalently $\lambda = -1$, is investigated in Ghosh and Nyquist[**22**] assuming $\eta = \psi(\mathbf{x}^\top \boldsymbol{\beta}) = \beta_0 + \mathbf{x}^\top \boldsymbol{\beta} = [\pi(\eta)/1 - \pi(\eta)]$. Furthermore, the model in (20) for $\gamma = 2$ or equivalently $\lambda = -1$, becomes the popular logistic regression model (Berkson ([**B5**], [**9**]), Cox([**15**],[**16**])) assuming $\eta = \psi(\mathbf{x}^\top \boldsymbol{\beta}) = e^{\beta_0 + \mathbf{x}^\top \boldsymbol{\beta}} = [\pi(\eta)/1 - \pi(\eta)]$. For the logistic regression model, *logit* $\pi(\eta) = log\,[\pi(\eta)/1 - \pi(\eta)] = \beta_0 + \mathbf{x}^\top \boldsymbol{\beta}$. Thus two different models, the model in Ghosh and Nyquist[**22**] as well as the logistic regression model, belong to the family of models in (20).

Assuming $\eta = \psi(\mathbf{x}^\top \boldsymbol{\beta})$, $\gamma = (\theta+1)/\theta$, and the Pareto cumulative distribution function for $\pi(\eta)$ as

$$\pi(\eta) = 1 - \left(\frac{a}{\eta}\right)^{\theta} = 1 - \left(\frac{\eta}{a}\right)^{-\theta}, \quad 0 < a \le \eta < \infty, \quad \theta > 0, \tag{28}$$

it can be seen that

$$\frac{d}{d\eta}\pi(\eta) = \frac{\theta}{a}\left(\frac{a}{\eta}\right)^{\theta+1} = \frac{\theta}{a}\left[\left(\frac{a}{\eta}\right)^{\theta}\right]^{\frac{\theta+1}{\theta}} = c\,(1-\pi(\eta))^{\gamma}.$$

Hence, $\pi(\eta)$ in (28) satisfies the differential equation in (1) for $c = (\theta/a)$ and $\gamma = (\theta+1)/\theta$, where $\gamma \ge 1$ and $c > 0$.

Consider now $\eta = \psi(\mathbf{x}^\top \boldsymbol{\beta})$, where $0 \le a \le \eta < b \le \infty$, a and b are real numbers, and $\psi(.)$ is a meaningful function. The $\pi(\eta)$ is an increasing function of η, $\pi(a) = 0$, $\pi(b) = 1$, $\pi(\eta)$ satisfies (1). When $\gamma \ne 1$, $u(\eta)$ in (24) satisfies $u(a) = 1$ and $u(b) = 0$. Hence

$$(1-\gamma)c = \frac{1}{b-a}, d = \frac{b}{b-a}, u(\eta) = \frac{b-\eta}{b-a} = 1 - \frac{\eta-a}{b-a},$$
$$\pi(\eta) = 1 - \left(1 - \frac{\eta-a}{b-a}\right)^{\frac{1}{1-\gamma}} = 1 - \left(1 - \frac{c(\eta-a)}{\lambda}\right)^{\lambda}.$$

Therefore, the above expression of $\pi(\eta)$ is exactly same as in (20) and consequently, (20) holds as well. When $\gamma = 1$, it follows from (1) that

$$\frac{d}{d\eta} log(1-\pi(\eta)) = (-c),$$

which has a general solution without satisfying two conditions $\pi(a) = 0$ and $\pi(b) = 1$,

$$1 - \pi(\eta) = ue^{-c\eta+v}.$$

The condition $\pi(a) = 0$ implies that $ue^{v} = e^{ca}$ and therefore $1 - \pi(\eta) = e^{-c(\eta-a)}$ which is the expression of $1-\pi(\eta)$ in (17), when $b = \infty$. For the extreme situations where $b = \infty$ or $c = \infty$, the condition $\pi(b) = 1$ holds. The situation $c = \infty$ does not provide a meaningful interpretation of the differential equation in (1).

5. Exponential integral function and α-Chernoff divergence

DEFINITION 5. The incomplete gamma function $\Gamma(w, x)$ is defined as

$$\Gamma(w,x) = \int_x^{\infty} t^{w-1} e^{-t} dt, \qquad x > 0, w \ge 0. \tag{29}$$

DEFINITION 6. The exponential integral function $E_n(s, x)$ of order n is defined as

$$E_n(s,x) = \int_x^{\infty} t^{-n} e^{-st} dt, \qquad x > 0, n \ge 0, s \ge 0. \tag{30}$$

When $x = 1$, the exponential integral function $E_n(s, 1)$ is denoted by $E_n(s)$. Taking $s = 1$ and $n = 1$ in $E_n(s, x)$ and $w = 0$ in $\Gamma(w, x)$, it follows that

$$E_1(1,x) = \Gamma(0,x) = \int_x^{\infty} t^{-1} e^{-t} dt = \int_x^{\infty} \frac{e^{-t}}{t} dt. \tag{31}$$

For $x = 1$ in (31),

$$E_1(1) = E_1(1,1) = \Gamma(0,1) = \int_1^\infty \frac{e^{-t}}{t} dt. \tag{32}$$

DEFINITION 7. Let $f(t)$ be defined for $t \geq 0$. The Laplace transformation of $f(t)$, denoted by $\mathcal{L}(f(t))$ or by $L(s)$, is defined as

$$\mathcal{L}(f(t)) = L(s) = \int_0^\infty f(t) e^{-st} dt, \tag{33}$$

provided the integral is convergent.

For $f(t) = 1/(1+t)$, it can be seen that

$$\begin{aligned} \int_0^\infty \frac{e^{-st}}{t+1} dt = e^s E_1(s) \qquad &= e^s \int_1^\infty \frac{e^{-st}}{t} dt, \\ \int_0^\infty \frac{e^{-\frac{t}{2}}}{t+1} dt = e^{\frac{1}{2}} E_1\left(\frac{1}{2}\right) \quad &= e^{\frac{1}{2}} \int_1^\infty \frac{e^{-\frac{t}{2}}}{t} dt. \end{aligned} \tag{34}$$

EXAMPLE 3. For two probability distributions: Lomax and exponential in Table 1, having the probability densities

$$f^{(1)}(\eta) = \frac{1}{(1+\eta)^2} \quad I(\eta \geq 0), f^{(2)}(\eta) = \quad e^{-\eta} \quad I(\eta \geq 0), \tag{35}$$

it can be seen from (11), (12), and (34) that

$$\begin{aligned} C_\alpha(f^{(1)}, f^{(2)}) &= -log\left(\int_0^\infty \frac{e^{-\alpha\eta}}{(1+\eta)^{2(1-\alpha)}} d\eta\right) \\ &= -log\left(e^\alpha \int_1^\infty \frac{e^{-\alpha\eta}}{\eta^{2(1-\alpha)}} d\eta\right) \\ &= -log\left(e^\alpha E_{2(1-\alpha)}(\alpha)\right), \\ B(f^{(1)}, f^{(2)}) &= -log\left(\int_0^\infty \frac{e^{-\frac{\eta}{2}}}{1+\eta} d\eta\right) \\ &= -log\left(e^{\frac{1}{2}} \int_1^\infty \frac{e^{-\frac{\eta}{2}}}{\eta} d\eta\right) \\ &= -log\left(e^{\frac{1}{2}} E_1\left(\frac{1}{2}\right)\right). \end{aligned} \tag{36}$$

Both the codes "$> expint\_E1(0.5, scale = FALSE)$" and "$> expint.E1$ $(0.5, deriv = 0)$" in the R console, give the same value $E_1\left(\frac{1}{2}\right) = 0.5597736$ and $B(f^{(1)}, f^{(2)}) = 0.08022287 = C_{\frac{1}{2}}(f^{(1)}, f^{(2)})$. It can be seen that

$$\begin{aligned} E_2(0) &= \int_1^\infty \frac{1}{t^2} dt = 1, \\ E_0(1) &= \int_1^\infty e^{-t} dt = \frac{1}{e}. \end{aligned}$$

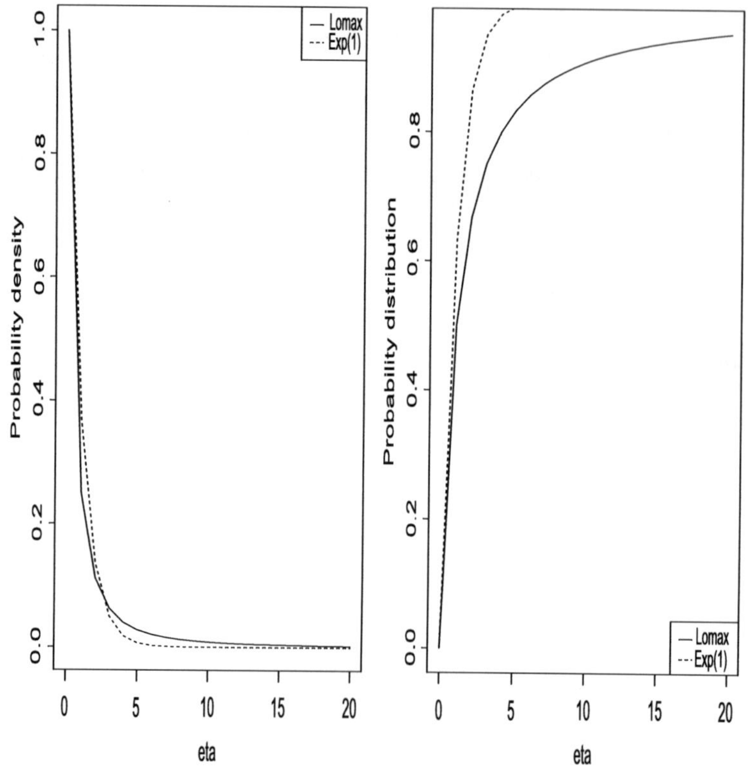

FIGURE 5. Plot of $f^{(1)}(\eta) = Lomax$ and $f^{(2)}(\eta) = Exp(1)$ against $\eta \in [0, 20]$

Hence the values of α-*Chernoff divergence measure* $C_\alpha(f^{(1)}, f^{(2)})$ for $\alpha = 0, 1/2$, and $\alpha = 1$ are

$$\begin{aligned}
C_0(f^{(1)}, f^{(2)}) &= -log(E_2(0)) = -log(1) = 0,\\
C_{\frac{1}{2}}(f^{(1)}, f^{(2)}) &= 0.08022287,\\
C_1(f^{(1)}, f^{(2)}) &= -log(eE_0(1)) = -log(ee^{-1}) = -log(1) = 0.
\end{aligned}$$

6. Second family of models

Denote $c(\eta)$ as

$$c(\eta) = \frac{f(\eta)}{(1 - F(\eta))^\gamma} = \frac{1}{m(\eta)(1 - F(\eta))^{\gamma-1}}. \tag{37}$$

When $\gamma = 1$, it follows from (37)

$$c(\eta) = \frac{1}{m(\eta)}.$$

Also, from (2), (3) and (37), the $\pi(\eta)$ $(= F(\eta))$ for the second family of models satisfies

$$\begin{aligned} \frac{d}{d\eta}\pi(\eta) = f(\eta) &= (1-F(\eta))^{\gamma}\frac{f(\eta)}{(1-F(\eta))^{\gamma}} \\ &= c(\eta)(1-F(\eta))^{\gamma} \\ &= c(\eta)(1-\pi(\eta))^{\gamma}. \end{aligned} \tag{38}$$

When $c(\eta) = c$, where c is a constant which does not depend on η, the equation (38) becomes exactly equal to the equation (1). For $0 < \eta < \infty$, the function $1 - F(\eta)$ is called the survival function and the $c(\eta)$, for $\gamma = 1$, is called the hazard function (Cox ([**15**], [**16**]) which is the inverse Mills' ratio (Mills[**M**]).

Consider the Weibull distribution with

$$F(\eta) = 1 - e^{-(\theta\eta)^{\delta}}, f(\eta) = \delta\theta^{\delta}\eta^{\delta-1}e^{-(\theta\eta)^{\delta}}, \qquad 0 \le \eta < \infty, \theta > 0, \delta > 0, \tag{39}$$

where θ and δ are constants that do not depend on η. The expression of $c(\eta)$ in (37) by using the expressions of $F(\eta)$ and $f(\eta)$ in (39), can be written as

$$c(\eta) = \frac{\delta\theta^{\delta}\eta^{\delta-1}e^{-(\theta\eta)^{\delta}}}{\left(e^{-(\theta\eta)^{\delta}}\right)^{\gamma}}, \tag{40}$$

and the expression of $c(\eta)$ in (40) for $\delta = 1$ becomes

$$c(\eta) = \theta e^{-\theta\eta(1-\gamma)}. \tag{41}$$

When $\delta = 1$, the Weibull distribution in (39) becomes the exponential distribution with

$$F(\eta) = 1 - e^{-\theta\eta}, f(\eta) = \theta e^{-\theta\eta}, \qquad 0 \le \eta < \infty, \theta > 0, \tag{42}$$

and $c(\eta)$ in (41). For $\gamma = 1$, the expression of $c(\eta)$ in (41) becomes $c(\eta) = c = \theta$, a constant independent of η. Consequently, the exponential model in (42) belongs to the family of models satisfying (1) for $\gamma = 1$.

For the logistic distribution with

$$F(\eta) = \frac{1}{1+e^{-\frac{\eta-s}{t}}}, \quad f(\eta) = \frac{e^{-\frac{\eta-s}{t}}}{t(1+e^{-\frac{\eta-s}{t}})^2}, \quad t > 0, -\infty < s, \eta < \infty, \tag{43}$$

it can be seen

$$f(\eta) = \frac{1}{t}F(\eta)(1-F(\eta)). \tag{44}$$

By using (43) and (44), the $c(\eta)$ in (37) can be expressed as

$$c(\eta) = \frac{F(\eta)}{t\ \ (1-F(\eta))^{\gamma-1}}. \tag{45}$$

When $\gamma = 2$, it follows from (37) that the $c(\eta)$ in (45) is

$$c(\eta) = \frac{F(\eta)}{t\ \ (1-F(\eta))} = \frac{1}{t}e^{\frac{\eta-s}{t}}. \tag{46}$$

For the logistic distribution, the (37) holds when $\gamma = 2$ and therefore

$$\frac{d}{d\eta}\pi(\eta) = f(\eta) = c(\eta)(1-\pi(\eta))^2 = c(\eta)(1-F(\eta))^2 = \frac{1}{t}F(\eta)(1-F(\eta)). \tag{47}$$

The Pareto distribution has

$$F(\eta) = 1 - \left(\frac{a}{\eta}\right)^{\theta}, \quad f(\eta) = \frac{\theta a^{\theta}}{\eta^{\theta+1}}, \tag{48}$$

where $a \leq \eta < \infty, a > 0, \theta > 0$. Hence

$$c(\eta) = \frac{f(\eta)}{(1 - F(\eta))^{\gamma}} = \theta a^{\theta(1-\gamma)} \eta^{\theta(\gamma-1)-1}. \tag{49}$$

Choosing $\gamma = (1/\theta) + 1$, $c(\eta) = (\theta/a)$ is a constant which does not depend on η and consequently, the Pareto model in (48) belongs to the family of models satisfying (1) for $\gamma = (1/\theta) + 1$.

Consider m distribution functions $F_i(\eta)$, $i = 1, \ldots, m$, satisfying

$$\frac{d}{d\eta} F_i(\eta) = k_i(1 - F_i(\eta)), \qquad i = 1, \ldots, m, \tag{50}$$

where k_i, $i = 1, \ldots, m$, are positive constants. Define

$$G(\eta) = \sum_{i=1}^{m} p_i F_i(\eta), \qquad p_i > 0, \quad i = 1, \ldots, m, \qquad \sum_{i=1}^{m} p_i = 1. \tag{51}$$

THEOREM 3. *A necessary and sufficient condition for $G(\eta)$ in (51) to satisfy the equation*

$$\frac{d}{d\eta} G(\eta) = k(1 - G(\eta)),$$

for a positive constant k is that

$$\sum_{i=1}^{m} p_i k_i - \sum_{i=1}^{m} p_i k_i F_i(\eta) = k\left(1 - \sum_{i=1}^{m} p_i F_i(\eta)\right).$$

7. Interpretations, explanations and applications

This section presents interpretations, explanations and applications of two classes of models presented in the earlier sections. The hazard rate (or failure rate) $h(\eta)$ (Barlow, Marshall and Proschan[**6**], Barlow and Proschan[**7**]) is defined as

$$h(\eta) = \frac{f(\eta)}{1 - F(\eta)} = -\frac{d}{d\eta} log(1 - F(\eta)).$$

It follows from (37) that

$$h(\eta) = c(\eta)(1 - F(\eta))^{\gamma-1} = c(\eta)(1 - \pi(\eta))^{\gamma-1}. \tag{52}$$

When $c(\eta) = c$, the equation (52) becomes

$$h(\eta) = c(1 - F(\eta))^{\gamma-1} = c(1 - \pi(\eta))^{\gamma-1} = \frac{1}{m(\eta)}. \tag{53}$$

where $m(\eta)$ is the Mills' ratio, defined in Section 1. When $c(\eta) = c$ and $\gamma > 1$, $h(\eta)$ in (53) is a monotonically increasing function of $(1 - F(\eta))$. Hence, $h(\eta)$ is a monotonically decreasing (meaning non-increasing) function of $F(\eta)$ or η. Consequently, the distribution $F(\eta)$ is said to have a decreasing hazard rate (DHR) or a decreasing failure rate (DFR) (Barlow, Marshall and Proschan[**6**], Barlow and Proschan[**7**]). When $c(\eta) = c$ and $\gamma < 1$, $h(\eta)$ in (53) is a monotonically increasing function of η and the distribution $F(\eta)$ is said to have an increasing hazard rate

(IHR) or an increasing failure rate (IFR) (Barlow, Marshall and Proschan[**6**], Barlow and Proschan[**7**]). When $c(\eta) = c$ and $\gamma = 1$, $h(\eta)$ in (53) is a flat function of η.

THEOREM 4 (Barlow, Marshall and Proschan [**6**]). *If* $F_i(\eta)$ *has a decreasing hazard rate,* $i = 1, \ldots, m$, *then* $G(\eta)$ *in* (51) *has a decreasing hazard rate.*

Proschan([**P2**]) demonstrated based on the pooled data on the times of successive failures of the air conditioning system of a fleet jet airlines, that the life distribution had an apparent decreasing failure rate. The detailed analysis showed that the failure distribution for each airplane separately was exponential with a different failure rate. Using Theorem 4, a mixture of exponential distributions each having a non-increasing failure rate, has a non-increasing failure rate. The apparent decreasing failure rate of the pooled air-conditioning life distribution was satisfactorily explained by Theorem 4.

Singh and Maddala [**34**] defined a process with the rate of decay $dF(\eta)/d\eta$ or $f(\eta)$ or $d\pi(\eta)/d(\eta)$ in (38) introducing "memory" when $c(\eta) \neq c$ and not introducing "memory" or "memoryless" when $c(\eta) = c$ for describing the size distribution of incomes. In this sense, the differential equation in (1) is for a process that does not introduce "memory" but the differential equation in (38) is for a process that does introduce "memory". Note that this "memoryless" is different from the popular condition

$$(1 - F(\eta + \delta)) = (1 - F(\eta))(1 - F(\delta)).$$

References

[1] R. R. Bahadur, *Sufficiency and statistical decision functions*, Ann. Math. Statistics **25** (1954), 423–462, DOI 10.1214/aoms/1177728715. MR63630

[2] Edward W. Barankin and Melvin Katz Jr., *Sufficient statistics of minimal dimension*, Sankhyā **21** (1959), 217–246. MR115235

[3] Edward W. Barankin, *Application to exponential families of the solution of the minimal dimensionality problem for sufficient statistics* (English, with French summary), Bull. Inst. Internat. Statist. **38** (1961), 141–150. MR150894

[4] Edward W. Barankin and Ashok P. Maitra, *Generalization of the Fisher-Darmois-Koopman-Pitman theorem on sufficient statistics*, Sankhyā Ser. A **25** (1963), 217–244. MR171342

[5] Ole Barndorff-Nielsen, *Information and exponential families in statistical theory*, John Wiley & Sons, Ltd., Chichester, 1978. Wiley Series in Probability and Mathematical Statistics. MR489333

[6] Richard E. Barlow, Albert W. Marshall, and Frank Proschan, *Properties of probability distributions with monotone hazard rate*, Ann. Math. Statist. **34** (1963), 375–389, DOI 10.1214/aoms/1177704147. MR171328

[7] Richard E. Barlow and Frank Proschan, *Mathematical theory of reliability*, With contributions by Larry C. Hunter. The SIAM Series in Applied Mathematics, John Wiley & Sons, Inc., New York-London-Sydney, 1965. MR0195566

[8] D. Basu, *On statistics independent of a complete sufficient statistic*, Sankhyā **15** (1955), 377–380, DOI 10.1007/978-1-4419-5825-9_14. MR74745

[B5] J. Berkson, *Maximum likelihood and minimum* χ^2 *estimates of the logistic function*, Journal of the American Statistical Association, **50**, 130−162, 1955.

[9] Joseph Berkson, *Tables for the maximum likelihood estimate of the logistic function*, Biometrics **13** (1957), 28–34, DOI 10.2307/3001900. MR123387

[10] A. Bhattacharyya, *On a measure of divergence between two statistical populations defined by their probability distributions*, Bull. Calcutta Math. Soc. **35** (1943), 99–109. MR10358

[11] A. Bhattacharyya, *On a measure of divergence between two multinomial populations*, Sankhyā **7** (1946), 401–406. MR18387

[12] Lawrence D. Brown, *Fundamentals of statistical exponential families with applications in statistical decision theory*, Institute of Mathematical Statistics Lecture Notes—Monograph Series, vol. 9, Institute of Mathematical Statistics, Hayward, CA, 1986. MR882001

[13] Herman Chernoff, *A measure of asymptotic efficiency for tests of a hypothesis based on the sum of observations*, Ann. Math. Statistics **23** (1952), 493–507, DOI 10.1214/aoms/1177729330. MR57518

[14] Herman Chernoff, *Large-sample theory: parametric case*, Ann. Math. Statist. **27** (1956), 1–22, DOI 10.1214/aoms/1177728347. MR76245

[15] D. R. Cox, *The regression analysis of binary sequences*, J. Roy. Statist. Soc. Ser. B **20** (1958), 215–242. MR99097

[16] D. R. Cox, *Regression models and life-tables*, J. Roy. Statist. Soc. Ser. B **34** (1972), 187–220. MR341758

[C5] D. R. Cox, *Analysis of Binary Data*, Chapman & Hall, London, 1977.

[17] D. R. Cox and D. Oakes, *Analysis of survival data*, Monographs on Statistics and Applied Probability, Chapman & Hall, London, 1984. MR751780

[18] Georges Darmois, *Sur certaines lois de probabilité* (French), C. R. Acad. Sci. Paris **222** (1946), 164–165. MR15729

[19] E. B. Dynkin, *Necessary and sufficient statistics for a family of probability distributions* (Russian), Uspehi Matem. Nauk (N.S.) **6** (1951), no. 1(41), 68–90. MR0041376

[20] Bradley Efron, *Defining the curvature of a statistical problem (with applications to second order efficiency)*, Ann. Statist. **3** (1975), no. 6, 1189–1242. MR428531

[F1] R. A. Fisher, *On the Mathematical Foundations of Theoretical Statistics*, Philosophical Transactions of the Royal Society, London, **222**, 309−368, 1922.

[21] D. A. S. Fraser, *On sufficiency and the exponential family*, J. Roy. Statist. Soc. Ser. B **25** (1963), 115–123. MR173345

[22] Subir Ghosh and Hans Nyquist, *Model fitting and optimal design for a class of binary response models*, J. Statist. Plann. Inference **179** (2016), 22–35, DOI 10.1016/j.jspi.2016.07.001. MR3550877

[23] I. R. Goodman, Ronald P. S. Mahler, and Hung T. Nguyen, *Mathematics of data fusion*, Theory and Decision Library. Series B: Mathematical and Statistical Methods, vol. 37, Kluwer Academic Publishers Group, Dordrecht, 1997, DOI 10.1007/978-94-015-8929-1. MR1635258

[24] Paul R. Halmos and L. J. Savage, *Application of the Radon-Nikodym theorem to the theory of sufficient statistics*, Ann. Math. Statistics **20** (1949), 225–241, DOI 10.1214/aoms/1177730032. MR30730

[25] Bent Jørgensen, *The theory of dispersion models*, Monographs on Statistics and Applied Probability, vol. 76, Chapman & Hall, London, 1997. MR1462891

[K1] Kailath, T., *The divergence and Bhattacharyya distance measures in signal selection*, IEEE Trans. Commun., **15**, 52−60, 1967.

[26] John D. Kalbfleisch and Ross L. Prentice, *The statistical analysis of failure time data*, John Wiley and Sons, New York-Chichester-Brisbane, 1980. Wiley Series in Probability and Mathematical Statistics. MR570114

[27] B. O. Koopman, *On distributions admitting a sufficient statistic*, Trans. Amer. Math. Soc. **39** (1936), no. 3, 399–409, DOI 10.2307/1989758. MR1501854

[28] S. Kullback and R. A. Leibler, *On information and sufficiency*, Ann. Math. Statistics **22** (1951), 79–86, DOI 10.1214/aoms/1177729694. MR39968

[29] E. L. Lehmann and Henry Scheffé, *Completeness, similar regions, and unbiased estimation. I*, Sankhyā **10** (1950), 305–340, DOI 10.1007/978-1-4614-1412-4_23. MR39201

[30] E. L. Lehmann, *An interpretation of completeness and Basu's theorem*, J. Amer. Statist. Assoc. **76** (1981), no. 374, 335–340. MR624335

[L2] K. S. Lomax, *Business Failures: Another Example of the Analysis of Failure Data*, Journal of the American Statistical Association, **49**, 847−852, 1954.

[31] P. McCullagh and J. A. Nelder, *Generalized linear models*, Monographs on Statistics and Applied Probability, Chapman & Hall, London, 1989. Second edition [of MR0727836], DOI 10.1007/978-1-4899-3242-6. MR3223057

[M] J. P. Mills, *Table of the ratio : Area to bounding ordinate, for any portion of normal curve*, Biometrika, **18**, 395−400, 1926.

[32] R. W. M. Wedderburn, *Quasi-likelihood functions, generalized linear models, and the Gauss-Newton method*, Biometrika **61** (1974), 439–447, DOI 10.1093/biomet/61.3.439. MR375592

[N] J. Neyman, *Su un teorema concernente le cosiddette statistiche sufficienti*, Inst. Ital. Atti Giorn., **6**, 320–334, 1935.

[P1] E. J. G. Pitman, *Sufficient statistics and intrinsic accuracy*, Proceedings of the Cambridge Philosophical Society **32**, 567–579, 1936.

[P2] F. Proschan, *Theoretical explanation of observed decreasing failure rate*, Technometrics. **5**, 375–383, 1963.

[33] C. Radhakrishna Rao, *Information and the accuracy attainable in the estimation of statistical parameters*, Bull. Calcutta Math. Soc. **37** (1945), 81–91. MR15748

[34] Kajal Lahiri and Peter C. B. Phillips, *Obituary: G. S. Maddala, 1933–1999*, Econometric Theory **15** (1999), no. 4, 639–641, DOI 10.1017/S0266466699154082. MR1717971

[S1] C. G. Small, *Expansions and Asymptotics for Statistics*, Chapman & Hall/CRC, Taylor & Francis Group, Boca Raton, Florida, 2010.

[35] Rolf Sundberg, *Statistical modelling by exponential families*, Institute of Mathematical Statistics Textbooks, vol. 12, Cambridge University Press, Cambridge, 2019, DOI 10.1017/9781108604574. MR3969949

Department of Statistics, University of California, Riverside, California 92521
Email address: subir.ghosh@ucr.edu

Department of Statistics, Stockholm University, SE-106 91 Stockholm, Sweden
Email address: Hans.Nyquist@stat.su.se

Contemporary Mathematics
Volume **774**, 2021
https://doi.org/10.1090/conm/774/15568

Nonlinear parabolic equations with Robin boundary conditions and Hardy-Leray type inequalities

Gisèle Ruiz Goldstein, Jerome A. Goldstein, Ismail Kömbe, and Reyhan Tellioğlu Balekoğlu

Dedicated to M. M. Rao, our mathematical father, grandfather, and great grandfather

ABSTRACT. We are primarily concerned with the absence of positive solutions of the following problem,

$$\begin{cases} \frac{\partial u}{\partial t} = \Delta(u^m) + V(x)u^m + \lambda u^q & \text{in} \quad \Omega \times (0,T), \\ u(x,0) = u_0(x) \geq 0 & \text{in} \quad \Omega, \\ \frac{\partial u}{\partial \nu} = \beta(x)u & \text{on } \partial\Omega \times (0,T), \end{cases}$$

where $0 < m < 1$, $V \in L^1_{loc}(\Omega)$, $\beta \in L^1_{loc}(\partial\Omega)$, $\lambda \in \mathbb{R}$, $q > 0$, $\Omega \subset \mathbb{R}^N$ is a bounded open subset of $\mathbb{R}^N$ with smooth boundary $\partial\Omega$, and $\frac{\partial u}{\partial \nu}$ is the outer normal derivative of u on $\partial\Omega$. Moreover, we also present some new sharp Hardy and Leray type inequalities with remainder terms that provide us concrete potentials to use in the partial differential equation of our interest.

1. Introduction

The main goal of this paper is to study nonexistence of positive solutions in the sense of distributions for the following nonlinear problem with Robin boundary condition,

$$(1.1) \qquad \begin{cases} \frac{\partial u}{\partial t} = \Delta(u^m) + V(x)u^m + \lambda u^q & \text{in} \quad \Omega \times (0,T), \\ u(x,0) = u_0(x) \geq 0 & \text{in} \quad \Omega, \\ \frac{\partial u}{\partial \nu} = \beta(x)u & \text{on} \quad \partial\Omega \times (0,T), \end{cases}$$

where $0 < m < 1$, $V \in L^1_{loc}(\Omega)$, $\beta \in L^1_{loc}(\partial\Omega)$, $\lambda \in \mathbb{R}$, $q > 0$, $\Omega \subset \mathbb{R}^N$ is a bounded open subset of $\mathbb{R}^N$ with smooth boundary $\partial\Omega$, and $\frac{\partial u}{\partial \nu}$ is the outer normal derivative of u on $\partial\Omega$. Let us provide some motivation for investigating problems of the form (1.1).

Linear problems. If we omit λu^q and take $m = 1$, then the problem (1.1) reduces to the linear heat equation with a potential. In this direction, a significant result has been given by Baras and Goldstein [**BG**]. They considered the linear heat problem with the inverse square potential,

---

2020 *Mathematics Subject Classification.* Primary 35K10, 35K15, 35K55; Secondary 26D10, 46E35.

Key words and phrases. Critical exponents, Hardy-Leray inequalities, Robin boundary conditions, positive solutions, nonexistence.

$$(1.2)\qquad \begin{cases} \frac{\partial u}{\partial t} = \Delta u + \frac{c}{|x|^2}u & \text{in} \quad \Omega \times (0,T), \\ u(x,t) = 0 & \text{on} \quad \partial\Omega \times (0,T), \\ u(x,0) = u_0(x) \geq 0 & \text{in} \quad \Omega, \end{cases}$$

where $\Omega \subset \mathbb{R}^N$ is a bounded domain with smooth boundary $\partial\Omega$ and $0 \in \Omega$. They proved that Cauchy-Dirichlet problem (1.2) has no nonnegative solutions in the sense of distributions except $u \equiv 0$ if $c > (\frac{N-2}{2})^2$, and positive weak solutions do exist if $c \leq (\frac{N-2}{2})^2$. The critical constant $C^*(N) = (\frac{N-2}{2})^2$ is the best constant in Hardy's inequality,

$$\int_{\mathbb{R}^N} |\nabla\phi(x)|^2 dx \geq \left(\frac{N-2}{2}\right)^2 \int_{\mathbb{R}^N} \frac{|\phi(x)|^2}{|x|^2} dx,$$

valid for all $\phi \in C_c^1(\mathbb{R}^N)$ if $N \geq 3$ and all $\phi \in C_c^1(\mathbb{R}^N \backslash \{0\})$ if $N = 1, 2$.

Obviously, the phenomenon of existence and nonexistence is caused by the singular potential $\frac{c}{|x|^2}$, which is controlled by Hardy's inequality together with its best constant. The results and ideas of the pioneering paper [**BG**] generated a new direction of nonexistence theory for linear parabolic equations, and we refer the reader to the articles [**CM**], [**GZ1**], [**GK1**], [**GZ2**], [**K**] and the references therein.

Nonlinear problems. The nonlinear partial differential equation

$$\frac{\partial u}{\partial t} = \Delta u^m, \qquad u = u(x,t),$$

is the famous heat equation for $m = 1$, the porous medium equation for $m > 1$, the fast diffusion equation for $0 < m < 1$, and is usually called very fast diffusion equation for $m < 0$. These problems for positive solutions arise in many applications in the fields of mechanics, physics, biology and have been studied by several authors on account of their physical and mathematical interest. We refer the reader to the monographs of Vazquez [**V1, V2**] and Daskalopoulos and Kenig [**DK**] and references quoted therein for more information.

In the classical paper [**F**], Fujita studied the following Cauchy problem for the semilinear heat equation

$$(1.3)\qquad \begin{cases} u_t = \Delta u + u^q & \text{in} \quad \mathbb{R}^N \times (0,\infty), \\ u(x,0) = u_0(x) \geq 0 & \text{in} \quad \mathbb{R}^N, \end{cases}$$

where $q > 1$ and $u_0(x)$ is a bounded nonnegative continuous function. He proved that

(i) if $1 < q < 1 + \frac{2}{N}$, then the problem (1.3) has no positive global in time solutions;

(ii) if $q > 1 + \frac{2}{N}$, then the problem (1.3) has a positive global solution for some initial values u_0, small enough in some sense.

The number $q^* = 1 + \frac{2}{N}$ is often called the critical Fujita exponent. The statement (i) also holds for the critical case $q = q^*$, which was proved later by Hayakawa [**H**] and Weissler [**W**]. The result of Fujita [**F**] has been extended and generalized in various directions. For instance, Qi [**Q**] considered the following fast diffusion problem,

$$\begin{cases} u_t = \Delta u^m + u^q & \text{in} \quad \mathbb{R}^N \times (0,\infty), \\ u(x,0) = u_0(x) \geq 0 & \text{in} \quad \mathbb{R}^N, \end{cases} \tag{1.4}$$

where $0 < m < 1$, $q > 1$ and obtained the following results.

(i) If $1 < q < m + \frac{2}{N}$, then the problem (1.4) has no global positive solutions;

(ii) If $q > m + \frac{2}{N}$, then the problem (1.4) has some global positive solutions.

Thus the critical Fujita exponent $q_m^* = m + \frac{2}{N}$ is the cut off point for existence of global positive solution for the Cauchy problem (1.4). We refer the reader to the survey papers by Deng and Levine [**DL**] and Levine [**Le**] for a good account of related works.

On the other hand, Goldstein and Kömbe [**GK2**] investigated the nonexistence of positive solutions for the following nonlinear equation,

$$\begin{cases} \frac{\partial u}{\partial t} = \Delta(u^m) + V(x)u^m & \text{in} \quad \Omega \times (0,T), \\ u(x,0) = u_0(x) \geq 0 & \text{in} \quad \Omega, \\ u(x,t) = 0 & \text{on} \quad \partial\Omega \times (0,T), \end{cases} \tag{1.5}$$

where $0 < m < 1$, $V \in L^1_{loc}(\Omega)$ and Ω is a bounded domain with smooth boundary in $\mathbb{R}^N$. Using the method of Cabré and Martel [**CM**], they found that the nonexistence of positive solutions of problem (1.5) is largely determined by the size of infimum of the spectrum of the symmetric operator $S = -\Delta - V$ on $L^2(\Omega)$ which is

$$\sigma_{\inf} = \inf_{0 \not\equiv \phi \in C_c^\infty(\Omega)} \frac{\int_\Omega |\nabla\phi|^2 dx - \int_\Omega V|\phi|^2 dx}{\int_\Omega |\phi|^2 dx}. \tag{1.6}$$

It is now clear that while the critical Fujita exponent q_m^* determines the existence and nonexistence of the positive solutions in problem (1.4), the bottom of spectrum $\sigma_{\inf}$ plays the similar role in problem (1.5). Therefore it would be important to investigate the nature of interactions of these two fundamental factors in a unified problem. One of the main goals of this paper is to address the questions we have proposed above. We note that our model problem (1.1) unifies the problems (1.4) and (1.5), and generalizes to the Robin boundary condition. Even though there is a vast literature on these type of problems (with or without potential and source) with Dirichlet boundary condition, the literature regarding these problems with Robin boundary condition is not as rich. Furthermore, according to our knowledge, the model problem (1.1) seems to have never been investigated.

On the other hand, the importance of Hardy type inequalities has been known so far in the study of spectral theory and partial differential equations when dealing with the Schrödinger operators $S = -\Delta - V$ for some potentials V. In this line of research, our second main goal is to find new sharp Hardy-Leray type inequalities which have singularities at the origin and boundary.

The rest of this paper is organized as follows. In Section 2, we study problem (1.1). In Section 3, we first study Hardy and Leray type inequalities with remainder terms. In Section 4 and Section 5, we present various corollaries of Theorem 2.2 with the help of Sobolev trace, Hardy and Leray type inequalities.

Before proceeding to the main results of this paper, we define positive solutions in the following sense.

DEFINITION 1.1. By a *positive local solution continuous* off of $\mathcal{K}$, we mean

(1) $\mathcal{K}$ is a closed Lebesgue null subset of Ω,
(2) $u : [0,T) \longrightarrow L^1(\Omega)$ is continuous for some $T > 0$,
(3) $(x,t) \longrightarrow u(x,t) \in C((\Omega \setminus \mathcal{K}) \times (0,T))$,
(4) $u(x,t) > 0$ on $(\Omega \setminus \mathcal{K}) \times (0,T)$,
(5) $\lim_{t\to 0} u(.,t) = u_0$ in the sense of distributions,
(6) $\nabla u \in L^2_{loc}(\Omega \setminus \mathcal{K})$, and u is a solution in the sense of distributions of the PDE (1.1).

REMARK 1.2. If $0 < a < b < T$ and $\mathcal{K}_o$ is a compact subset of $\Omega \setminus \mathcal{K}$, then $u(x,t) \geq \epsilon_1 > 0$ for $(x,t) \in \mathcal{K}_o \times [a,b]$ for some $\epsilon_1 > 0$. We can weaken (3), (4) to be

(3)' $u(x,t)$ is positive and locally bounded on $(\Omega \setminus \mathcal{K}) \times (0,T)$,
(4)' $\frac{1}{u(x,t)}$ is locally bounded on $(\Omega \setminus \mathcal{K}) \times (0,T)$.

If a solution satisfies (1), (2), (3)', (4)', (5), and (6) then we call it a " general positive local solution off of $\mathcal{K}$ ". This is more general than a positive local solution continuous off of $\mathcal{K}$. If $\mathcal{K} = \emptyset$, we simply call u "general positive local solution".

2. Main result

Before we state and prove our main result, we first recall the following weighted Sobolev interpolation inequality, which plays an important role in our proof.

LEMMA 2.1. *Let Ω be a bounded open subset of $\mathbb{R}^N$ with C^1 boundary, $N \geq 3$ and $M(x) \in L^{N/2}(\Omega)$. Then for each $\epsilon > 0$, there exists a positive constant $C(\epsilon)$ such that*

$$\Big|\int_\Omega M(x)\phi^2 dx\Big| \leq \frac{\epsilon}{2(1-\epsilon)}\int_\Omega |\nabla\phi|^2 dx + C(\epsilon)\int_\Omega \phi^2 dx.$$

for all $\phi \in W^{1,2}(\Omega)$.

PROOF. The proof is similar to the proof of Proposition A.1 in [**GK2**]. The only difference is that we use the Sobolev inequality for functions in $W^{1,2}(\Omega)$ instead of $W^{1,2}_0(\Omega)$. □

We are now ready to state the main theorem of this section.

THEOREM 2.2. *Let $N \geq 3$, $\frac{N-2}{N} \leq m < 1$ and $m < q \leq m + \frac{2}{N}$. Let $\beta(x) \in L^1_{loc}(\partial\Omega)$ be a nonnegative function and $V(x) \in L^1_{loc}(\Omega \setminus \mathcal{K})$ where $\mathcal{K}$ is a closed Lebesgue null subset of Ω. If*

$$\inf_{0\not\equiv\phi\in C^\infty(\Omega\setminus\mathcal{K})} \frac{\int_\Omega |\nabla\phi|^2 dx - (1-\epsilon)\int_\Omega V\phi^2 dx - m(1-\epsilon)\int_{\partial\Omega}\beta\phi^2 ds}{\int_\Omega \phi^2 dx} = -\infty$$

for some $\epsilon > 0$, then the problem (1.1) *has no general positive local solution off of $\mathcal{K}$.*

PROOF. The proof is by contradiction. Given any $T > 0$, let $u : [0,T) \longrightarrow L^1(\Omega)$ be a general positive local solution to (1.1) in $(\Omega \setminus \mathcal{K}) \times (0,T)$ with $u_0 \geq 0$ but not identically zero.

Multiply both sides of (1.1) by the test function ϕ^2/u^m and integrate over Ω, where $\phi \in C^\infty(\Omega \setminus \mathcal{K})$,

$$\begin{aligned} \frac{1}{1-m}\frac{d}{dt}\int_\Omega u^{1-m}\phi^2 dx = &\int_\Omega \Delta u^m \,(\frac{\phi^2}{u^m})dx + \int_\Omega V(x)\phi^2(x)dx \\ &+ \int_\Omega \lambda u^{q-m}\phi^2(x)dx. \end{aligned} \tag{2.1}$$

Integration by parts gives

$$\begin{aligned} \frac{1}{1-m}\frac{d}{dt}\int_\Omega u^{1-m}\phi^2 dx = &\int_\Omega (m^2\phi^2\frac{|\nabla u|^2}{u^2} - 2m\frac{\phi}{u}\nabla u\cdot\nabla\phi)dx \\ &+ \int_{\partial\Omega} m\beta\phi^2 ds + \int_\Omega V(x)\phi^2(x)dx \\ &+ \int_\Omega \lambda u^{q-m}\phi^2(x)dx, \end{aligned} \tag{2.2}$$

where ds denotes the $(N-1)$ dimensional surface measure on $\partial\Omega$.

A direct computation shows that

$$\int_\Omega (m^2\frac{\phi^2}{u^2}|\nabla u|^2 - 2m\frac{\phi}{u}\nabla u\cdot\nabla\phi)dx \geq -\int_\Omega |\nabla\phi|^2 dx. \tag{2.3}$$

Substituting (2.3) into (2.2), we obtain

$$\begin{aligned} \int_\Omega V(x)\phi^2 dx - \int_\Omega |\nabla\phi|^2 dx + \int_{\partial\Omega} m\beta\phi^2 ds \leq &\frac{1}{1-m}\frac{d}{dt}\int_\Omega u^{1-m}\phi^2 dx \\ &- \int_\Omega \lambda u^{q-m}\phi^2 dx. \end{aligned} \tag{2.4}$$

Integrating from t_1 to t_2 $(0 < t_1 < t_2 < T)$ yields

$$\begin{aligned} &\int_\Omega V(x)\phi^2(x)dx - \int_\Omega |\nabla\phi|^2 dx + \int_{\partial\Omega} m\beta\phi^2 ds \\ &\leq K_1\int_\Omega (u^{1-m}(x,t_2) - u^{1-m}(x,t_1))\phi^2 dx - \int_\Omega\int_{t_1}^{t_2} \lambda u^{q-m}\phi^2(x)dtdx, \end{aligned} \tag{2.5}$$

where

$$K_1 = \frac{1}{(1-m)(t_2-t_1)}.$$

We now focus our attention to the integrals of the right hand side in (2.5). Using Jensen's inequality for concave functions, we obtain

$$\int_\Omega \Big(u(x,t_i)\Big)^{\frac{(1-m)N}{2}} dx \leq C(|\Omega|)\Big(\int_\Omega u(x,t_i)dx\Big)^{\frac{(1-m)N}{2}} < \infty.$$

Therefore,

$$u^{1-m}(x,t_i) \in L^{N/2}(\Omega),$$

for $i = 1,2$, and the function is concave since $\frac{(1-m)N}{2} \leq 1$, which follows from the assumption $\frac{N-2}{N} \leq m$. Let $F(x) := \lambda\int_{t_1}^{t_2}\Big(u(x,t)\Big)^{q-m} dt$. Applying Jensen's inequality, we find that

$$F(x,t) \in L^{N/2}(\Omega).$$

By Lemma 2.1, we have

$$K_1 \int_\Omega (u^{1-m}(x,t_2) - u^{1-m}(x,t_1))\phi^2 dx - \int_\Omega \int_{t_1}^{t_2} \lambda u^{q-m}\phi^2(x) dt dx \le \frac{\epsilon}{(1-\epsilon)} \int_\Omega |\nabla\phi|^2 dx + C(\epsilon)\int_\Omega \phi^2 dx, \tag{2.6}$$

where $C(\epsilon)$ is a positive constant and $0<\epsilon<1$. Substituting (2.6) into (2.5) gives

$$\int_\Omega V(x)\phi^2(x)dx - \int_\Omega |\nabla\phi|^2 dx + m\int_{\partial\Omega} \beta\phi^2(x)ds \le \frac{\epsilon}{1-\epsilon}\int_\Omega |\nabla\phi|^2 dx + C(\epsilon)\int_\Omega \phi^2(x)dx. \tag{2.7}$$

We can rearrange (2.7) in the following way,

$$\frac{\int_\Omega |\nabla\phi|^2 dx - \int_\Omega (1-\epsilon)V(x)\phi^2(x)dx - m(1-\epsilon)\int_{\partial\Omega}\beta\phi^2 ds}{\int_\Omega \phi^2 dx} \ge -(1-\epsilon)C(\epsilon). \tag{2.8}$$

Therefore,

$$\inf_{0\not\equiv\phi\in C^\infty(\Omega\setminus\mathcal{K})} \frac{\int_\Omega |\nabla\phi|^2 dx - (1-\epsilon)\int_\Omega V(x)\phi^2(x)dx - m(1-\epsilon)\int_{\partial\Omega}\beta(x)\phi^2 ds}{\int_\Omega \phi^2 dx} > -\infty. \tag{2.9}$$

This contradicts our assumption of Theorem 2.2. The proof is now complete. □

REMARK 2.3. Even though our problem has been considered under the Robin boundary condition, we found the same upper bound for q as with Qi [**Q**]. On the other hand our lower bound is higher than his.

Sobolev Trace Inequality. Sobolev and Sobolev trace inequalities are among the most famous and useful functional inequalities in analysis and geometry. We now use the following Sobolev trace inequalities to control the boundary integral term in (2.9) in terms of the L^2 integrals of ϕ and $|\nabla\phi|$ over the domain Ω. The first one is the classical trace inequality, see [**A**] or [**P**].

LEMMA 2.4. *Let $N\ge 3$ and Ω be a bounded open subset of $\mathbb{R}^N$ with smooth boundary $\partial\Omega$. Then, for every $\phi\in W^{1,2}(\Omega)$, we have*

$$\Big(\int_{\partial\Omega} |\phi|^{\frac{2(N-1)}{N-2}} ds\Big)^{\frac{N-2}{N-1}} \le \frac{1}{S}\Big(\int_\Omega |\nabla\phi|^2 dx + \int_\Omega |\phi|^2 dx\Big),$$

for some constant $S>0$ depending on N and Ω.

Thanks to the continuity of the immersion $W^{1,p}(\Omega)\subset L^p(\partial\Omega)$, we have the following version of the Sobolev trace inequality (see also [**AB**] and [**CL**]).

LEMMA 2.5. *Let $N\ge 2$ and Ω be a bounded open subset of $\mathbb{R}^N$ with smooth boundary $\partial\Omega$. Then for every $\epsilon>0$ there exists a constant $C(\epsilon)>0$ such that*

$$\int_{\partial\Omega} |\phi|^2 ds \le \epsilon\int_\Omega |\nabla\phi|^2 dx + C(\epsilon)\int_\Omega |\phi|^2 dx,$$

for all $\phi\in W^{1,2}(\Omega)$.

As a consequence of Sobolev trace and Hölder inequalities, we derive the following weighted Sobolev trace inequality.

LEMMA 2.6. *(Weighted trace inequality) Let $N \geq 3$ and Ω be a bounded open subset of $\mathbb{R}^N$ with smooth boundary $\partial\Omega$. If $\beta(x) \in L^{N-1}(\partial\Omega)$, then for each $\epsilon > 0$, there exists a positive constant $\tilde{C}(\epsilon)$ such that*

$$\left| \int_{\partial\Omega} \beta(x)\phi^2 ds \right| \leq \epsilon \int_\Omega |\nabla\phi|^2 dx + \tilde{C}(\epsilon) \int_\Omega \phi^2 dx,$$

for all $\phi \in W^{1,2}(\Omega)$.

PROOF. Let $\beta_n(x)$ be the sequence defined by $\beta_n(x) = \min\{\beta(x), n\}$ for almost every $x \in \partial\Omega$ and $n \geq 1$. Then $\beta_n(x) \to \beta(x)$ as $n \to \infty$ and $|\beta_n(x)| \leq |\beta(x)|$ for almost every $x \in \partial\Omega$. By using Lebesgue's dominated convergence theorem, we have

$$\beta_n(x) \to \beta(x) \text{ in } L^{N-1}(\partial\Omega) \text{ as } n \to \infty. \tag{2.10}$$

Clearly, we have

$$\left| \int_{\partial\Omega} \beta\phi^2 ds \right| \leq \int_{\partial\Omega} |\beta - \beta_n|\phi^2 ds + n \int_{\partial\Omega} \phi^2 ds.$$

Using Hölder's inequality for the first integral on the right side yields

$$\left| \int_{\partial\Omega} \beta\phi^2 ds \right| \leq \Big(\int_{\partial\Omega} |\beta - \beta_n|^{N-1} ds \Big)^{\frac{1}{N-1}} \Big(\int_{\partial\Omega} |\phi|^{\frac{2(N-1)}{N-2}} ds \Big)^{\frac{N-2}{N-1}} + n \int_{\partial\Omega} \phi^2 ds.$$

Applying the classical trace inequality, we get

$$\begin{aligned} \left| \int_{\partial\Omega} \beta\phi^2 ds \right| \leq & \Big(\int_{\partial\Omega} |\beta - \beta_n|^{N-1} ds \Big)^{\frac{1}{N-1}} \Big(\frac{1}{S} \int_\Omega (|\nabla\phi|^2 + \phi^2) dx \Big) \\ & + n \int_{\partial\Omega} \phi^2 ds. \end{aligned} \tag{2.11}$$

Due to the limit (2.10), for every given $\epsilon \in (0,1)$, there is a $n(\epsilon) \geq 1$ such that

$$\Big(\int_{\partial\Omega} |\beta - \beta_n|^{N-1} ds \Big)^{\frac{1}{N-1}} \leq S\frac{\epsilon}{2} \qquad \text{for} \quad n \geq n(\epsilon). \tag{2.12}$$

Fix $n \geq n(\epsilon)$. Substituting (2.12) into (2.11) gives

$$\left| \int_{\partial\Omega} \beta\phi^2 ds \right| \leq \frac{\epsilon}{2} \int_\Omega (|\nabla\phi|^2 + \phi^2) dx + n \int_{\partial\Omega} \phi^2 ds. \tag{2.13}$$

By Lemma 2.5, we have

$$n \int_{\partial\Omega} \phi^2 ds \leq \frac{\epsilon}{2} \int_\Omega |\nabla\phi|^2 dx + C(\epsilon) \int_\Omega \phi^2 dx \tag{2.14}$$

and we can choose $C(\epsilon, n)$ to depend only on epsilon since $n = n(\epsilon)$. Substituting (2.14) into (2.13) gives the desired inequality

$$\left| \int_{\partial\Omega} \beta(x)\phi^2 ds \right| \leq \epsilon \int_\Omega |\nabla\phi|^2 dx + \tilde{C}(\epsilon) \int_\Omega \phi^2 dx.$$

□

An immediate consequence of the previous result is contained in the following remark.

REMARK 2.7. Note that if $\beta \in L^{N-1}(\partial\Omega)$ then the equation (2.9) reduces to

$$\inf_{0\not\equiv\phi\in C^\infty(\Omega\backslash\mathcal{K})} \frac{\int_\Omega |\nabla\phi|^2 dx - \int_\Omega (1-\epsilon)V\phi^2 dx}{\int_\Omega \phi^2 dx} > -\infty, \tag{2.15}$$

and we use this result frequently in Section 4 and Section 5.

REMARK 2.8. Note that the kinetic energy $\int_\Omega |\nabla\phi|^2 dx$, the potential energy $\int_\Omega V\phi^2 dx$ and the quantity $\int_{\partial\Omega} \beta\phi^2 ds$ are in the competition in the bottom of the spectrum (2.9) and one could expect that the bottom of the spectrum (2.9) can be $-\infty$. In fact, this depends on the choices of potential V and weight function β. Our interest here is to consider only the critical potentials, which are related to sharp Hardy and Leray type inequalities for the Dirichlet-Laplacian. On the other hand, we should mention that there have been some interesting developments regarding Hardy type inequalities for the Robin-Laplacian [**KL**], [**EKL**].

3. Improved Hardy type inequalities and applications

Let Ω be a bounded domain in $\mathbb{R}^N$ with $0 \in \Omega$. The classical Hardy inequality involving the distance to the origin states that

$$\int_\Omega |\nabla\phi|^2 dx \geq \left(\frac{N-2}{2}\right)^2 \int_\Omega \frac{|\phi|^2}{|x|^2} dx, \tag{3.1}$$

where $\phi \in C_c^\infty(\Omega)$ and $N \geq 3$. Here the constant $(\frac{N-2}{2})^2$ is sharp, in the sense that

$$\inf_{0\not\equiv\phi\in C_c^\infty(\Omega)} \frac{\int_\Omega |\nabla\phi|^2 dx}{\int_\Omega \frac{|\phi|^2}{|x|^2} dx} = \left(\frac{N-2}{2}\right)^2.$$

It is clear that this form of Hardy's inequality (3.1) fails when $N = 2$. However, there is another version of the Hardy inequality. In [**L**], Leray presented the following integral inequality, which has singularity at both the center and boundary of the two dimensional unit ball,

$$\int_{B_1} |\nabla\phi|^2 dx \geq \frac{1}{4} \int_{B_1} \frac{|\phi|^2}{|x|^2 \ln(\frac{1}{|x|})^2} dx, \tag{3.2}$$

where B_1 is the unit ball in $\mathbb{R}^2$ centered at the origin and $\phi \in C_c^\infty(B_1)$. In [**AS**], Adimurthi and Sandeep proved that the constant $\frac{1}{4}$ is sharp,

$$\inf_{0\not\equiv\phi\in C_c^\infty(B_1)} \frac{\int_{B_1} |\nabla\phi|^2 dx}{\int_{B_1} \frac{|\phi|^2}{|x|^2 \ln(\frac{1}{|x|})^2} dx} = \frac{1}{4}.$$

It is natural to ask whether the Hardy and Leray inequalities given above can be unified into one sharp inequality for $\Omega \subset \mathbb{R}^N$ and $N \geq 3$,

$$\int_\Omega |\nabla\phi|^2 dx \geq H \int_\Omega \frac{|\phi|^2}{|x|^2} dx + L \int_\Omega \frac{|\phi|^2}{|x|^2 \ln(\frac{1}{|x|})^2} dx, \tag{3.3}$$

where H and L are positive constants. The first affirmative answer in this direction with the sharp constant $H = (\frac{N-2}{2})^2$ and some $L > 0$ was given by Adimurthi, Chaudhuri and Ramaswamy [**ACR**]. On the other hand, Wang and Willem [**WW**] obtained both sharp constants $H = (\frac{N-2}{2})^2$ and $L = \frac{1}{4}$.

Our first goal in this section is to prove a new sharp Leray inequality with a remainder term on the N-dimensional unit ball centered at the origin. More precisely, we have the following theorem.

THEOREM 3.1. *Let $N \geq 3$ and $B_1 \subset \mathbb{R}^N$ be the N-dimensional open unit ball centered at the origin. Then the following inequality holds,*

$$\int_{B_1} |\nabla\phi|^2 dx \geq \frac{1}{4}\int_{B_1} \frac{\phi^2}{|x|^2 \ln^2(\frac{1}{|x|})} dx + \frac{N-2}{2}\int_{B_1} \frac{\phi^2}{|x|^2 \ln(\frac{1}{|x|})} dx \tag{3.4}$$

for all $\phi \in C_c^\infty(B_1)$, and the constant $\frac{1}{4}$ is sharp.

PROOF. Let $v(x) = -\ln(\ln(\frac{1}{|x|}))$. A direct computation shows that

$$\Delta v = \frac{N-2}{|x|^2 \ln(\frac{1}{|x|})} + \frac{1}{|x|^2 \ln^2(\frac{1}{|x|})}. \tag{3.5}$$

Multiplying both sides of (3.5) by ϕ^2 and integrating over Ω, we obtain

$$(N-2)\int_{B_1} \frac{\phi^2}{|x|^2 \ln(\frac{1}{|x|})} dx + \int_{B_1} \frac{\phi^2}{|x|^2 \ln^2(\frac{1}{|x|})} dx = -2\int_{B_1} \phi \nabla v \cdot \nabla\phi dx \tag{3.6}$$

since $\phi = 0$ on and near ∂B_1 by hypothesis. Applying Young's inequality, we have

$$-2\int_{B_1} \phi\nabla v \cdot \nabla\phi dx \leq 2\epsilon \int_{B_1} |\nabla\phi|^2 dx + \frac{1}{2\epsilon}\int_{B_1} |\phi|^2 |\nabla v|^2 dx, \tag{3.7}$$

where $|\nabla v|^2 = \frac{1}{|x|^2 \ln^2(\frac{1}{|x|})}$ and $\epsilon > 0$ will be chosen later. Combining (3.7) and (3.6) we get

$$\int_{B_1} |\nabla\phi|^2 dx \geq \left(\frac{1}{2\epsilon} - \frac{1}{4\epsilon^2}\right)\int_{B_1} \frac{\phi^2}{|x|^2 \ln^2(\frac{1}{|x|})} dx + \frac{N-2}{2\epsilon}\int_{B_1} \frac{\phi^2}{|x|^2 \ln(\frac{1}{|x|})} dx.$$

Observe that the function $f(\epsilon) = \frac{1}{2\epsilon} - \frac{1}{4\epsilon^2}$ attains the maximum for $\epsilon = 1$ and this maximum is equal to $\frac{1}{4}$. Therefore we obtain the desired inequality

$$\int_{B_1} |\nabla\phi|^2 dx \geq \frac{1}{4}\int_{B_1} \frac{\phi^2}{|x|^2 \ln^2(\frac{1}{|x|})} dx + \frac{N-2}{2}\int_{B_1} \frac{\phi^2}{|x|^2 \ln(\frac{1}{|x|})} dx.$$

□

Note that the technique used in Theorem 3.1 gives us a certain Leray inequality with a remainder term. Therefore it is natural to investigate a general inequality that allows us to find different Hardy and Leray type inequalities with remainder terms.

Now, using the same technique as in [**KO**] and relaxing some of assumptions on the weight function, we obtain the following result.

THEOREM 3.2. *Let Ω be a bounded domain with smooth boundary in $\mathbb{R}^N, N \geq 3$ with $0 \in \Omega$. Let ρ and δ be nonnegative functions on Ω such that $-\Delta\rho \geq 0$ and $-div(\rho\nabla\delta) \geq 0$ in the sense of distributions. Then we have*

$$\int_\Omega |\nabla\phi|^2 dx \geq \frac{1}{4}\int_\Omega \frac{|\nabla\rho|^2}{\rho^2}\phi^2 dx - \frac{1}{2}\int_\Omega \frac{\Delta\rho}{\rho}\phi^2 dx + \frac{1}{4}\int_\Omega \frac{|\nabla\delta|^2}{\delta^2}\phi^2 dx \tag{3.8}$$

for all $\phi \in C_c^\infty(\Omega)$.

PROOF. Let $\phi \in C_c^\infty(\Omega)$ and define $\psi = \rho^{-\frac{1}{2}}\phi$. A direct calculation shows that

$$|\nabla\phi|^2 = \frac{1}{4}\frac{|\nabla\rho|^2}{\rho}\psi^2 + \psi\nabla\rho\cdot\nabla\psi + \rho|\nabla\psi|^2. \tag{3.9}$$

Then integration by parts (i.e., the divergence theorem) gives

$$\int_\Omega |\nabla\phi|^2 dx = \frac{1}{4}\int_\Omega \frac{|\nabla\rho|^2}{\rho^2}\phi^2 dx - \frac{1}{2}\int_\Omega \frac{\Delta\rho}{\rho}\phi^2 dx + \int_\Omega \rho|\nabla\psi|^2 dx. \tag{3.10}$$

We now focus on the last term on the right-hand side of (3.10). Let us define a new function $\varphi(x) := \delta(x)^{-1/2}\psi(x)$ where $0 < \delta(x) \in C^2(\Omega)$. It is clear that

$$|\nabla\psi|^2 = \frac{1}{4}\frac{\varphi^2}{\delta}|\nabla\delta|^2 + \varphi\nabla\delta\cdot\nabla\varphi + \delta|\nabla\varphi|^2.$$

Therefore,

$$\begin{aligned}\int_\Omega \rho|\nabla\psi|^2 dx &\geq \frac{1}{4}\int_\Omega \rho\frac{\varphi^2}{\delta}|\nabla\delta|^2 dx + \int_\Omega \rho\varphi\nabla\delta\cdot\nabla\varphi dx \\ &= \frac{1}{4}\int_\Omega \rho\frac{|\nabla\delta|^2}{\delta^2}\psi^2 dx - \frac{1}{2}\int_\Omega div(\rho\nabla\delta)\frac{\psi^2}{\delta}dx.\end{aligned}$$

Since $-div(\rho\nabla\delta) \geq 0$ and $\psi^2 = \frac{\phi^2}{\rho}$ then we get

$$\int_\Omega \rho|\nabla\psi|^2 dx \geq \frac{1}{4}\int_\Omega \frac{|\nabla\delta|^2}{\delta^2}\phi^2 dx. \tag{3.11}$$

Substituting (3.11) into (3.10) gives the desired inequality (3.8),

$$\int_\Omega |\nabla\phi|^2 dx \geq \frac{1}{4}\int_\Omega \frac{|\nabla\rho|^2}{\rho^2}\phi^2 dx - \frac{1}{2}\int_\Omega \frac{\Delta\rho}{\rho}\phi^2 dx + \frac{1}{4}\int_\Omega \frac{|\nabla\delta|^2}{\delta^2}\phi^2 dx.$$

□

Before giving the application of Theorem 3.2 below, we should mention that Hardy type inequalities involving both the distance to the boundary and the distance to the origin have been studied by Filippas, Moschini Tertikas [**FMT**] and Avkhadiev and Laptev [**AL**].

We now present various Hardy and Leray type inequalities with remainder terms, which can be obtained after suitable choices of weight functions ρ and δ in Theorem 3.2.

In our first example, the choices

$$\rho = |x|^{2-N} \quad \text{and} \quad \delta = \ln(\frac{1}{|x|})$$

give the following sharp Hardy-Leray inequality (3.3) obtained by J. Wang and M. Willem [**WW**].

COROLLARY 3.3. *Let $N \geq 3$ and $B_1 \subset \mathbb{R}^N$ be the N-dimensional unit ball centered at the origin. Then for all $\phi \in C_c^\infty(B_1)$, we have*

$$\int_\Omega |\nabla\phi|^2 dx \geq \left(\frac{N-2}{2}\right)^2 \int_\Omega \frac{|\phi|^2}{|x|^2}dx + \frac{1}{4}\int_\Omega \frac{|\phi|^2}{|x|^2\ln(\frac{1}{|x|})^2}dx. \tag{3.12}$$

In [**T**], Tidblom obtained the following Hardy type inequality with a non-radial remainder term,

$$\int_\Omega |\nabla\phi|^2 dx \geq \frac{1}{4}\int_\Omega \frac{|\phi|^2}{|x|^2}dx + \Big(\frac{2n-1}{4n^2}\Big)\int_\Omega \big(\frac{1}{x_1^2}+\frac{1}{x_2^2}+\cdots+\frac{1}{x_n^2}\big)|\phi|^2 dx, \tag{3.13}$$

where $\Omega = \{x_1 \geq 0, x_2 \geq 0, \dots, x_n \geq 0\}$ and $\phi \in C_c^\infty(\Omega)$.

The result of Tidblom raises the question of whether we can achieve other Hardy-Leray-type inequalities with the same non-radial remainder term. Following the line of this question and the suitable choice of the weight functions in Theorem 3.2 gives the following corollaries.

For instance, let us consider the pair

$$\rho = |x|^{2-n} \quad \text{and} \quad \delta = x_1 \cdots x_n > 0.$$

Then we have the following Hardy inequality with non-radial remainder term.

COROLLARY 3.4. *Let Ω be a bounded domain with smooth boundary in $\mathbb{R}^N$, $N \geq 3$ with $0 \in \Omega$. Then for all $\phi \in C_c^\infty(\Omega)$, we have*

$$\int_\Omega |\nabla\phi|^2 dx \geq \Big(\frac{N-2}{2}\Big)^2\int_\Omega \frac{|\phi|^2}{|x|^2}dx + \frac{1}{4}\int_\Omega \big(\frac{1}{x_1^2}+\frac{1}{x_2^2}+\cdots+\frac{1}{x_n^2}\big)|\phi|^2 dx. \tag{3.14}$$

On the other hand, by making the choices

$$\rho = \ln(\frac{1}{|x|}) \quad \text{and} \quad \delta = x_1 \cdots x_n > 0,$$

we obtain the following Hardy-Leray type inequality with remainders.

COROLLARY 3.5. *Let $B_1 \subset \mathbb{R}^N$ be the unit ball centered at the origin, $N \geq 3$. Then for all $\phi \in C_c^\infty(B_1)$, we have*

$$\begin{aligned}\int_{B_1} |\nabla\phi|^2 dx \geq &\frac{1}{4}\int_{B_1} \frac{\phi^2}{|x|^2\ln^2(\frac{1}{|x|})}dx + \frac{N-2}{2}\int_{B_1}\frac{\phi^2}{|x|^2\ln(\frac{1}{|x|})}dx\\ &+\frac{1}{4}\int_{B_1}\big(\frac{1}{x_1^2}+\frac{1}{x_2^2}+\cdots+\frac{1}{x_n^2}\big)|\phi|^2 dx.\end{aligned} \tag{3.15}$$

Finally, setting the pair of parameters as

$$\rho = \frac{1-|x|}{|x|} \quad \text{and} \quad \delta = x_1 \cdots x_n > 0$$

gives the another sharp form of the Leray type inequality with remainder terms.

COROLLARY 3.6. *Let $N \geq 3$ and $B_1 \subset \mathbb{R}^N$ be the N-dimensional unit ball centered at the origin. Then for all $\phi \in C_c^\infty(B_1)$, we have*

$$\begin{aligned}\int_{B_1} |\nabla\phi|^2 dx \geq &\frac{1}{4}\int_{B_1} \frac{\phi^2}{|x|^2(1-|x|)^2}dx + \frac{N-3}{2}\int_{B_1}\frac{\phi^2}{|x|^2(1-|x|)}dx\\ &+\frac{1}{4}\int_{B_1}\big(\frac{1}{x_1^2}+\frac{1}{x_2^2}+\cdots+\frac{1}{x_n^2}\big)|\phi|^2 dx.\end{aligned} \tag{3.16}$$

Moreover the constant $\frac{1}{4}$ in front of the first integral on the right hand side of (3.16) *is sharp.*

REMARK 3.7. To show that constant $\frac{1}{4}$ is sharp, we consider the family of functions

$$\phi_\epsilon(x) = \begin{cases} \left(\frac{|x|}{1-|x|}\right)^{\frac{1+\epsilon}{2}} & \text{if } 0 \le |x| \le \frac{1}{2}, \\ \left(\frac{1-|x|}{|x|}\right)^{\frac{1+\epsilon}{2}} & \text{if } \frac{1}{2} \le |x| \le 1, \end{cases}$$

and pass to the limit as $\epsilon \to 0$.

4. Applications

We now present various corollaries of Theorem 2.2 with the help of Sobolev trace, Hardy and Leray type inequalities.

In our first result below, we consider the positive radial potential V.

COROLLARY 3.8. *Let $0 \in \Omega$, $N \ge 3$, $V(x) = \frac{c}{|x|^2}$ and $\beta \in L^{N-1}(\partial\Omega)$. Then the problem* (1.1) *has no general positive local solution off of $\mathcal{K} = \{0\}$ if $c > (\frac{N-2}{2})^2$ and $\frac{N-2}{N} \le m < 1$.*

Secondly, as a sign changing potential, we consider the highly singular, oscillating potential.

COROLLARY 3.9. *Let $0 \in \Omega$, $N \ge 3$, $V(x) = \frac{c}{|x|^2} + \frac{\delta}{|x|^2}\sin(\frac{1}{|x|^\alpha})$ where $c > 0$, $\alpha > 0$, $\delta \in \mathbb{R}\backslash\{0\}$ and $\beta \in L^{N-1}(\partial\Omega)$. Then the problem* (1.1) *has no general positive local solution off of $\mathcal{K} = \{0\}$ if $c > (\frac{N-2}{2})^2$ and $\frac{N-2}{N} \le m < 1$.*

To prove Corollary 3.8 and Corollary 3.9, we use the same family of test functions ϕ_ϵ used in the proof of Corollary 3.2 in [**GK2**] . □

REMARK 3.10. Note that the potential in Corollary 3.9 has very large positive and negative oscillating parts, in particular, it oscillates wildly, but important cancellations occur between the positive and negative parts in the quadratic form. As a result, the nonexistence of positive solutions only depends on the size of c.

In the following corollary, we consider a potential that has singularities at the center and on boundary of the unit ball $B_1 \subset \mathbb{R}^N$.

COROLLARY 3.11. *Let $N \ge 3$, $V(x) = \frac{c}{|x|^2 \ln^2(\frac{1}{|x|})}$ and $\beta \in L^{N-1}(\partial B_1)$. Then the problem* (1.1) *has no general positive local solution off of $\mathcal{K} = \{0\} \cup \partial B_1$ if $c > \frac{1}{4}$ and $\frac{N-2}{N} \le m < 1$.*

PROOF. In order to show

$$\inf_{0 \not\equiv \phi \in C^1(\Omega\backslash\mathcal{K})} \frac{\int_\Omega |\nabla\phi|^2 dx - \int_\Omega (1-\epsilon) V(x)\phi^2 dx}{\int_\Omega \phi^2 dx} = -\infty, \tag{3.17}$$

we use the following test function among others. Let $\phi(x) = \varphi_\epsilon(r)$ be the radial function ($r = |x|$) defined by

$$\varphi_\epsilon(r) = \begin{cases} (\ln(\frac{1}{a}))^{\frac{1+\epsilon}{2}} & \text{if } 0 \le r \le a, \\ (\ln(\frac{1}{r}))^{\frac{1+\epsilon}{2}} & \text{if } a \le r \le 1, \end{cases} \tag{3.18}$$

where $\epsilon > 0$ and $r = |x|$.

A direct computation shows that

$$\int_{B_1} |\nabla\phi|^2 dx = \omega_N \Big(\frac{1+\epsilon}{2}\Big)^2 \int_a^1 r^{N-3}(\ln(\frac{1}{r}))^{\epsilon-1} dr, \tag{3.19}$$

where ω_N is the surface area of the $(N-1)$ dimensional unit sphere.

Similarly we get

$$\int_{B_1} \frac{|\phi|^2}{r^2(\ln(\frac{1}{r}))^2} dx = \omega_N \Big(\ln(\frac{1}{a}))^{1+\epsilon} \int_0^a \frac{r^{N-3}}{(\ln(\frac{1}{r}))^2} dr + \int_a^1 r^{N-3}(\ln(\frac{1}{r}))^{\epsilon-1} dr \Big). \tag{3.20}$$

Since the first integral on the right hand side of (3.20) is finite, we write

$$\int_{B_1} \frac{|\phi|^2}{r^2(\ln(\frac{1}{r}))^2} dx = \omega_N \int_a^1 r^{N-3}(\ln(\frac{1}{r}))^{\epsilon-1} dr + C_1. \tag{3.21}$$

It is clear that

$$\int_{B_1} |\phi|^2 dx \geq \omega_N \Big(\frac{a^N}{N} (\ln \frac{1}{a})^{1+\epsilon} \Big) = C_2. \tag{3.22}$$

Substituting (3.19), (3.21) and (3.22) into the Rayleigh quotient gives

$$\begin{aligned} \mathcal{R} &= \frac{\int_\Omega |\nabla\phi|^2 dx - \int_\Omega (1-\epsilon_1)\alpha V(x)|\phi|^2 dx}{\int_\Omega |\phi|^2 dx} \\ &\leq \frac{\omega_N \Big(\frac{(1+\epsilon)^2}{4} - c(1-\epsilon) \Big) \int_a^1 r^{N-3}(\ln(\frac{1}{r}))^{\epsilon-1} dr - C_1}{C_2}. \end{aligned} \tag{3.23}$$

Now, letting ϵ go to 0, we get

$$\frac{1}{4}(1+\epsilon)^2 - c(1-\epsilon) < 0 \quad \text{and} \quad \lim_{\epsilon\to 0} \int_a^1 r^{N-3}(\ln(\frac{1}{r}))^{\epsilon-1} dr = +\infty.$$

Hence,

$$\inf_{0\not\equiv\phi\in C^\infty(\Omega\setminus\mathcal{K})} \frac{\int_\Omega |\nabla\phi|^2 dx - \int_\Omega (1-\epsilon)V(x)\phi^2 dx}{\int_\Omega \phi^2 dx} = -\infty. \tag{3.24}$$

The proof of Corollary (3.11) is now complete. □

Another result in this direction is the following.

COROLLARY 3.12. *Let $N \geq 3$, $V(x) = \frac{c}{|x|^2(1-|x|)^2}$ and $\beta \in L^{N-1}(\partial B_1)$. Then the problem* (1.1) *has no general positive local solution off of $\mathcal{K} = \{0\}\cup\partial B_1$ if $c > \frac{1}{4}$ and $\frac{N-2}{N} \leq m < 1$.*

To prove Corollary(3.12), we use the same family of test functions ϕ_ϵ used in the proof of Corollary (3.6). □

5. The one and two-dimensional cases

We now present some one and two dimensional results. Since the proofs in each case are similar to the proof of Theorem 2.2, we will state them without proof.

THEOREM 3.13. *Let $N = 2$, $\frac{1}{2} \leq m < 1$, $m < q \leq \frac{1}{2} + m$, $\beta(x) \in L^1_{loc}(\partial\Omega)$ and $V(x) \in L^1_{loc}(\Omega \setminus \mathcal{K})$ where $\mathcal{K}$ is a closed Lebesgue null subset of $\bar{\Omega}$. If*

$$\inf_{0 \not\equiv \phi \in C^\infty(\Omega \setminus \mathcal{K})} \frac{\int_\Omega |\nabla\phi|^2 dx - \int_\Omega (1-\epsilon) V \phi^2 dx - m(1-\epsilon) \int_{\partial\Omega} \beta \phi^2 ds}{\int_\Omega \phi^2 dx} = -\infty$$

for some $\epsilon > 0$, then problem (1.1) has no general positive local solution off of $\mathcal{K}$.

THEOREM 3.14. *Let $N = 1$, $0 < m < 1$, $m < q \leq m + 1$, $\beta \in \mathbb{R} \setminus \{0\}$ and $V(x) \in L^1_{loc}(\Omega \setminus \mathcal{K})$ where $\mathcal{K}$ is a closed Lebesgue null subset of $\bar{\Omega}$; we could also take $\Omega = (0, r)$ for $r > 0$. If*

$$\inf_{0 \not\equiv \phi \in C^\infty(\Omega \setminus \mathcal{K})} \frac{\int_\Omega |\nabla\phi|^2 dx - \int_\Omega (1-\epsilon) V \phi^2 dx}{\int_\Omega \phi^2 dx} = -\infty$$

for some $\epsilon > 0$, then problem (1.1) has no general positive local solution off of $\mathcal{K}$.

NOTE 3.15. As in the application of Theorem 2.2, some applications of Theorem 3.13 and Theorem 3.14 can be given with the help of Sobolev trace, Hardy and Leray type inequalities.

References

[A] Robert A. Adams, *Sobolev spaces*, Academic Press [A subsidiary of Harcourt Brace Jovanovich, Publishers], New York-London, 1975. Pure and Applied Mathematics, Vol. 65. MR0450957

[ACR] Adimurthi, Nirmalendu Chaudhuri, and Mythily Ramaswamy, *An improved Hardy-Sobolev inequality and its application*, Proc. Amer. Math. Soc. **130** (2002), no. 2, 489–505, DOI 10.1090/S0002-9939-01-06132-9. MR1862130

[AS] Adimurthi and K. Sandeep, *Existence and non-existence of the first eigenvalue of the perturbed Hardy-Sobolev operator*, Proc. Roy. Soc. Edinburgh Sect. A **132** (2002), no. 5, 1021–1043, DOI 10.1017/S0308210500001992. MR1938711

[AB] G. A. Afrouzi and K. J. Brown, *On principal eigenvalues for boundary value problems with indefinite weight and Robin boundary conditions*, Proc. Amer. Math. Soc. **127** (1999), no. 1, 125–130, DOI 10.1090/S0002-9939-99-04561-X. MR1469392

[AL] Ari Laptev (ed.), *Around the research of Vladimir Maz'ya. II*, International Mathematical Series (New York), vol. 12, Springer, New York; Tamara Rozhkovskaya Publisher, Novosibirsk, 2010. Partial differential equations. MR2664211

[BG] Pierre Baras and Jerome A. Goldstein, *The heat equation with a singular potential*, Trans. Amer. Math. Soc. **284** (1984), no. 1, 121–139, DOI 10.2307/1999277. MR742415

[CM] Xavier Cabré and Yvan Martel, *Existence versus explosion instantanée pour des équations de la chaleur linéaires avec potentiel singulier* (French, with English and French summaries), C. R. Acad. Sci. Paris Sér. I Math. **329** (1999), no. 11, 973–978, DOI 10.1016/S0764-4442(00)88588-2. MR1733904

[CL] Mabel Cuesta and Liamidi Leadi, *Weighted eigenvalue problems for quasilinear elliptic operators with mixed Robin-Dirichlet boundary conditions*, J. Math. Anal. Appl. **422** (2015), no. 1, 1–26, DOI 10.1016/j.jmaa.2014.08.015. MR3263445

[DK] Panagiota Daskalopoulos and Carlos E. Kenig, *Degenerate diffusions*, EMS Tracts in Mathematics, vol. 1, European Mathematical Society (EMS), Zürich, 2007. Initial value problems and local regularity theory, DOI 10.4171/033. MR2338118

[DL] Keng Deng and Howard A. Levine, *The role of critical exponents in blow-up theorems: the sequel*, J. Math. Anal. Appl. **243** (2000), no. 1, 85–126, DOI 10.1006/jmaa.1999.6663. MR1742850

[EKL] Tomas Ekholm, Hynek Kovařík, and Ari Laptev, *Hardy inequalities for p-Laplacians with Robin boundary conditions*, Nonlinear Anal. **128** (2015), 365–379, DOI 10.1016/j.na.2015.08.013. MR3399533

[FMT] Stathis Filippas, Luisa Moschini, and Achilles Tertikas, *Sharp two-sided heat kernel estimates for critical Schrödinger operators on bounded domains*, Comm. Math. Phys. **273** (2007), no. 1, 237–281, DOI 10.1007/s00220-007-0253-z. MR2308757

[F] Hiroshi Fujita, *On the blowing up of solutions of the Cauchy problem for* $u_t = \Delta u + u^{1+\alpha}$, J. Fac. Sci. Univ. Tokyo Sect. I **13** (1966), 109–124 (1966). MR214914

[GK1] Jerome A. Goldstein and Ismail Kombe, *Instantaneous blow up*, Advances in differential equations and mathematical physics (Birmingham, AL, 2002), Contemp. Math., vol. 327, Amer. Math. Soc., Providence, RI, 2003, pp. 141–150, DOI 10.1090/conm/327/05810. MR1991537

[GK2] Jerome A. Goldstein and Ismail Kombe, *Nonlinear degenerate prabolic equations with singular lower-order term*, Adv. Differential Equations **8** (2003), no. 10, 1153–1192. MR2016679

[GZ1] Jerome A. Goldstein and Qi S. Zhang, *On a degenerate heat equation with a singular potential*, J. Funct. Anal. **186** (2001), no. 2, 342–359, DOI 10.1006/jfan.2001.3792. MR1864826

[GZ2] Jerome A. Goldstein and Qi S. Zhang, *Linear parabolic equations with strong singular potentials*, Trans. Amer. Math. Soc. **355** (2003), no. 1, 197–211, DOI 10.1090/S0002-9947-02-03057-X. MR1928085

[H] Kantaro Hayakawa, *On nonexistence of global solutions of some semilinear parabolic differential equations*, Proc. Japan Acad. **49** (1973), 503–505. MR338569

[K] Ismail Kombe, *The linear heat equation with highly oscillating potential*, Proc. Amer. Math. Soc. **132** (2004), no. 9, 2683–2691, DOI 10.1090/S0002-9939-04-07392-7. MR2054795

[KL] Hynek Kovařík and Ari Laptev, *Hardy inequalities for Robin Laplacians*, J. Funct. Anal. **262** (2012), no. 12, 4972–4985, DOI 10.1016/j.jfa.2012.03.021. MR2916058

[KO] Ismail Kombe and Murad Özaydin, *Hardy-Poincaré, Rellich and uncertainty principle inequalities on Riemannian manifolds*, Trans. Amer. Math. Soc. **365** (2013), no. 10, 5035–5050, DOI 10.1090/S0002-9947-2013-05763-7. MR3074365

[L] Jean Leray, *Étude de diverses équations intégrales non linéaires et de quelques problèmes que pose l'hydrodynamique* (French), NUMDAM, [place of publication not identified], 1933. MR3533002

[Le] Howard A. Levine, *The role of critical exponents in blowup theorems*, SIAM Rev. **32** (1990), no. 2, 262–288, DOI 10.1137/1032046. MR1056055

[P] Augusto C. Ponce, *Elliptic PDEs, measures and capacities*, EMS Tracts in Mathematics, vol. 23, European Mathematical Society (EMS), Zürich, 2016. From the Poisson equations to nonlinear Thomas-Fermi problems, DOI 10.4171/140. MR3675703

[Q] Yuan-Wei Qi, *On the equation* $u_t = \Delta u^\alpha + u^\beta$, Proc. Roy. Soc. Edinburgh Sect. A **123** (1993), no. 2, 373–390, DOI 10.1017/S0308210500025750. MR1215421

[T] J. Tidblom, *L^p Hardy inequalities in general domains*, Research Reports in Mathematics Stockholm University no. 4, `http://www2.math.su.se/reports/2003/4/2003-4.pdf`, 2003.

[V1] Juan Luis Vázquez, *The porous medium equation*, Oxford Mathematical Monographs, The Clarendon Press, Oxford University Press, Oxford, 2007. Mathematical theory. MR2286292

[V2] Juan Luis Vázquez, *Smoothing and decay estimates for nonlinear diffusion equations*, Oxford Lecture Series in Mathematics and its Applications, vol. 33, Oxford University Press, Oxford, 2006. Equations of porous medium type, DOI 10.1093/acprof:oso/9780199202973.001.0001. MR2282669

[WW] Zhi-Qiang Wang and Michel Willem, *Caffarelli-Kohn-Nirenberg inequalities with remainder terms*, J. Funct. Anal. **203** (2003), no. 2, 550–568, DOI 10.1016/S0022-1236(03)00017-X. MR2003359

[W] Fred B. Weissler, *Existence and nonexistence of global solutions for a semilinear heat equation*, Israel J. Math. **38** (1981), no. 1-2, 29–40, DOI 10.1007/BF02761845. MR599472

Department of Mathematical Sciences, University of Memphis, Dunn Hall 373, Memphis, Tennessee 38152
Email address: ggoldste@memphis.edu

Department of Mathematical Sciences, University of Memphis, Dunn Hall 373, Memphis, Tennessee 38152
Email address: jgoldste@memphis.edu

Department of Mathematics, Faculty of Humanities and Social Sciences, Istanbul Commerce University, Beyoglu, Istanbul, Turkey
Email address: ikombe@ticaret.edu.tr

Department of Mathematics, Faculty of Humanities and Social Sciences, Istanbul Commerce University, Beyoglu, Istanbul, Turkey
Email address: reyhan.tellioglu@istanbulticaret.edu.tr

Contemporary Mathematics
Volume **774**, 2021
https://doi.org/10.1090/conm/774/15569

Banach space valued weak second order stochastic processes

Yûichirô Kakihara

This paper is dedicated to Professor M. M. Rao on the occasion of his 90th birthday.

ABSTRACT. Banach space valued stochastic processes of weak second order on a locally compact abelian group G is considered. These processes are recognized as operator valued processes on G. More fully, letting $\mathfrak{U}$ be a Banach space and $\mathfrak{H}$ a Hilbert space, we study $B(\mathfrak{U},\mathfrak{H})$-valued processes. Since $B(\mathfrak{U},\mathfrak{H})$ has a $B(\mathfrak{U},\mathfrak{U}^*)$-valued gramian, every $B(\mathfrak{U},\mathfrak{H})$-valued process has a $B(\mathfrak{U},\mathfrak{U}^*)$-valued covariance function. Using this property we can define operator stationarity, operator harmonizability and operator V-boundedness for $B(\mathfrak{U},\mathfrak{H})$-valued processes, in addition to scalar ones. Interrelations among these processes are obtained together with the operator stationary dilation.

1. Introduction

We are interested in Banach space valued second order stochastic processes. So let $\mathfrak{X}$ be a Banach space and $(\Omega,\mathfrak{F},\mu)$ be a probability measure space, and consider $\mathfrak{X}$-valued random variables on Ω. Let $L^2(\Omega)=L^2(\Omega,\mu)$ be the L^2-space on $(\Omega,\mathfrak{F},\mu)$. Also let $L^2(\Omega\,;\mathfrak{X})$ be the Banach space of all $\mathfrak{X}$-valued strong random variables x on Ω such that

$$\int_\Omega \|x(\omega)\|_{\mathfrak{X}}^2\,\mu(d\omega)<\infty,$$

where $\|\cdot\|_{\mathfrak{X}}$ is the norm in $\mathfrak{X}$. Each $x\in L^2(\Omega\,;\mathfrak{X})$ is said to be of **strong second order**. If $x:\Omega\to\mathfrak{X}$ is weakly measurable such that $x^*(x(\cdot))\in L^2(\Omega)$ for $x^*\in\mathfrak{X}^*$, then it is said to be of **weak second order**, where $\mathfrak{X}^*$ is the adjoint space of $\mathfrak{X}$ consisting of all bounded conjugate linear functionals on $\mathfrak{X}$. The usual dual space is denoted by $\mathfrak{X}'$, so that $\mathfrak{X}^*=\{\overline{x'}:x'\in\mathfrak{X}'\}$.

We need some terminologies on stochastic processes on a locally compact abelian group G. $\{x(t)\}$ is called an $\mathfrak{X}$-valued **strong second order stochastic process** on G if $x(t)=x(t,\cdot)\in L^2(\Omega\,;\mathfrak{X})$ for $t\in G$. $\{x(t)\}$ is called an $\mathfrak{X}$-valued **weak second order stochastic process** on G if $x(t)=x(t,\cdot)$ is weakly measurable and $x^*(x(t,\cdot))\in L^2(\Omega)$ for $t\in G$ and $x^*\in\mathfrak{X}^*$. If "2" is replaced by "p" with $1\le p<\infty$, then we can define $\mathfrak{X}$-valued **weak or strong pth order stochastic processes**.

2020 *Mathematics Subject Classification.* Primary 60G10; Secondary 46E25.

Key words and phrases. Banach space valued stochastic processes, gramian, orthogonaly scattered measures, $\mathfrak{U}$-operator semivariation.

For Banach spaces $\mathfrak{U}$ and $\mathfrak{V}$ let $B(\mathfrak{U},\mathfrak{V})$ be the set of all bounded linear operators from $\mathfrak{U}$ to $\mathfrak{V}$. Gangolli **[6]** considered $B(\mathfrak{U},\mathfrak{V})$-valued processes when $\mathfrak{U}$ and $\mathfrak{V}$ are Hilbert spaces (see also Makagon and Salehi **[11]**). Loynes **[8, 9]** started a theory of VH-spaces and LVH-spaces, which is an abstraction of $B(\mathfrak{U},\mathfrak{H})$ type spaces and considered processes with values in an LVH-space in **[10]**. On the other hand, the study of Banach space valued stochastic processes is initiated by Chobanyan **[1–3]**, and Chobanyan and Weron **[4]** laid the foundation for the theory of stationary such processes (see also Miamee **[12, 13]**). We shall follow the lines given in **[4]** and **[13]**.

The following proposition is a basic fact connecting an $\mathfrak{X}$-valued random variable of weak pth order and a bounded linear operator between $\mathfrak{X}^*$ and $L^p(\Omega)$ obtained by Chobanyan and Weron **[4]**. Here, we denote the duality pair of $\mathfrak{X}$ and $\mathfrak{X}^*$ by $x^*(x) = \langle x, x^*\rangle$ for $x \in \mathfrak{X}$ and $x^* \in \mathfrak{X}^*$.

PROPOSITION 1.1. *Let $1 \le p < \infty$ and assume that $x(\cdot) : \Omega \to \mathfrak{X}$ is of weak pth order. Define an operator $T_x : \mathfrak{X}^* \to L^p(\Omega)$ by*

$$T_x x^*(\cdot) = x^*\big(x(\cdot)\big) = \big\langle x(\cdot), x^*\big\rangle, \qquad x^* \in \mathfrak{X}^*. \tag{1.1}$$

Then, T_x is a bounded linear operator, i.e., $T_x \in B(\mathfrak{X}^*, L^p(\Omega))$.

It follows from Proposition 1.1 that if $\{x(t)\}$ is an $\mathfrak{X}$-valued stochastic process of weak second order, then there corresponds a $B(\mathfrak{X}^*, L^2(\Omega))$-valued process $\{T_{x(t)}\}$ given by (1.1). Writing $\mathfrak{U} = \mathfrak{X}^*$ and $\mathfrak{H} = L^2(\Omega)$ we can consider $B(\mathfrak{U},\mathfrak{H})$-valued processes as models for Banach space valued weak second order stochastic processes.

In this paper, we shall define operator stationarity, operator harmonizability, operator V-boundedness and operator stationary dilations for $B(\mathfrak{U},\mathfrak{H})$-valued processes on G, and examine interrelations among these concepts. This paper contains some new results as well as old ones, which serves as a review of Banach space valued stochastic processes. Also scalar stationarity, harmonizability and V-boundedness are defined and the connection to operator ones are considered.

Here are the contents of this paper. In Section 2, we shall explore the structure of the spaces $B(\mathfrak{U},\mathfrak{H})$ and $B(\mathfrak{U},\mathfrak{U}^*)$, and note that $B(\mathfrak{U},\mathfrak{H})$ is a right $B(\mathfrak{U})$-module and has a $B(\mathfrak{U},\mathfrak{U}^*)$-valued gramian. In Section 3, we study $B(\mathfrak{U},\mathfrak{H})$-valued measures, which is the basis of representing above mentioned processes. In Section 4, $B(\mathfrak{U},\mathfrak{U}^*)$-valued measures and bimeasures are examined to represent the covariance functions of processes. Finally in Section 5, we deal with $B(\mathfrak{U},\mathfrak{H})$-valued processes on a locally compact abelian group as models for Banach space valued second order stochastic processes of weak second order. Three types of processes mentioned above are considered with the interrelations among them. Hilbert space valued strong second order processes, i.e., $L^2(\Omega\,;\mathfrak{H})$-valued processes, are fully explained in Kakihara **[7]** and we shall use the notations used there as well as some results.

2. The spaces $B(\mathfrak{U},\mathfrak{H})$ and $B(\mathfrak{U},\mathfrak{U}^*)$

The structure of the spaces $X = B(\mathfrak{U},\mathfrak{H})$ and $B(\mathfrak{U},\mathfrak{U}^*)$ will be clarified in this section, where $\mathfrak{U}$ is a Banach space and $\mathfrak{H}$ is a Hilbert space. First, we note that X is a right $B(\mathfrak{U})$-module and a left $B(\mathfrak{H})$-module. That is,

$$x \in X,\ a \in B(\mathfrak{U}),\ p \in B(\mathfrak{H}) \implies xa,\ px \in X.$$

Second, we consider a mapping $[\cdot,\cdot] : X \times X \to B(\mathfrak{U},\mathfrak{U}^*)$ defined by

$$[x, y] = y^*x, \qquad x, y \in X. \tag{2.1}$$

The mapping $[\cdot,\cdot]$ defined by (2.1) is called a **gramian** on X. If $\mathfrak{U}$ is reflexive, then $B(\mathfrak{U},\mathfrak{U}^*)$ is closed under taking the adjoint, i.e., if $U \in B(\mathfrak{U},\mathfrak{U}^*)$, then $U^* \in B(\mathfrak{U},\mathfrak{U}^*)$. However, if $\mathfrak{U}$ is not reflexive, we consider the domain of the operator $U^* : \mathfrak{U}^{**} \to \mathfrak{U}^*$ be restricted to $\mathfrak{U}$, so that we can write $U^* \in B(\mathfrak{U},\mathfrak{U}^*)$.

For an operator $T \in X = B(\mathfrak{U},\mathfrak{H})$ the **covariance operator** Γ of T is defined by $\Gamma = T^*T : \mathfrak{U} \to \mathfrak{U}^*$, so that

$$(\Gamma u)(v) = \langle v, \Gamma u\rangle = \langle v, T^*Tu\rangle = (Tu, Tv)_{\mathfrak{H}}, \qquad u, v \in \mathfrak{U},$$

where $(\cdot,\cdot)_{\mathfrak{H}}$ is the inner product in $\mathfrak{H}$. Then the following properties of Γ are easily verified.

(1) $\Gamma \in B(\mathfrak{U},\mathfrak{U}^*)$.
(2) Γ is **hermitian**, i.e., $(\Gamma u)(v) = \overline{(\Gamma v)(u)}$ for $u, v \in \mathfrak{U}$.
(3) Γ is **nonnegative**, i.e., it is hermitian and $(\Gamma u)(u) \geq 0$ for $u \in \mathfrak{U}$. In this case, we write $\Gamma \geq 0$.

Let $B^+(\mathfrak{U},\mathfrak{U}^*)$ denote the set of all hermitian and nonnegative operators in $B(\mathfrak{U},\mathfrak{U}^*)$. Then, we have the following basic lemma.

LEMMA 2.1. *Let $X = B(\mathfrak{U},\mathfrak{H})$. Then, the gramian $[\cdot,\cdot]$ on X defined by* (2.1) *satisfies the following properties, where $x, y, z \in X$, $a, b \in B(\mathfrak{U})$ and $p \in B(\mathfrak{H})$.*

(1) $[x,x] \geq 0$.
(2) $[x,x] = 0$ *if and only if* $x = 0$.
(3) $[x, y+z] = [x,y] + [x,z]$.
(4) $[xa, y] = [x,y]a,\ \ [x, yb] = b^*[x,y]$.
(5) $[px, y] = [x, p^*y]$.
(6) $[x,y]^* = [y,x]$.

PROOF. (1) For any $u \in \mathfrak{U}$ we have

$$\begin{aligned}\big([x,x]u\big)(u) &= \big\langle u, [x,x]u\big\rangle = \big\langle u, (x^*x)u\big\rangle \\ &= (xu, xu)_{\mathfrak{H}} = \|xu\|_{\mathfrak{H}}^2 \geq 0, \end{aligned} \tag{2.2}$$

where $\|\cdot\|_{\mathfrak{H}}$ is the norm in $\mathfrak{H}$.

(2) follows from (2.2) and (3) is obvious.

(4) The first one is seen from $[xa, y] = y^*(xa) = (y^*x)a = [x,y]a$. The other one is similarly shown.

(5) is checked as $[px, y] = y^*(px) = (p^*y)^*x = [x, p^*y]$ and (6) as $[x,y]^* = (y^*x)^* = x^*y^{**} = x^*y = [y,x]$. □

One of the important properties of the gramian is positive definiteness. We examine $B(\mathfrak{U},\mathfrak{U}^*)$-valued positive definite kernels.

DEFINITION 2.2. Let Λ be any nonempty set and $\Gamma : \Lambda \times \Lambda \to B(\mathfrak{U},\mathfrak{U}^*)$. Then, Γ is said to be **positive definite** or a **positive definite kernel** if for any $n \in \mathbb{N}$, $\lambda_1, \ldots, \lambda_n \in \Lambda$ and $a_1, \ldots, a_n \in B(\mathfrak{U})$ it holds that

$$\sum_{i,j=1}^n a_j^*\Gamma(\lambda_i,\lambda_j)a_i \geq 0,$$

i.e., $\sum_{i,j=1}^n (a_j^*\Gamma(\lambda_i,\lambda_j)a_i u)(u) \geq 0$ for any $u \in \mathfrak{U}$.

We can introduce a reproducing kernel Hilbert space for a $B(\mathfrak{U},\mathfrak{U}^*)$-valued positive definite kernel as in Miamee and Salehi [**14**] as follows.

DEFINITION 2.3. Let $\Gamma : \Lambda \times \Lambda \to B(\mathfrak{U}, \mathfrak{U}^*)$ be a positive definite kernel and $\mathfrak{H}$ be a Hilbert space consisting of $\mathfrak{U}^*$-valued functions on Λ. Then, $\mathfrak{H}$ is said to be a **reproducing kernel Hilbert space** (**RKHS**) of Γ if the following conditions are satisfied:

(a) $\Gamma(\lambda, \cdot)u \in \mathfrak{H}$ for each $\lambda \in \Lambda$ and $u \in \mathfrak{U}$;
(b) $\langle u, \varphi(\lambda)\rangle = (\varphi(\cdot), \Gamma(\lambda, \cdot)u)_{\mathfrak{H}}$ for each $\lambda \in \Lambda$, $u \in \mathfrak{U}$ and $\varphi \in \mathfrak{H}$.

The existence of a RKHS for each $B(\mathfrak{U}, \mathfrak{U}^*)$-valued positive definite kernel was shown in Miamee and Salehi [**14**].

PROPOSITION 2.4. *Every positive definite kernel* $\Gamma : \Lambda \times \Lambda \to B(\mathfrak{U}, \mathfrak{U}^*)$ *admits a unique RKHS* $\mathfrak{H}_\Gamma$ *consisting of* $\mathfrak{U}^*$*-valued functions on* Λ.

A connection between a $B(\mathfrak{U}, \mathfrak{U}^*)$-valued positive definite kernel and a $B(\mathfrak{U}, \mathfrak{H})$-valued function is given through a RKHS.

COROLLARY 2.5. *Let* $\Gamma : \Lambda \times \Lambda \to B(\mathfrak{U}, \mathfrak{U}^*)$ *be a positive definite kernel and* $\mathfrak{H}$ *be its RKHS consisting of* $\mathfrak{U}^*$*-valued functions on* Λ. *Then there exists a* $B(\mathfrak{U}, \mathfrak{H})$*-valued function* $T(\cdot)$ *on* Λ *such that* $\Gamma(\lambda, \mu) = T(\mu)^*T(\lambda)$ *for* $\lambda, \mu \in \Lambda$.

The idea of the RKHS can be applied to create a space of the type $B(\mathfrak{U}, \mathfrak{H})$ from a right $B(\mathfrak{U})$-module with a $B(\mathfrak{U}, \mathfrak{U}^*)$-valued gramian like function.

COROLLARY 2.6. *Let* $\mathfrak{U}$ *be a Banach space and* X *a right* $B(\mathfrak{U})$*-module with a mapping* $[\cdot, \cdot] : X \times X \to B(\mathfrak{U}, \mathfrak{U}^*)$ *such that for* $x, y, z \in X$ *and* $a \in B(\mathfrak{U})$

(1) $[x, x] \geq 0$;
(2) $[x, x] = 0$ *if and only if* $x = 0$;
(3) $[x, y + z] = [x, y] + [x, z]$;
(4) $[xa, y] = [x, y]a$;
(5) $[x, y]^* = [y, x]$.

Then, there exists a Hilbert space $\mathfrak{H}$ *such that* $X = B(\mathfrak{U}, \mathfrak{H})$ *and* $[x, y] = y^*x$ *for* $x, y \in X$, *i.e.,* $[\cdot, \cdot]$ *is a gramian on* X. *Moreover, the Hilbert space* $\mathfrak{H}$ *is unique in the sense that if* $\mathfrak{K}$ *is another Hilbert space such that* $X = B(\mathfrak{U}, \mathfrak{K})$, *then* $\mathfrak{H}$ *and* $\mathfrak{K}$ *are unitarily isomorphic.*

PROOF. Let $\Gamma(x, y) = [x, y]$ for $x, y \in X$. Then, it is seen from the properties (1) – (5) that Γ is a positive definite kernel. Then, there is a RKHS $\mathfrak{H}$ of Γ by Proposition 2.4. In view of Corollary 2.5 it is not hard to see that $X = B(\mathfrak{U}, \mathfrak{H})$ and $\mathfrak{H}$ is unique within unitary equivalence. □

The operator norm in $X = B(\mathfrak{U}, \mathfrak{H})$ is denoted by $\|\cdot\|_X$. Then, the following lemma is obtained.

LEMMA 2.7. *Let* $X = B(\mathfrak{U}, \mathfrak{H})$. *Then, for* $x, y \in X$, $a \in B(\mathfrak{U})$ *and* $p \in B(\mathfrak{H})$ *we have the following.*

(1) $\|[x, y]\| \leq \|x\|_X \|y\|_X$.
(2) $\|[x, x]\| = \|x\|_X^2$.
(3) $\|x\|_X = \sup\{\|[x, y]\| : \|y\|_X \leq 1\}$.
(4) $\|xa\|_X \leq \|x\|_X \|a\|$.
(5) $\|px\|_X \leq \|p\| \|x\|_X$.

Here, the norm $\|\cdot\|$ *is taken in the respective space such as* $B(\mathfrak{U}, \mathfrak{U}^*)$, $B(\mathfrak{U})$ *and* $B(\mathfrak{H})$.

PROOF. (1), (4) and (5) are obvious.

(2) By (1) we have $\|[x,x]\| \leq \|x\|_X^2$. To see the opposite inequality, observe that

$$\begin{aligned}\big\|[x,x]\big\| = \|x^*x\| &= \sup_{\|u\|_{\mathfrak{U}}\leq 1} \|x^*xu\|_{\mathfrak{U}^*} = \sup_{\|u\|_{\mathfrak{U}}\leq 1}\ \sup_{\|v\|_{\mathfrak{U}}\leq 1} \big|\langle v, x^*xu\rangle\big| \\ &= \sup_{\|u\|_{\mathfrak{U}}\leq 1,\, \|v\|_{\mathfrak{U}}\leq 1} \big|(xu,xv)_{\mathfrak{H}}\big| \geq \sup_{\|u\|_{\mathfrak{U}}\leq 1} \|xu\|_{\mathfrak{H}}^2 = \|x\|_X^2,\end{aligned}$$

where $\|\cdot\|_{\mathfrak{U}}$ and $\|\cdot\|_{\mathfrak{U}^*}$ are the norms in $\mathfrak{U}$ and $\mathfrak{U}^*$, respectively. Hence, (2) is proved.

(3) follows from (1) and (2). □

3. $B(\mathfrak{U},\mathfrak{H})$-valued measures

Let $\mathfrak{U}$ be a Banach space and $\mathfrak{H}$ be a Hilbert space. We shall consider $X = B(\mathfrak{U},\mathfrak{H})$-valued measures on a measurable space $(\Theta,\mathfrak{A})$, where $\mathfrak{A}$ is a σ-algebra of subsets of Θ. Let $ca(\mathfrak{A},X)$ denote the set of all X-valued **countably additive** measures in the norm $\|\cdot\|_X$ on $\mathfrak{A}$. In addition to countably additive measures we need to study finitely additive and weakly countably additive measures. In fact, weakly countably additive measures will be used to represent X-valued operator stationary and harmonizable processes later. So let $fa(\mathfrak{A},X)$ be the set of all X-valued **finitely additive** measures on $\mathfrak{A}$. Similarly let $wca(\mathfrak{A},X)$ be the set of all X-valued **weakly countably additive** measures on $\mathfrak{A}$, where $\mathbb{C}$ is the set of all complex numbers. That is, $\xi \in wca(\mathfrak{A},X)$ if $(\xi(\cdot)u,\phi)_{\mathfrak{H}} \in ca(\mathfrak{A},\mathbb{C})$ for every $u \in \mathfrak{U}$ and $\phi \in \mathfrak{H}$. In this case, $\xi \in wca(\mathfrak{A},X)$ if and only if it is **strongly countably additive**, i.e., $\xi(\cdot)u \in ca(\mathfrak{A},\mathfrak{H})$ for every $u \in \mathfrak{U}$ by the Orlicz-Pettis Theorem (cf. Diestel and Uhl [**5**, p. 22]). We can define variations and semivariations for X-valued measures as follows.

DEFINITION 3.1. Let $\xi \in fa(\mathfrak{A},X)$ and $A \in \mathfrak{A}$. Let $\Pi(A)$ denote the set of all finite measurable partitions of A.

(1) The **variation** of ξ at A is defined by

$$|\xi|(A) = \sup\left\{\sum_{\Delta\in\pi} \big\|\xi(\Delta)\big\|_X : \pi \in \Pi(A)\right\}.$$

Let $vca(\mathfrak{A},X)$ denote the set of all X-valued countably additive measures $\xi \in ca(\mathfrak{A},X)$ of bounded variation, i.e., $|\xi|(\Theta) < \infty$.

(2) The **semivariation** of ξ at A is defined by

$$\|\xi\|(A) = \sup\left\{\bigg\|\sum_{\Delta\in\pi} \alpha_\Delta \xi(\Delta)\bigg\|_X : \alpha_\Delta \in \mathbb{C},\ |\alpha_\Delta| \leq 1,\ \Delta \in \pi \in \Pi(A)\right\}.$$

(3) The **$\mathfrak{U}$-operator semivariation** of ξ at A is defined by

$$\|\xi\|_{\mathfrak{U},o}(A) = \sup\left\{\bigg\|\sum_{\Delta\in\pi} \xi(\Delta)a_\Delta\bigg\|_X : a_\Delta \in B(\mathfrak{U}),\ \|a_\Delta\| \leq 1,\ \Delta \in \pi \in \Pi(A)\right\}.$$

We shall use the following notation.

$$\begin{aligned} bfa(\mathfrak{A},X) &= \big\{\xi \in fa(\mathfrak{A},X) : \|\xi\|_{\mathfrak{U},o}(\Theta) < \infty\big\}, \\ bwca(\mathfrak{A},X) &= \big\{\xi \in wca(\mathfrak{A},X) : \|\xi\|_{\mathfrak{U},o}(\Theta) < \infty\big\}, \\ bca(\mathfrak{A},X) &= \big\{\xi \in ca(\mathfrak{A},X) : \|\xi\|_{\mathfrak{U},o}(\Theta) < \infty\big\}. \end{aligned}$$

(4) The **second order variation** of ξ at A is defined by

$$|\xi|_2(A) = \sup\left\{\left(\sum_{\Delta\in\pi}\|\xi(\Delta)\|_X^2\right)^{\frac{1}{2}} : \pi \in \Pi(A)\right\}.$$

(5) The **strong semivariation** of ξ for A is defined by

$$\|\xi\|_s(A) = \sup\left\{\left\|\sum_{\Delta\in\pi}\xi(\Delta)u_\Delta\right\|_{\mathfrak{H}} : \|u_\Delta\|_{\mathfrak{U}} \le 1,\ \Delta\in\pi\in\Pi(A)\right\}.$$

The following lemma gives basic relations among the above notions.

LEMMA 3.2. *For $\xi \in fa(\mathfrak{A}, X)$ the following statements are true.*

(1) $\|\xi(A)\|_X \le \|\xi\|(A) \le \|\xi\|_s(A) = \|\xi\|_{\mathfrak{U},o}(A) \le |\xi|(A)$ *for* $A \in \mathfrak{A}$.
(2) $\|\xi(A)\|_X \le |\xi|_2(A) \le \|\xi\|_s(A)$ *for* $A \in \mathfrak{A}$.
(3) $\|\xi\|(A) = \sup\left\{|\langle\xi(\cdot), x^*\rangle|(A) : x^* \in B(\mathfrak{U},\mathfrak{H})^*,\ \|x^*\| \le 1\right\}$ *for* $A \in \mathfrak{A}$.
(4) $\|\xi\|_{\mathfrak{U},o}(A) \le \sup\left\{|[\xi(\cdot), x]|(A) : x \in X,\ \|x\|_X \le 1\right\}$ *for* $A \in \mathfrak{A}$.

PROOF. (1) We only need to show $\|\xi\|_s(A) = \|\xi\|_{\mathfrak{U},o}(A)$ for $A \in \mathfrak{A}$.

To see $\|\xi\|_s(A) \le \|\xi\|_{\mathfrak{U},o}(A)$ let $\pi \in \Pi(A)$ and $u_\Delta \in \mathfrak{U}$ with $\|u_\Delta\|_{\mathfrak{U}} \le 1$ for $\Delta \in \pi$. Choose $u \in \mathfrak{U}$ with $\|u\|_{\mathfrak{U}} \le 1$ and $a_\Delta \in B(\mathfrak{U})$ such that $a_\Delta u = u_\Delta$ and $\|a_\Delta\| \le 1$ for $\Delta \in \pi$. Then we have

$$\begin{aligned}\left\|\sum_{\Delta\in\pi}\xi(\Delta)u_\Delta\right\|_{\mathfrak{H}} &= \left\|\sum_{\Delta\in\pi}\xi(\Delta)a_\Delta u\right\|_{\mathfrak{H}} \\ &\le \left\|\sum_{\Delta\in\pi}\xi(\Delta)a_\Delta\right\|_X \le \|\xi\|_{\mathfrak{U},o}(A).\end{aligned}$$

Hence, $\|\xi\|_s(A) \le \|\xi\|_{\mathfrak{U},o}(A)$.

To show the opposite inequality let $\pi \in \Pi(A)$ and $a_\Delta \in B(\mathfrak{U})$ with $\|a_\Delta\| \le 1$ for $\Delta \in \pi$ be given. For any $\varepsilon > 0$ choose $u \in \mathfrak{U}$ such that $\|u\|_{\mathfrak{U}} \le 1$ and

$$\left\|\sum_{\Delta\in\pi}\xi(\Delta)a_\Delta u\right\|_{\mathfrak{H}} > \left\|\sum_{\Delta\in\pi}\xi(\Delta)a_\Delta\right\|_X - \varepsilon.$$

Letting $u_\Delta = a_\Delta u$ for $\Delta \in \pi$ we see that $\|u_\Delta\|_{\mathfrak{U}} \le 1$ for $\Delta \in \pi$ and

$$\left\|\sum_{\Delta\in\pi}\xi(\Delta)a_\Delta\right\|_X < \left\|\sum_{\Delta\in\pi}\xi(\Delta)u_\Delta\right\|_{\mathfrak{H}} + \varepsilon \le \|\xi\|_s(A) + \varepsilon.$$

Since $\varepsilon > 0$ is arbitrary we conclude that $\|\xi\|_{\mathfrak{U},o}(A) \le \|\xi\|_s(A)$.

(2) Let $\{A_1, \dots, A_n\} \in \Pi(A)$ and $\varepsilon > 0$ be given. Choose $u_i \in \mathfrak{U}$ such that $\|u_i\|_{\mathfrak{U}} \le 1$ and $\|\xi(A_i)u_i\|_{\mathfrak{H}}^2 > \|\xi(A_i)\|_X^2 - \frac{\varepsilon}{n}$, $1 \le i \le n$. Letting $\{r_j(t) : j \in \mathbb{N}\}$ be the Rademacher system in $L^2([0,1], dt)$, where dt is the Lebesgue measure and r_j is defined by

$$r_j(t) = \left\{\begin{array}{ll} 1, & t \in \left[\frac{k}{2^j}, \frac{k+1}{2^j}\right),\ k = 0, 2, 4, \dots, 2^{j-1} \\ -1, & \text{otherwise}\end{array}\right\}.$$

Note that $\{r_j : j \in \mathbb{N}\}$ is an orthonormal set in $L^2([0,1], dt)$. Now we see that

$$\begin{aligned}\int_0^1 \left\| \sum_{j=1}^n \xi(A_j) u_j r_j(t) \right\|_{\mathfrak{H}}^2 dt &= \int_0^1 \sum_{i,j=1}^n \big(\xi(A_i)u_i, \xi(A_j)u_j\big)_{\mathfrak{H}} r_i(t) r_j(t)\, dt \\ &= \sum_{i,j=1}^n \big(\xi(A_i)u_i, \xi(A_j)u_j\big)_{\mathfrak{H}} \int_0^1 r_i(t) r_j(t)\, dt \\ &= \sum_{j=1}^n \big\|\xi(A_j)u_j\big\|_{\mathfrak{H}}^2 \\ &> \sum_{j=1}^n \big\|\xi(A_j)\big\|_X^2 - \varepsilon\end{aligned}$$

and

$$\int_0^1 \left\| \sum_{j=1}^n \xi(A_j) u_j r_j(t) \right\|_{\mathfrak{H}}^2 dt \le \int_0^1 \|\xi\|_s(A)^2\, dt = \|\xi\|_s(A)^2.$$

It follows that $|\xi|_2(A) \le \|\xi\|_s(A)$.

(3) is well-known. See, e.g., Diestel and Uhl [**5**, pp. 3–4].

(4) Let α be the RHS (right hand side) of the inequality in (4). Let $\pi \in \Pi(A)$ and $a_\Delta \in B(\mathfrak{U})$ with $\|a_\Delta\| \le 1$ for $\Delta \in \pi$. It follows from Lemma 2.7(3) that

$$\begin{aligned}\left\| \sum_{\Delta\in\pi} \xi(\Delta) a_\Delta \right\|_X &= \sup\left\{ \left\| \left[\sum_{\Delta\in\pi} \xi(\Delta) a_\Delta, x \right] \right\| : x \in X,\ \|x\|_X \le 1 \right\} \\ &= \sup\left\{ \left\| \sum_{\Delta\in\pi} [\xi(\Delta), x] a_\Delta \right\| : x \in X,\ \|x\|_X \le 1 \right\} \\ &\le \sup\left\{ \sum_{\Delta\in\pi} \big\| [\xi(\Delta), x] \big\| : x \in X,\ \|x\|_X \le 1 \right\} \\ &\le \sup\Big\{ \big|[\xi(\cdot), x]\big|(A) : x \in X,\ \|x\|_X \le 1 \Big\} = \alpha.\end{aligned}$$

This implies $\|\xi\|_{\mathfrak{U},o}(A) \le \alpha$. □

We shall use gramian orthogonally scattered measures to represent X-valued (operator) stationary processes. Here is the definition.

DEFINITION 3.3. An X-valued measure $\xi \in fa(\mathfrak{A}, X)$ is said to be **gramian orthogonally scattered** if $[\xi(A), \xi(B)] = 0$ for disjoint $A, B \in \mathfrak{A}$. Let us use the following notations.

$$\begin{aligned} fagos(\mathfrak{A}, X) &= \big\{\xi \in fa(\mathfrak{A}, X) : \xi \text{ is gramian orthogonally scattered}\big\}, \\ wcagos(\mathfrak{A}, X) &= \big\{\xi \in wca(\mathfrak{A}, X) : \xi \text{ is gramian orthogonally scattered}\big\}, \\ cagos(\mathfrak{A}, X) &= \big\{\xi \in ca(\mathfrak{A}, X) : \xi \text{ is gramian orthogonally scattered}\big\}. \end{aligned}$$

Some properties of a gramian orthogonally scattered measure are given below.

LEMMA 3.4. *If $\xi \in fagos(\mathfrak{A}, X)$ is an X-valued finitely additive gramian orthogonally scattered measure, then $\|\xi\|_{\mathfrak{U},o}(A) = \|\xi\|_s(A) = |\xi|_2(A)$.*

PROOF. In view of Lemma 3.2 we only need to show $\|\xi\|_{\mathfrak{U},o}(A) \le |\xi|_2(A)$ for $A \in \mathfrak{A}$. Let $A \in \mathfrak{A}$, $\pi \in \Pi(A)$ and $a_\Delta \in B(\mathfrak{U})$ with $\|a_\Delta\| \le 1$ for $\Delta \in \pi$. Then,

$$\begin{aligned}
\Big\| \sum_{\Delta\in\pi} \xi(\Delta)a_\Delta \Big\|_X^2 &= \Big\| \Big[\sum_{\Delta\in\pi} \xi(\Delta)a_\Delta, \sum_{\Delta'\in\pi} \xi(\Delta')a_{\Delta'} \Big] \Big\| \\
&= \Big\| \sum_{\Delta,\Delta'\in\pi} a_{\Delta'}^* \big[\xi(\Delta), \xi(\Delta')\big] a_\Delta \Big\| \\
&= \Big\| \sum_{\Delta\in\pi} a_\Delta^* \big[\xi(\Delta), \xi(\Delta)\big] a_\Delta \Big\| \\
&\le \sum_{\Delta\in\pi} \Big\| a_\Delta^* \big[\xi(\Delta), \xi(\Delta)\big] a_\Delta \Big\| \\
&\le \sum_{\Delta\in\pi} \Big\| \big[\xi(\Delta), \xi(\Delta)\big] \Big\| \\
&= \sum_{\Delta\in\pi} \|\xi(\Delta)\|_X^2 \\
&\le |\xi|_2^2(A).
\end{aligned}$$

Hence, $\|\xi\|_{\mathfrak{U},o}(A) \le |\xi|_2(A)$. □

An $\mathfrak{H}$-valued measure $\zeta \in ca(\mathfrak{A}, \mathfrak{H})$ is said to be **orthogonally scattered** if $(\zeta(A), \zeta(B))_{\mathfrak{H}} = 0$ for disjoint $A, B \in \mathfrak{A}$, denoted $\zeta \in caos(\mathfrak{A}, \mathfrak{H})$. The next lemma gives a necessary and sufficient condition for an X-valued measure to be gramian orthogonally scattered.

LEMMA 3.5. *For an X-valued weakly countably additive measure $\xi \in wca(\mathfrak{A}, X)$ and $u \in \mathfrak{U}$ let $\xi_u \in ca(\mathfrak{A}, \mathfrak{H})$ be defined by $\xi_u(\cdot) = \xi(\cdot)u$. Then, ξ is gramian orthogonally scattered if and only if ξ_u is orthogonally scattered for every $u \in \mathfrak{U}$, i.e., $\xi_u \in caos(\mathfrak{A}, \mathfrak{H})$.*

PROOF. Let $\xi \in wca(\mathfrak{A}, X)$, $u \in \mathfrak{U}$ and $A, B \in \mathfrak{A}$. Then we see that

$$\begin{aligned}
\big(\xi_u(A), \xi_u(B)\big)_{\mathfrak{H}} &= \big(\xi(A)u, \xi(B)u\big)_{\mathfrak{H}} \\
&= \langle u, \xi(B)^*\xi(A)u \rangle \\
&= \Big\langle u, \big[\xi(A), \xi(B)\big]u \Big\rangle.
\end{aligned}$$

Hence it follows that if ξ is gramian orthogonally scattered, then ξ_u is orthogonally scattered for every $u \in \mathfrak{U}$.

The converse is obtained as follows. Let $A, B \in \mathfrak{A}$ and observe that

$$\begin{aligned}
\big[\xi(A), \xi(B)\big] = 0 &\iff \Big\langle u, \big[\xi(A), \xi(B)\big]v \Big\rangle = 0, \quad u, v \in \mathfrak{U} \\
&\iff \Big\langle u, \big[\xi(A), \xi(B)\big]u \Big\rangle = 0, \quad u \in \mathfrak{U}, \quad \text{by polarization}, \\
&\iff \big(\xi_u(A), \xi_u(B)\big)_{\mathfrak{H}} = 0, \quad u \in \mathfrak{U}.
\end{aligned}$$

Thus, if ξ_u, $u \in \mathfrak{U}$ are orthogonally scattered, then ξ is gramian orthogonally scattered. □

In the above lemma, we can say that $\xi \in wca(\mathfrak{A}, X)$ is gramian orthogonally scattered if and only if ξ_u and ξ_v are **biorthogonally scattered** for $u, v \in \mathfrak{U}$, i.e., $(\xi_u(A), \xi_v(B))_{\mathfrak{H}} = 0$ for disjoint $A, B \in \mathfrak{A}$.

Some examples of X-valued measures are given below.

EXAMPLE 3.6. Consider a measurable space $(\mathbb{R}, \mathfrak{B})$ where $\mathfrak{B}$ is the Borel σ-algebra of $\mathbb{R}$. Let $\mathfrak{U} = \ell^1$, so that $\mathfrak{U}^* = \ell^\infty$. Write a standard basis of ℓ^1 by $\{e_k : k \in \mathbb{N}\}$, i.e., $e_k = (\alpha_1, \alpha_2, \ldots)$ with $\alpha_k = 1$ and $\alpha_j = 0$ for $j \neq k$. Then, $e_k^* = e_k$, $k \in \mathbb{N}$ serve as a set of bounded linear functionals on ℓ^1 such that $\langle e_j, e_k^* \rangle = \delta_{jk}$ for $j, k \in \mathbb{N}$. Assume that $\dim \mathfrak{H} = \infty$ and let $\{\phi_k : k \in \mathbb{N}\}$ be an orthonormal basis in $\mathfrak{H}$. Note that the algebraic tensor product $\ell^\infty \odot \mathfrak{H}$ is a subset of $X = B(\ell^1, \mathfrak{H})$ by the identification $(u^* \otimes \phi)u = \langle u, u^* \rangle \phi$ for $u \in \ell^1$, $u^* \in \ell^\infty$ and $\phi \in \mathfrak{H}$.

(1) Define ξ by

$$\xi(A) = \sum_{k \in A \cap \mathbb{N}} \frac{1}{k^2} e_k^* \otimes \phi_k, \qquad A \in \mathfrak{B}.$$

Then we see that ξ is well-defined, countably additive in the operator norm and it holds that

$$|\xi|(\mathbb{R}) = \sum_{k=1}^{\infty} \left\| \frac{1}{k^2} e_k^* \otimes \phi_k \right\| = \sum_{k=1}^{\infty} \frac{1}{k^2} < \infty,$$

so that $\xi \in vca(\mathfrak{B}, X)$. Also we see that ξ is gramian orthogonally scattered, $\xi \in cagos(\mathfrak{B}, X)$, since for $u \in \ell^1$ and disjoint $A, B \in \mathfrak{B}$ it holds that

$$\begin{aligned}
\big(\xi(A)u, \xi(B)u\big)_{\mathfrak{H}} &= \left(\sum_{i \in A \cap \mathbb{N}} \frac{1}{i^2} (e_i^* \otimes \phi_i)u, \sum_{j \in B \cap \mathbb{N}} \frac{1}{j^2} (e_j^* \otimes \phi_j)u \right)_{\mathfrak{H}} \\
&= \left(\sum_{i \in A \cap \mathbb{N}} \frac{1}{i^2} \langle u, e_i^* \rangle \phi_i, \sum_{j \in B \cap \mathbb{N}} \frac{1}{j^2} \langle u, e_j^* \rangle \phi_j \right)_{\mathfrak{H}} \\
&= \sum_{i \in A \cap \mathbb{N}} \sum_{j \in B \cap \mathbb{N}} \frac{1}{i^2 j^2} \langle u, e_i^* \rangle \overline{\langle u, e_j^* \rangle} (\phi_i, \phi_j)_{\mathfrak{H}} = 0
\end{aligned}$$

and Lemma 3.5 applies.

(2) Define ξ by

$$\xi(A) = \sum_{k \in A \cap \mathbb{N}} \frac{1}{k} e_k^* \otimes \phi_k, \qquad A \in \mathfrak{B}.$$

Then we see that ξ is well-defined and $\xi \notin vca(\mathfrak{B}, X)$ since

$$|\xi|(\mathbb{R}) = \sum_{k=1}^{\infty} \frac{1}{k} = \infty.$$

ξ is gramian orthogonally scattered, which is seen from the computation in (1). Moreover, we can see that $\xi \in bca(\mathfrak{B}, X)$. In fact, for any $\pi \in \Pi(\mathbb{R})$ take $u_\Delta \in \ell^1$

with $\|u_\Delta\|_1 \le 1$ for $\Delta \in \pi$ and see that

$$\begin{aligned}\Big\|\sum_{\Delta\in\pi}\xi(\Delta)u_\Delta\Big\|_{\mathfrak{H}}^2 &= \Big\|\sum_{\Delta\in\pi}\sum_{k\in\Delta\cap\mathbb{N}}\frac{1}{k}\langle u_\Delta, e_k^*\rangle\phi_k\Big\|_{\mathfrak{H}}^2 \\ &= \sum_{\Delta\in\pi}\sum_{k\in\Delta\cap\mathbb{N}}\frac{1}{k^2}\big|\langle u_\Delta, e_k^*\rangle\big|^2\|\phi_k\|_{\mathfrak{H}}^2 \\ &\le \sum_{k=1}^{\infty}\frac{1}{k^2} < \infty.\end{aligned}$$

Hence, $\|\xi\|_s(\mathbb{R}) < \infty$. Thus, $\xi \in bca(\mathfrak{B}, X) \cap cagos(\mathfrak{B}, X)$.

(3) Define ξ by

$$\xi(A) = \sum_{k\in A\cap\mathbb{N}}\frac{1}{k^2}e_k^*\otimes(\phi_k+\phi_{k+1}), \qquad A \in \mathfrak{B}.$$

Then we see that ξ is well-defined, $\xi \in vca(\mathfrak{B}, X)$, but ξ is not gramian orthogonally scattered.

(4) Let $E(\cdot)$ be a (weakly) countably additive spectral measure in $\mathfrak{H}$ and $S \in B(\ell^1, \mathfrak{H})$. Then, $\xi(\cdot) = E(\cdot)S \in wcagos(\mathfrak{B}, X)$ but $\notin ca(\mathfrak{B}, X)$ since

$$\big[\xi(A), \xi(B)\big] = \big(E(B)S\big)^* E(A)S = S^*E(A\cap B)S, \qquad A, B \in \mathfrak{B}.$$

(5) Define $E(\cdot)$ on $\mathfrak{H}$ and $S : \ell^1 \to \mathfrak{H}$ respectively by

$$\begin{aligned}E(A) &= \sum_{k\in A\cap\mathbb{N}}\phi_k\otimes\overline{\phi}_k, \qquad A \in \mathfrak{B}, \\ Se_k &= \phi_k, k \in \mathbb{N},\end{aligned}$$

where $(\phi_k\otimes\overline{\phi}_k)\phi = (\phi, \phi_k)_{\mathfrak{H}}\,\phi_k$ for $\phi \in \mathfrak{H}$ (cf. Schatten [**15**]). Hence, we see $S \in B(\ell^1, \mathfrak{H})$ and $\xi(\cdot) = E(\cdot)S \in wcagos(\mathfrak{B}, B(\ell^1, \mathfrak{H}))$ by (4) above. Moreover, for $\pi_n = \{\Delta_0, \Delta_1, \ldots, \Delta_{n+1}\} \in \Pi(\mathbb{R})$, where $\Delta_0 = (-\infty, 0]$, $\Delta_1 = (0, 1]$, $\Delta_2 = (1, 2], \ldots, \Delta_n = (n-1, n]$, $\Delta_{n+1} = (n, \infty)$, it holds that

$$\Big\|\sum_{i=0}^{n+1}\xi(\Delta_i)e_i\Big\|_{\mathfrak{H}} = \Big\|\sum_{k=1}^{n}(\phi_k\otimes\overline{\phi}_k)\phi_k\Big\|_{\mathfrak{H}} = \sqrt{n} \to \infty$$

as $n \to \infty$, so that $\|\xi\|_s(\mathbb{R}) = \|\xi\|_{\mathfrak{U},o}(\mathbb{R}) = \infty$ or $\xi \notin bwca(\mathfrak{B}, B(\ell^1, \mathfrak{H}))$.

(6) Define $E(\cdot)$ as in (5) and $S : \ell^1 \to \mathfrak{H}$ by

$$Se_k = \phi_k + \phi_{k+1}, \qquad k \in \mathbb{N}.$$

Then, we see that $\xi(\cdot) = E(\cdot)S \in wca(\mathfrak{B}, X)$, $\notin bwca(\mathfrak{B}, X)$ and $\notin wcagos(\mathfrak{B}, X)$.

To describe (operator) stationary dilations of X-valued processes we have to introduce gramian orthogonally scattered dilation of X-valued measures.

DEFINITION 3.7. An X-valued finitely additive measure $\xi \in fa(\mathfrak{A}, X)$ is said to have a **gramian orthogonally scattered dilation** if there exist a Hilbert space $\mathfrak{K}$ containing $\mathfrak{H}$ as a closed subspace and a $Y = B(\mathfrak{U}, \mathfrak{K})$-valued finitely additive gramian orthogonally scattered measure $\eta \in fagos(\mathfrak{A}, B(\mathfrak{U}, \mathfrak{K}))$ such that $\xi = P\eta$, where $P : \mathfrak{K} \to \mathfrak{H}$ is the orthogonal projection. The triple $\{\eta, Y, P\}$ is also called a **gramian orthogonally scattered dilation** of ξ. When ξ is weakly countably additive or countably additive, so is the corresponding η.

To consider gramian orthogonally scattered dilation we need some preparations. Let $L^0(\Theta\,;\mathfrak{U})$ be the set of all $\mathfrak{U}$-valued $\mathfrak{A}$-simple functions on Θ and $\zeta \in fa(\mathfrak{A}, \mathfrak{U}^*)$, the set of all $\mathfrak{U}^*$-valued finitely additive measures on $\mathfrak{A}$. For $\varphi = \sum_{j=1}^n u_j 1_{A_j} \in L^0(\Theta\,;\mathfrak{U})$, where $u_1, \dots, u_n \in \mathfrak{U}$ and $\{A_1, \dots, A_n\} \in \Pi(\Theta)$, we define the integral of φ with respect to ζ by

$$\int_\Theta \langle \varphi, d\zeta \rangle = \sum_{j=1}^n \langle u_j, \zeta(A_j) \rangle.$$

We also define two kinds of norms for φ as follows:

$$\|\varphi\|_\infty = \sup_{t\in\Theta} \big\|\varphi(t)\big\|_{\mathfrak{U}}, \tag{3.1}$$

$$\|\varphi\|_* = \sup\left\{\left|\int_\Theta \langle \varphi, d\zeta\rangle\right| : \zeta \in fa(\mathfrak{A}, \mathfrak{U}^*),\ \|\zeta\|(\Theta) \le 1\right\}. \tag{3.2}$$

Now let $X = B(\mathfrak{U}, \mathfrak{H})$ and consider an X-valued finitely additive measure $\xi \in fa(\mathfrak{A}, X)$. For $\varphi = \sum_{j=1}^n u_j 1_{A_j} \in L^0(\Theta\,;\mathfrak{U})$ the integral of φ with respect to ξ is defined by

$$\int_\Theta d\xi\, \varphi = \sum_{j=1}^n \xi(A_j) u_j \in \mathfrak{H}.$$

Define an operator $S_\xi : L^0(\Theta\,;\mathfrak{U}) \to \mathfrak{H}$ by

$$S_\xi \varphi = \int_\Theta d\xi\, \varphi, \qquad \varphi \in L^0(\Theta\,;\mathfrak{U}). \tag{3.3}$$

Let $\mathfrak{V}$ be another Banach space. We need the following lemma on the semi-variation of a $B(\mathfrak{U}, \mathfrak{V})$-valued measure, which is a generalization of Makagon and Salehi [**11**, 2.9 Lemma]. The proof is similar, so it is omitted.

LEMMA 3.8. *If $\xi \in fa(\mathfrak{A}, B(\mathfrak{U}, \mathfrak{V}))$ and $A \in \mathfrak{A}$, it holds that*

$$\begin{aligned}\|\xi\|(A) &= \sup\left\{\big\|\xi(\cdot)u\big\|(A) : u \in \mathfrak{U},\ \|u\|_{\mathfrak{U}} \le 1\right\} \\ &= \sup\left\{\big\|\xi(\cdot)^* v^*\big\|(A) : v^* \in \mathfrak{V}^*,\ \|v^*\|_{\mathfrak{V}^*} \le 1\right\},\end{aligned}$$

where $\|\cdot\|_{\mathfrak{V}^*}$ is the norm in $\mathfrak{V}^*$.

The following lemma follows from Miamee [**13**, p. 844], which is a generalization of Makagon and Salehi [**11**, p. 263].

LEMMA 3.9. *Let $\xi \in fa(\mathfrak{A}, X)$ and $S_\xi : (L^0(\Theta\,;\mathfrak{U}), \|\cdot\|_*) \to \mathfrak{H}$ be given by* (3.3). *Then, it holds that*

$$\|S_\xi \varphi\|_{\mathfrak{H}} = \left\|\int_\Theta d\xi\, \varphi\right\|_{\mathfrak{H}} \le \|\varphi\|_* \|\xi\|(\Theta), \qquad \varphi \in L^0(\Theta\,;\mathfrak{U})$$

and $\|S_\xi\| = \|\xi\|(\Theta)$, where $\|\cdot\|_*$ is defined by (3.2).

We have defined gramian orthogonally scattered dilation for an X-valued measure in Definition 3.7. We need spectral dilations together with 2-majorants for such measures. As we know the space $B(\mathfrak{U}, \mathfrak{H})$ has a $B(\mathfrak{U}, \mathfrak{U}^*)$-valued gramian $[\cdot, \cdot]$. When the Hilbert space $\mathfrak{H}$ is replaced by another Hilbert space $\mathfrak{K}$, then the space $Y = B(\mathfrak{U}, \mathfrak{K})$ also has a $B(\mathfrak{U}, \mathfrak{U}^*)$-valued gramian.

DEFINITION 3.10. Let $X = B(\mathfrak{U}, \mathfrak{H})$.

(1) $E \in fa(\mathfrak{A}, B(\mathfrak{H}))$ is said to be a **finitely additive spectral measure** in $\mathfrak{H}$ if $E(A)$ is an orthogonal projection in $\mathfrak{H}$ for $A \in \mathfrak{A}$, $E(\Theta) = \mathbf{1}$ and $E(A)E(B) = E(A \cap B)$ for $A, B \in \mathfrak{A}$, $\mathbf{1}$ being the identity operator on $\mathfrak{H}$.

(2) $E \in wca(\mathfrak{A}, B(\mathfrak{H}))$ is said to be a **weakly countably additive spectral measure** in $\mathfrak{H}$ if the conditions in (1) above hold.

(3) $\xi \in fa(\mathfrak{A}, X)$ (respectively $wca(\mathfrak{A}, X)$) is said to have a **finitely additive** (respectively **weakly countably additive**) **spectral dilation** if there exist a Hilbert space $\mathfrak{K}$, a finitely additive (respectively weakly countably additive) spectral measure $E(\cdot)$ in $\mathfrak{K}$ and bounded operators $R \in B(\mathfrak{U}, \mathfrak{K})$ and $S \in B(\mathfrak{K}, \mathfrak{H})$ such that $\xi(\cdot) = SE(\cdot)R$.

(4) $F \in fa(\mathfrak{A}, B(\mathfrak{U}, \mathfrak{U}^*))$ is said to be **weak* countably additive** if $\langle v, F(\cdot)u\rangle \in ca(\mathfrak{A}, \mathbb{C})$ for any $u, v \in \mathfrak{U}$. Let $w^*ca(\mathfrak{A}, B(\mathfrak{U}, \mathfrak{U}^*))$ denote the set of all $B(\mathfrak{U}, \mathfrak{U}^*)$-valued weak* countably additive measures.

(5) $\xi \in fa(\mathfrak{A}, B(\mathfrak{U}, \mathfrak{H}))$ (respectively $wca(\mathfrak{A}, B(\mathfrak{U}, \mathfrak{H})))$ is said to have a **finitely additive** (respectively **weak* countably additive**) **2-majorant** if there exists an $F \in fa(\mathfrak{A}, B^+(\mathfrak{U}, \mathfrak{U}^*))$ (respectively $w^*ca(\mathfrak{A}, B^+(\mathfrak{U}, \mathfrak{U}^*)))$ such that for any $n \in \mathbb{N}$, $u_1, \dots, u_n \in \mathfrak{U}$ and $A_1, \dots, A_n \in \mathfrak{A}$ it holds that

$$\left\| \sum_{j=1}^{n} \xi(A_j)u_j \right\|_{\mathfrak{H}}^2 \le \sum_{j=1}^{n}\sum_{k=1}^{n} \langle u_k, F(A_j \cap A_k)u_j \rangle. \tag{3.4}$$

REMARK 3.11. (1) For an $F \in fa(\mathfrak{A}, B^+(\mathfrak{U}, \mathfrak{U}^*))$ consider the following condition: for any $n \in \mathbb{N}$, $u_1, \dots, u_n \in \mathfrak{U}$ and $\{A_1, \dots, A_n\} \in \Pi(\Theta)$ it holds that

$$\left\| \sum_{j=1}^{n} \xi(A_j)u_j \right\|_{\mathfrak{H}}^2 \le \sum_{j=1}^{n} \langle u_j, F(A_j)u_j \rangle. \tag{3.5}$$

We can show that two conditions (3.4) and (3.5) are equivalent.

(2) Let $\xi \in fa(\mathfrak{A}, B(\mathfrak{U}, \mathfrak{H}))$ have a finitely additive gramian orthogonally scattered dilation $\eta \in fagos(\mathfrak{A}, B(\mathfrak{U}, \mathfrak{K}))$ for some Hilbert space $\mathfrak{K}$, so that $\xi(\cdot) = P\eta(\cdot)$, $P : \mathfrak{K} \to \mathfrak{H}$ being the orthogonal projection. Here, $fagos(\mathfrak{A}, B(\mathfrak{U}, \mathfrak{K}))$ denotes the set of all $B(\mathfrak{U}, \mathfrak{K})$-valued finitely additive gramian orthogonally scattered measures. Then, (3.4) holds with $F \in fa(\mathfrak{A}, B^+(\mathfrak{U}, \mathfrak{U}^*))$ given by $F(A \cap B) = \eta(B)^*\eta(A)$ for $A, B \in \mathfrak{A}$. Hence, F is a 2-majorant for ξ. The weakly countably additive case is similarly proved.

(3) We defined weak* countable additivity for $B(\mathfrak{U}, \mathfrak{U}^*)$-valued measures. In Section 4, we shall define weak and strong countable additivities for these measures and see the equivalence among them.

The following theorem clarifies the relations among spectral dilation, gramian orthogonally scattered dilation and 2-majorants.

THEOREM 3.12. *Let $\xi \in fa(\mathfrak{A}, B(\mathfrak{U}, \mathfrak{H}))$ (respectively $wca(\mathfrak{A}, B(\mathfrak{U}, \mathfrak{H})))$. Then the following conditions are equivalent.*

(1) *ξ has a finitely additive (respectively weak* countably additive) 2-majorant.*
(2) *ξ has a finitely additive (respectively weakly countably additive) gramian orthogonally scattered dilation.*
(3) *ξ has a finitely additive (respectively weakly countably additive) spectral dilation.*

PROOF. We shall prove the finitely additive case. The weakly countably additive case is similarly shown.

(1) $\Leftrightarrow$ (3) was shown by Miamee [**13**]. So we shall prove (1) $\Rightarrow$ (2) $\Rightarrow$ (3).

(1) $\Rightarrow$ (2). Let us suppose that ξ has a finitely additive 2-majorant $F \in fa(\mathfrak{A}, B^+(\mathfrak{U}, \mathfrak{U}^*))$. Define $\Gamma : \mathfrak{A} \times \mathfrak{A} \to B(\mathfrak{U}, \mathfrak{U}^*)$ by

$$\Gamma(A, B) = F(A \cap B) - \xi(B)^* \xi(A), \qquad A, B \in \mathfrak{A}.$$

Then, Γ is a positive definite kernel since for any $n \in \mathbb{N}$, $A_1, \ldots, A_n \in \mathfrak{A}$ and $u_1, \ldots, u_n \in \mathfrak{U}$ we have that

$$\begin{aligned}
\sum_{j,k=1}^{n} \langle u_k, \Gamma(A_j, A_k) u_j \rangle &= \sum_{j,k=1}^{n} \Big\langle u_k, \big[F(A_j \cap A_k) - \xi(A_k)^* \xi(A_j)\big] u_j \Big\rangle \\
&= \sum_{j,k=1}^{n} \langle u_k, F(A_j \cap A_k) u_j \rangle - \sum_{j,k=1}^{n} \langle u_k, \xi(A_k)^* \xi(A_j) u_j \rangle \\
&= \sum_{j,k=1}^{n} \langle u_k, F(A_j \cap A_k) u_j \rangle - \sum_{j,k=1}^{n} \big(\xi(A_j) u_j, \xi(A_k) u_k\big)_{\mathfrak{H}} \\
&= \sum_{j,k=1}^{n} \langle u_k, F(A_j \cap A_k) u_j \rangle - \Big\| \sum_{j=1}^{n} \xi(A_j) u_j \Big\|_{\mathfrak{H}}^2 \ge 0
\end{aligned}$$

by (3.4). It follows from Proposition 2.4 that there exists a RKHS $\mathfrak{K}_1$ of Γ consisting of $\mathfrak{U}^*$-valued functions on $\mathfrak{A}$. Let $\zeta : \mathfrak{A} \to B(\mathfrak{U}, \mathfrak{K}_1)$ be defined by $\zeta(A) = \Gamma(A, \cdot)$ for $A \in \mathfrak{A}$. Then, for $u, v \in \mathfrak{U}$ and $A, B \in \mathfrak{A}$ we see that

$$\begin{aligned}
\langle u, \Gamma(A, B) v \rangle &= \big(\Gamma(A, \cdot) v, \Gamma(B, \cdot) u\big)_{\mathfrak{K}_1} \\
&= \big(\zeta(A) v, \zeta(B) u\big)_{\mathfrak{K}_1} \\
&= \langle u, \zeta(B)^* \zeta(A) v \rangle.
\end{aligned}$$

Hence we conclude $\Gamma(A, B) = \zeta(B)^* \zeta(A)$ for $A, B \in \mathfrak{A}$. It is easy to check that $\zeta \in fa(\mathfrak{A}, B(\mathfrak{U}, \mathfrak{K}_1))$. Now let $\mathfrak{K} = \mathfrak{H} \oplus \mathfrak{K}_1$ and define $\eta : \mathfrak{A} \to B(\mathfrak{U}, \mathfrak{K})$ by $\eta(\cdot) = \xi(\cdot) \oplus \zeta(\cdot)$. That is,

$$\eta(A) u = \big(\xi(A) u, \zeta(A) u\big) \in \mathfrak{K}, \qquad u \in \mathfrak{U},\ A \in \mathfrak{A}.$$

Thus, it follows that for $A, B \in \mathfrak{A}$ and $u, v \in \mathfrak{U}$

$$\begin{aligned}
\langle u, \eta(B)^* \eta(A) v \rangle &= \big(\eta(A) v, \eta(B) u\big)_{\mathfrak{K}} \\
&= \big(\xi(A) v, \xi(B) u\big)_{\mathfrak{H}} + \big(\zeta(A) v, \zeta(B) u\big)_{\mathfrak{K}_1} \\
&= \langle u, \xi(B)^* \xi(A) v \rangle + \langle u, \zeta(B)^* \zeta(A) v \rangle \\
&= \Big\langle u, \big[\xi(B)^* \xi(A) + \Gamma(A, B)\big] v \Big\rangle \\
&= \langle u, F(A \cap B) v \rangle,
\end{aligned}$$

so that $F(A \cap B) = \eta(B)^* \eta(A)$ for $A, B \in \mathfrak{A}$. Consequently η is gramian orthogonally scattered, i.e., $\eta \in fagos(\mathfrak{A}, B(\mathfrak{U}, \mathfrak{K}))$.

Finally, let $P : \mathfrak{K} \to \mathfrak{H} \oplus \{0\} \simeq \mathfrak{H}$ be the orthogonal projection. Then we see that $\xi = P\eta$ and ξ has a gramian orthogonally scattered dilation.

(2) $\Rightarrow$ (3). Assume that $\xi \in fa(\mathfrak{A}, B(\mathfrak{U}, \mathfrak{H}))$ has a gramian orthogonally scattered dilation $\eta \in fagos(\mathfrak{A}, B(\mathfrak{U}, \mathfrak{K}))$ for some Hilbert space $\mathfrak{K}$ containing $\mathfrak{H}$ as a

closed subspace such that $\xi(\cdot) = P\eta(\cdot)$, where $P : \mathfrak{K} \to \mathfrak{H}$ is the orthogonal projection. We can suppose that

$$\mathfrak{H} = \mathfrak{S}_0\{\xi(A)u : A \in \mathfrak{A},\ u \in \mathfrak{U}\},$$
$$\mathfrak{K} = \mathfrak{S}_0\{\eta(A)u : A \in \mathfrak{A},\ u \in \mathfrak{U}\},$$

where $\mathfrak{S}\{\cdot\}$ is a closed subspace of $\mathfrak{H}$ or $\mathfrak{K}$ generated by the set $\{\cdot\}$. For each $A \in \mathfrak{A}$ let $E(A)$ be the orthogonal projection of $\mathfrak{K}$ onto the closed subspace $\mathfrak{K}(A) = \mathfrak{S}_0\{\eta(A \cap B)u : B \in \mathfrak{A},\ u \in \mathfrak{U}\}$. Then we see that $E(\cdot)$ is a finitely additive spectral measure in $\mathfrak{K}$ such that $\eta(A) = E(A)\eta(\Theta)$ for $A \in \mathfrak{A}$. Hence, $\xi(\cdot) = P\eta(\cdot) = PE(\cdot)\eta(\Theta)$, i.e., ξ has a finitely additive spectral dilation. □

REMARK 3.13. In Definition 3.1(5), for $\xi \in fa(\mathfrak{A}, B(\mathfrak{U}, \mathfrak{H}))$ and $A \in \mathfrak{A}$, the strong semivariation $\|\xi\|_s(A)$ was defined and it holds that

$$\begin{aligned}\|\xi\|_s(A) &= \sup\left\{\left\|\sum_{\Delta\in\pi} \xi(\Delta)u_\Delta\right\|_{\mathfrak{H}} : \|u_\Delta\|_{\mathfrak{U}} \le 1,\ \Delta \in \pi \in \Pi(A)\right\}\\ &= \sup\left\{\left\|S_\xi(1_A\varphi)\right\|_{\mathfrak{H}} : \varphi \in L^0(\Theta\,;\mathfrak{U}),\ \|\varphi\|_\infty \le 1\right\},\end{aligned}$$

where $S_\xi : (L^0(\Theta\,;\mathfrak{U}), \|\cdot\|_\infty) \to \mathfrak{H}$ is given by (3.3) and $\|\cdot\|_\infty$ by (3.1). Hence $\|\xi\|_s(\Theta) = \|S_\xi\|$, the operator norm of S_ξ.

A sufficient condition for dilation was given by Miamee [**13**] as follows.

THEOREM 3.14. *If $\xi \in fa(\mathfrak{A}, B(\mathfrak{U}, \mathfrak{H}))$ is such that $\|\xi\|_s(\Theta) < \infty$, then ξ is dilatable, i.e., it has a finitely additive spectral dilation and a finitely additive gramian orthogonally scattered dilation. If $\xi \in wca(\mathfrak{A}, B(\mathfrak{U}, \mathfrak{H}))$ and $\|\xi\|_s(\Theta) < \infty$, then it has weakly countably additive spectral and gramian orthogonally scattered dilations.*

To finish this section let us state Riesz type theorems for an operator on a space of vector-valued continuous functions. Let $\mathfrak{U}$ be a Banach space, Θ be a locally compact Hausdorff space, $\mathfrak{A}$ be its Borel σ-algebra and $C_0(\Theta\,;B(\mathfrak{U}))$ be the Banach space of all $B(\mathfrak{U})$-valued norm continuous functions on Θ vanishing at infinity with the sup-norm $\|\cdot\|_\infty$. Let $\mathfrak{H}$ be a Hilbert space and $X = B(\mathfrak{U}, \mathfrak{H})$, and consider an integral representation of an operator $T : C_0(\Theta\,;B(\mathfrak{U})) \to X$. Recall that an X-valued measure $\xi \in ca(\mathfrak{A}, X)$ is **regular** if for any $A \in \mathfrak{A}$ and $\varepsilon > 0$ there exist an open set O and a compact set C in Θ such that $C \subseteq A \subseteq O$ and $\|\xi\|(O\backslash C) < \varepsilon$. $rca(\mathfrak{A}, X)$ etc denote the set of all regular measures in the respective space.

Also we need to recall the integration of $\Phi \in C_0(\Theta\,;B(\mathfrak{U}))$ with respect to an X-valued measure $\xi \in bfa(\mathfrak{A}, X)$ of bounded $\mathfrak{U}$-operator semivariation. For $\Phi \in L^0(\Theta\,;B(\mathfrak{U}))$ we define the **ξ-sup-norm** $\|\Phi\|_{\infty,\xi}$ by

$$\|\Phi\|_{\infty,\xi} = \inf\left\{\alpha > 0 : \left[\|\Phi\| \ge \alpha\right] \text{ is } \xi\text{-null}\right\},$$

where $[\|\Phi\| \ge \alpha] = \{t \in \Theta : \|\Phi(t)\| \ge \alpha\} \subseteq \Theta$ and $L^0(\Theta\,;B(\mathfrak{U}))$ is the set of all $B(\mathfrak{U})$-valued $\mathfrak{A}$-simple functions on Θ. Then, for $\Phi = \sum_{i=1}^n a_i 1_{A_i} \in L^0(\Theta\,;B(\mathfrak{U}))$ where $a_1, \dots, a_n \in B(\mathfrak{U})$ and $\{A_1, \dots, A_n\} \in \Pi(\Theta)$ the integral of Φ with respect to ξ over $A \in \mathfrak{A}$ is defined by

$$\int_A d\xi\,\Phi = \sum_{i=1}^n \xi(A_i \cap A)a_i \in X.$$

Note that $\| \int_A d\xi\, \Phi \|_X \leq \|\xi\|_{\mathfrak{U},o}(A)\|\Phi\|_{\infty,\xi}$. For any $\Phi \in C_0(\Theta\,; B(\mathfrak{U}))$ there exists a sequence $\{\Phi_n\}_{n=1}^{\infty} \subset L^0(\Theta\,; B(\mathfrak{U}))$ such that $\|\Phi_n - \Phi\|_{\infty,\xi} \to 0$ as $n \to \infty$. Then, for $A \in \mathfrak{A}$, it holds that

$$\left\| \int_A d\xi\, \Phi_n - \int_A d\xi\, \Phi_m \right\|_X \leq \|\xi\|_{\mathfrak{U},o}(A)\|\Phi_n - \Phi_m\|_{\infty,\xi} \to 0$$

as $n, m \to \infty$. Hence, $\{\int_A d\xi \Phi_n\}_{n=1}^{\infty}$ is a Cauchy sequence in X, so that the integral of Φ with respect to ξ over $A \in \mathfrak{A}$ is defined by

$$\int_A d\xi\, \Phi = \lim_{n\to\infty} \int_A d\xi\, \Phi_n. \tag{3.6}$$

Let us consider the space

$$L^{\infty}(\xi\,; B(\mathfrak{U})) = \Big\{ \Phi : \Theta \to B(\mathfrak{U}),\ \exists \{\Phi_n\}_{n=1}^{\infty} \subset L^0(\Theta\,; B(\mathfrak{U})) \\ \text{such that } \|\Phi_n - \Phi\|_{\infty,\xi} \to 0 \Big\}. \tag{3.7}$$

Note that $(L^{\infty}(\Theta\,; B(\mathfrak{U})), \|\cdot\|_{\infty,\xi})$ is a Banach space, $C_0(\Theta\,; B(\mathfrak{U})) \subset L^{\infty}(\xi\,; B(\mathfrak{U}))$ and the integral $\int_A d\xi\, \Phi$ of $\Phi \in L^{\infty}(\xi\,; B(\mathfrak{U}))$ with respect to ξ over $A \in \mathfrak{A}$ is defined by (3.6).

Now a Riesz type theorem is stated as follows.

PROPOSITION 3.15. *Assume that Θ is a locally compact Hausdorff space, $\mathfrak{A}$ is its Borel σ-algebra and $X = B(\mathfrak{U}, \mathfrak{H})$. Let $T : C_0(\Theta\,; B(\mathfrak{U})) \to X$ be a weakly compact and bounded right $B(\mathfrak{U})$-module map. Then, there exists a unique X-valued regular countably additive measure $\xi \in rbca(\mathfrak{A}, X)$ of bounded $\mathfrak{U}$-operator semivariation such that*

$$T(\Phi) = \int_{\Theta} d\xi\, \Phi, \qquad \Phi \in C_0(\Theta\,; B(\mathfrak{U}))$$

and $\|T\| = \|\xi\|_{\mathfrak{U},o}(\Theta)$.

The proof is similar to that of Theorem III.6.7 in [**7**] and is omitted.

We need to consider the case where T is not necessarily weakly compact. In this case we have to deal with weakly countably additive measures as will be seen as follows. First we note that $\xi \in fa(\mathfrak{A}, X)$ is weakly countably additive if and only if $\xi(\cdot)u \in ca(\mathfrak{A}, \mathfrak{H})$ for every $u \in \mathfrak{U}$ due to Orlicz-Pettis Theorem. Recall that $wca(\mathfrak{A}, X)$ denotes the set of all X-valued weakly countably additive measures. We begin with lemmas.

LEMMA 3.16. *Let Θ be a locally compact Hausdorff space, $\mathfrak{A}$ be its Borel σ-algebra and $X = B(\mathfrak{U}, \mathfrak{H})$. If a mapping $T : C_0(\Theta) \to X$ is a bounded linear operator, then there exists a unique X-valued regular weakly countably additive measure $\xi \in rwca(\mathfrak{A}, X)$ such that*

$$T(\varphi) = \int_{\Theta} \varphi\, d\xi, \qquad \varphi \in C_0(\Theta)$$

and $\|T\| = \|\xi\|(\Theta)$. Here the integral is taken in the sense that

$$\big((T\varphi)u, \phi\big)_{\mathfrak{H}} = \int_{\Theta} \varphi \big(d\xi\, u, \phi\big)_{\mathfrak{H}} \tag{3.8}$$

for $\varphi \in C_0(\Theta)$, $u \in \mathfrak{U}$ and $\phi \in \mathfrak{H}$.

PROOF. For any $u \in \mathfrak{U}$ let $T_u : C_0(\Theta) \to \mathfrak{H}$ be defined by

$$T_u\varphi = (T\varphi)u, \qquad \varphi \in C_0(\Theta).$$

Then we see that for $\varphi \in C_0(\Theta)$

$$\begin{aligned}\|T_u\varphi\|_{\mathfrak{H}} = \big\|(T\varphi)u\big\|_{\mathfrak{H}} &\leq \|T\varphi\|_X\|u\|_{\mathfrak{U}}\\ &\leq \|T\|\|\varphi\|_\infty\|u\|_{\mathfrak{U}},\end{aligned}$$

so that T_u is bounded with $\|T_u\| \leq \|T\|\|u\|_{\mathfrak{U}}$. Since $\mathfrak{H}$ is a Hilbert space it follows from the Riesz Theorem (see, e.g., [**7**, p. 131]) that there exists a unique measure $\xi_u \in rca(\mathfrak{A}, \mathfrak{H})$ such that

$$T_u\varphi = \int_\Theta \varphi\,d\xi_u, \qquad \varphi \in C_0(\Theta)$$

and $\|T_u\| = \|\xi_u\|(\Theta)$. Define $\xi(\cdot)u = \xi_u(\cdot)$. Then, since ξ_u is regular, for $A \in \mathfrak{A}$ there exists a sequence $\{\varphi_n\}_{n=1}^\infty \subset C_0(\Theta)$ such that $\|\varphi_n\|_\infty \leq 1$ for $n \geq 1$ and $\|T_u\varphi_n - \xi_u(A)\|_{\mathfrak{H}} \to 0$ as $n \to \infty$. Hence we have that

$$\begin{aligned}\big\|\xi(A)u\big\|_{\mathfrak{H}} = \big\|\xi_u(A)\big\|_{\mathfrak{H}} &\leq \liminf_{n\to\infty} \|T_u\varphi_n\|_{\mathfrak{H}}\\ &\leq \liminf_{n\to\infty} \|T_u\|\|\varphi_n\|_\infty\\ &\leq \|T_u\| \leq \|T\|\|u\|_{\mathfrak{U}}.\end{aligned}$$

Consequently, ξ is $X = B(\mathfrak{U}, \mathfrak{H})$-valued, $\xi \in rwca(\mathfrak{A}, X)$ and (3.8) holds. □

Using the above lemma the following representation is obtained.

PROPOSITION 3.17. *Let Θ be a locally compact Hausdorff space, $\mathfrak{A}$ be its Borel σ-algebra, $X = B(\mathfrak{U}, \mathfrak{H})$ and $T : C_0(\Theta; B(\mathfrak{U})) \to X$ be a bounded right $B(\mathfrak{U})$-module map. Then there exists a unique regular weakly countably additive measure $\xi \in rwca(\mathfrak{A}, X)$ with $\|\xi\|_{\mathfrak{U},o}(\Theta) < \infty$ such that $T\Phi = \int_\Theta d\xi\,\Phi$ for $\Phi \in C_0(\Theta; B(\mathfrak{U}))$ and $\|T\| = \|\xi\|_{\mathfrak{U},o}(\Theta)$, where the integral is in the sense of*

$$((T\Phi)u, \phi)_{\mathfrak{H}} = \int_\Theta (d\xi\,\Phi u, \phi)_{\mathfrak{H}} \tag{3.9}$$

for $\Phi \in C_0(\Theta; B(\mathfrak{U}))$, $u \in \mathfrak{U}$ and $\phi \in \mathfrak{H}$.

PROOF. Consider the algebraic tensor product $C_0(\Theta) \odot B(\mathfrak{U})$. By Lemma 3.16 there exists a unique regular weakly countably additive measure $\xi \in rwca(\mathfrak{A}, X)$ such that

$$T(\varphi \otimes \mathbf{1}) = T\varphi = \int_\Theta \varphi\,d\xi, \qquad \varphi \in C_0(\Theta),$$

where $\mathbf{1}$ is the identity operator on $\mathfrak{U}$. Now for $\Phi = \sum_{i=1}^n \varphi_i \otimes a_i \in C_0(\Theta) \odot B(\mathfrak{U})$ it holds that

$$T\Phi = \int_\Theta d\xi\,\Phi \tag{3.10}$$

since T is a right $B(\mathfrak{U})$-module map. In a similar way as in the proof of [**7**, Theorem III.6.7 (p. 134)] we can show that ξ is of bounded $\mathfrak{U}$-operator semivariation. Hence (3.10) holds for every $\Phi \in C_0(\Theta; B(\mathfrak{U})) = C_0(\Theta) \otimes_\lambda B(\mathfrak{U})$ in the sense of (3.9) since $C_0(\Theta) \odot B(\mathfrak{U})$ is dense in $C_0(\Theta; B(\mathfrak{U}))$. Finally, $\|T\| = \|\xi\|_{\mathfrak{U},o}(\Theta)$ is shown in the same fashion as in the proof of Lemma III.6.6 in [**7**]. □

4. $B(\mathfrak{U}, \mathfrak{U}^*)$-valued measures and bimeasures

To describe $X = B(\mathfrak{U}, \mathfrak{H})$-valued processes we use X-valued weakly countably additive measures rather than countably additive (in $\|\cdot\|_X$) measures, where $\mathfrak{U}$ is a Banach space and $\mathfrak{H}$ is a Hilbert space. Since the covariance function of an X-valued process is a $B(\mathfrak{U}, \mathfrak{U}^*)$-valued function, to define stationary or harmonizable processes we need to consider weakly or weak* countably additive $B(\mathfrak{U}, \mathfrak{U}^*)$-valued measures and bimeasures. In this case, there are three types of countable additivity: strong, weak and weak* countable additivities. First we shall state that these notions are equivalent after giving some definitions and notations. As before let $(\Theta, \mathfrak{A})$ be a measurable space. The duality pair of $\mathfrak{U}$ and $\mathfrak{U}^*$ is denoted by $\langle u, u^* \rangle$ for $u \in \mathfrak{U}$ and $u^* \in \mathfrak{U}^*$, while that of $\mathfrak{U}^{**}$ and $\mathfrak{U}^*$ by $\langle u^{**}, u^* \rangle$ for $u^{**} \in \mathfrak{U}^{**}$ and $u^* \in \mathfrak{U}^*$.

DEFINITION 4.1. Let $fa(\mathfrak{A}, B(\mathfrak{U}, \mathfrak{U}^*))$ (respectively $ca(\mathfrak{A}, B(\mathfrak{U}, \mathfrak{U}^*))$) be the set of all $B(\mathfrak{U}, \mathfrak{U}^*)$-valued finitely additive (respectively countably additive in the operator norm) measures on $\mathfrak{A}$. Let $F \in fa(\mathfrak{A}, B(\mathfrak{U}, \mathfrak{U}^*))$.

(1) F is said to be **weak* countably additive** if $\langle u, F(\cdot)v \rangle \in ca(\mathfrak{A}, \mathbb{C})$ for any $u, v \in \mathfrak{U}$. Let $w^*ca(\mathfrak{A}, B(\mathfrak{U}, \mathfrak{U}^*))$ denote the set of all $B(\mathfrak{U}, \mathfrak{U}^*)$-valued weak* countably additive measures on $\mathfrak{A}$ (cf. Definition 3.10(4)).

(2) F is said to be **weakly countably additive** if $\langle u^{**}, F(\cdot)u \rangle \in ca(\mathfrak{A}, \mathbb{C})$ for any $u^{**} \in \mathfrak{U}^{**}$ and $u \in \mathfrak{U}$. Let $wca(\mathfrak{A}, B(\mathfrak{U}, \mathfrak{U}^*))$ denote the set of all $B(\mathfrak{U}, \mathfrak{U}^*)$-valued weakly countably additive measures on $\mathfrak{A}$.

(3) F is said to be **strongly countably additive** if $F(\cdot)u \in ca(\mathfrak{A}, \mathfrak{U}^*)$ for any $u \in \mathfrak{U}$. Let $sca(\mathfrak{A}, B(\mathfrak{U}, \mathfrak{U}^*))$ denote the set of all $B(\mathfrak{U}, \mathfrak{U}^*)$-valued strongly countably additive measures on $\mathfrak{A}$.

For $F \in fa(\mathfrak{A}, B(\mathfrak{U}, \mathfrak{U}^*))$ and $A \in \mathfrak{A}$ the variation $|F|(A)$ and the semivariation $\|F\|(A)$ are defined in a similar way as in Definition 3.1. Furthermore, the **$\mathfrak{U}$-operator semivariation** $\|F\|_{\mathfrak{U},o}(A)$ is given by

$$\|F\|_{\mathfrak{U},o}(A) = \sup \left\{ \left\| \sum_{\Delta \in \pi} F(\Delta) a_\Delta \right\| : a_\Delta \in B(\mathfrak{U}),\ \|a_\Delta\| \le 1,\ \Delta \in \pi \in \Pi(A) \right\}.$$

Let $bw^*ca(\mathfrak{A}, B(\mathfrak{U}, \mathfrak{U}^*))$ denote the set of all $F \in w^*ca(\mathfrak{A}, B(\mathfrak{U}, \mathfrak{U}^*))$ of bounded $\mathfrak{U}$-operator semivariation.

It follows from the above definition that

$$\begin{aligned} ca(\mathfrak{A}, B(\mathfrak{U}, \mathfrak{U}^*)) \subseteq sca(\mathfrak{A}, B(\mathfrak{U}, \mathfrak{U}^*)) &= wca(\mathfrak{A}, B(\mathfrak{U}, \mathfrak{U}^*)) \\ &\subseteq w^*ca(\mathfrak{A}, B(\mathfrak{U}, \mathfrak{U}^*)) \subseteq fa(\mathfrak{A}, B(\mathfrak{U}, \mathfrak{U}^*)). \end{aligned}$$

Miamee and Salehi [**14**] proved that strong, weak and weak* countable additivities are equivalent for $B^+(\mathfrak{U}, \mathfrak{U}^*)$-valued measures, which is stated as follows.

THEOREM 4.2. *If $F \in w^*ca(\mathfrak{A}, B^+(\mathfrak{U}, \mathfrak{U}^*))$, then $F \in sca(\mathfrak{A}, B^+(\mathfrak{U}, \mathfrak{U}^*))$.*

From now on we are going to deal with weak* countably additive $B(\mathfrak{U}, \mathfrak{U}^*)$-valued measures and bimeasures. Let $\mathfrak{A} \times \mathfrak{A} = \{A \times B : A, B \in \mathfrak{A}\}$. For a function M on $\mathfrak{A} \times \mathfrak{A}$ we denote the value of M at $A \times B$ exchangeably by $M(A \times B)$ or $M(A, B)$.

DEFINITION 4.3. (1) A function $M : \mathfrak{A} \times \mathfrak{A} \to B(\mathfrak{U}, \mathfrak{U}^*)$ is said to be a **countably additive operator bimeasure** if $M(A, \cdot)$, $M(\cdot, B) \in ca(\mathfrak{A}, B(\mathfrak{U}, \mathfrak{U}^*))$ for every A, $B \in \mathfrak{A}$. Let $\mathfrak{M} = \mathfrak{M}(\mathfrak{A} \times \mathfrak{A}; B(\mathfrak{U}, \mathfrak{U}^*))$ denote the set of all countably additive operator bimeasures on $\mathfrak{A} \times \mathfrak{A}$.

(2) A function $M : \mathfrak{A} \times \mathfrak{A} \to B(\mathfrak{U}, \mathfrak{U}^*)$ is said to be a **finitely additive operator bimeasure** if $M(A, \cdot),\ M(\cdot, B) \in fa(\mathfrak{A}, B(\mathfrak{U}, \mathfrak{U}^*))$ for every $A, B \in \mathfrak{A}$. Let $\mathfrak{M}^f = \mathfrak{M}^f(\mathfrak{A} \times \mathfrak{A}\,; B(\mathfrak{U}, \mathfrak{U}^*))$ denote the set of all finitely additive operator bimeasures on $\mathfrak{A} \times \mathfrak{A}$.

(3) A function $M : \mathfrak{A} \times \mathfrak{A} \to B(\mathfrak{U}, \mathfrak{U}^*)$ is said to be a **weak* countably additive operator bimeasure** if $M(A, \cdot),\ M(\cdot, B) \in w^*ca(\mathfrak{A}, B(\mathfrak{U}, \mathfrak{U}^*))$ for every $A, B \in \mathfrak{A}$. Let $\mathfrak{M}^{w^*} = \mathfrak{M}^{w^*}(\mathfrak{A} \times \mathfrak{A}\,; B(\mathfrak{U}, \mathfrak{U}^*))$ denote the set of all weak* countably additive operator bimeasures.

(4) A finitely additive operator bimeasure $M \in \mathfrak{M}^f$ is said to be **positive definite** if, for any $n \in \mathbb{N}$, $a_1, \dots, a_n \in B(\mathfrak{U})$ and $A_1, \dots, A_n \in \mathfrak{A}$, it holds that

$$\sum_{i,j=1}^{n} a_j^* M(A_i, A_j) a_i \geq 0.$$

(5) A scalar valued function $m : \mathfrak{A} \times \mathfrak{A} \to \mathbb{C}$ is said to be a **scalar bimeasure** if $m(A, \cdot),\ m(\cdot, B) \in ca(\mathfrak{A}, \mathbb{C})$ for every $A, B \in \mathfrak{A}$. Let $\mathbf{M} = \mathfrak{M}(\mathfrak{A} \times \mathfrak{A}\,; \mathbb{C})$ denote the set of all scalar bimeasures on $\mathfrak{A} \times \mathfrak{A}$. A scalar bimeasure $m \in \mathbf{M}$ is said to be **positive definite** if

$$\sum_{i,j=1}^{n} \alpha_i \overline{\alpha}_j m(A_i, A_j) \geq 0$$

for any $n \in \mathbb{N}$, $\alpha_1, \dots, \alpha_n \in \mathbb{C}$ and $A_1, \dots, A_n \in \mathfrak{A}$.

In the following we mainly consider positive definite operator or scalar bimeasures. As before let $X = B(\mathfrak{U}, \mathfrak{H})$ with $\mathfrak{H}$ a Hilbert space and a gramian $[x, y] = y^* x \in B(\mathfrak{U}, \mathfrak{U}^*)$ for $x, y \in X$. Let $\xi, \eta \in fa(\mathfrak{A}, X)$ be X-valued finitely additive measures and define $M_{\xi\eta}$ and M_ξ by

$$M_{\xi\eta}(A, B) = \big[\xi(A), \eta(B)\big] = \eta(B)^* \xi(A), \qquad A, B \in \mathfrak{A}, \tag{4.1}$$

$$M_\xi = M_{\xi\xi}. \tag{4.2}$$

Then it is easy to see that $M_{\xi\eta}$, $M_\xi \in \mathfrak{M}^f$. Especially, M_ξ is called the operator bimeasure induced by ξ and is positive definite. Also it is clear that if $\xi, \eta \in wca(\mathfrak{A}, X)$ (respectively $ca(\mathfrak{A}, X)$), then $M_{\xi\eta} \in \mathfrak{M}^{w^*}$ (respectively $\mathfrak{M}$).

DEFINITION 4.4. Let $M \in \mathfrak{M}^f$. The **variation**, **semivariation** and **$\mathfrak{U}$-operator semivariation** of M at $A \times B \in \mathfrak{A} \times \mathfrak{A}$ are respectively defined by

$$|M|(A, B) = \sup\left\{ \sum_{\Delta \in \pi, \Delta' \in \pi'} \|M(\Delta, \Delta')\| : \pi \in \Pi(A),\ \pi' \in \Pi(B) \right\},$$

$$\|M\|(A, B) = \sup\left\{ \left\| \sum_{\Delta \in \pi, \Delta' \in \pi'} \alpha_\Delta \overline{\beta}_{\Delta'} M(\Delta, \Delta') \right\| : \alpha_\Delta, \beta_{\Delta'} \in \mathbb{C}, \right.$$

$$\left. |\alpha_\Delta|, |\beta_{\Delta'}| \leq 1,\ \Delta \in \pi \in \Pi(A),\ \Delta' \in \pi' \in \Pi(B) \right\},$$

$$\|M\|_{\mathfrak{U},o}(A, B) = \sup\left\{ \left\| \sum_{\Delta \in \pi, \Delta' \in \pi'} b_{\Delta'}^* M(\Delta, \Delta') a_\Delta \right\| : a_\Delta, b_{\Delta'} \in B(\mathfrak{U}), \right.$$

$$\left. \|a_\Delta\|, \|b_{\Delta'}\| \leq 1,\ \Delta \in \pi \in \Pi(A),\ \Delta' \in \pi' \in \Pi(B) \right\}.$$

Here $\Pi(A)$ and $\Pi(B)$ are the sets of all finite measurable partitions of A and B, respectively. Let $\mathfrak{M}_b^f = \mathfrak{M}_b^f(\mathfrak{A} \times \mathfrak{A}\,; B(\mathfrak{U}, \mathfrak{U}^*))$ denote the set of all finitely additive operator bimeasures M of bounded $\mathfrak{U}$-operator semivariation, i.e., $\|M\|_{\mathfrak{U},o}(\Theta, \Theta) < \infty$. Similarly, we shall write $\mathfrak{M}_b^{w^*} = \{M \in \mathfrak{M}^{w^*} : \|M\|_{\mathfrak{U},o}(\Theta, \Theta) < \infty\}$ and $\mathfrak{M}_b = \{M \in \mathfrak{M} : \|M\|_{\mathfrak{U},o}(\Theta, \Theta) < \infty\}$.

The following lemma corresponds to Lemma III.1.19 in [**7**, p. 66] and the proof is routine.

LEMMA 4.5. *Let $X = B(\mathfrak{U}, \mathfrak{H})$, $\xi, \eta \in fa(\mathfrak{A}, X)$ and $M_\xi, M_\eta \in \mathfrak{M}^f$ be given by (4.2) and $M_{\xi\eta}$ be given by (4.1). Then the following statements are true.*

(1) $\|M_{\xi\eta}\|(A, B) \leq \|\xi\|(A)\|\eta\|(B)$ *for* $A, B \in \mathfrak{A}$.
(2) $\|M_{\xi\eta}\|_{\mathfrak{U},o}(A, B) \leq \|\xi\|_{\mathfrak{U},o}(A)\|\eta\|_{\mathfrak{U},o}(B)$ *for* $A, B \in \mathfrak{A}$.
(3) $\|M_\xi\|(A, A) = \|\xi\|(A)^2$ *for* $A \in \mathfrak{A}$.
(4) $\|\xi\|_{\mathfrak{U},o}(A)^2 = \|M_\xi\|_{\mathfrak{U},o}(A, A) \leq |M_\xi|(A, A) \leq |\xi|(A)^2$ *for* $A \in \mathfrak{A}$.

The following is an application of RKHS.

PROPOSITION 4.6. *For any $B(\mathfrak{U}, \mathfrak{U}^*)$-valued finitely additive positive definite bimeasure $M \in \mathfrak{M}^f$ there exist a Hilbert space $\mathfrak{H}$ and an $X = B(\mathfrak{U}, \mathfrak{H})$-valued finitely additive measure $\xi \in fa(\mathfrak{A}, X)$ such that $M = M_\xi$, given by (4.2). If M is of bounded $\mathfrak{U}$-operator semivariation, then so is ξ. If M is weak* countably additive, then ξ is weakly countably additive. Furthermore, if $M \in \mathfrak{M}$, then $\xi \in ca(\mathfrak{A}, X)$.*

PROOF. Let $M \in \mathfrak{M}^f$. Since $M(\cdot, \cdot) : \mathfrak{A} \times \mathfrak{A} \to B(\mathfrak{U}, \mathfrak{U}^*)$ is a positive definite kernel it follows from Proposition 2.4 and Corollary 2.5 that there exists a Hilbert space $\mathfrak{H}$ and an $X = B(\mathfrak{U}, \mathfrak{H})$-valued function ξ such that $M = M_\xi$. The finite additivity of ξ is easily checked, so that $\xi \in fa(\mathfrak{A}, \mathfrak{U})$.

If M is of bounded $\mathfrak{U}$-operator semivariation, then $\|\xi\|_{\mathfrak{U},o}(\Theta) < \infty$ follows from Lemma 4.5(4). Other two statements are almost obvious. □

It follows from the above proposition that given a positive definite operator bimeasure M we can find some $X = B(\mathfrak{U}, \mathfrak{H})$-valued measure ξ such that $M = M_\xi$, so that the integration of $B(\mathfrak{U})$-valued functions with respect to M can be defined using ξ-integrations. More fully, let $\xi \in bfa(\mathfrak{A}, X)$ and consider $M = M_\xi$. Also let $\Phi, \Psi \in L^\infty(\xi\,; B(\mathfrak{U}))$ (cf. (3.7)) and $A, B \in \mathfrak{A}$. Then, the integral of (Φ, Ψ) with respect to M over $A \times B$ is defined by

$$\begin{aligned}
\iint_{A\times B} (\Phi, \Psi)\, dM &= \iint_{A\times B} \Psi^*\, dM\, \Phi \\
&= \left[\int_A d\xi\, \Phi, \int_B d\xi\, \Psi \right] \\
&= \left(\int_B d\xi\, \Psi \right)^* \left(\int_A d\xi\, \Phi \right),
\end{aligned} \tag{4.3}$$

where $\int_A d\xi\, \Phi$ and $\int_B d\eta\, \Psi$ are defined by (3.6). We may define a gramian in $L^\infty(\xi\,; B(\mathfrak{U}))$ by

$$[\Phi, \Psi]_M = \iint_{\Theta\times\Theta} (\Phi, \Psi)\, dM, \qquad \Phi, \Psi \in L^\infty(\Theta\,; B(\mathfrak{U})). \tag{4.4}$$

If $\xi \in bwca(\mathfrak{A}, X)$, then the integral (4.3) is defined in the sense that

$$\left\langle u, \iint_{A\times B} (\Phi, \Psi)\, dM\, v \right\rangle = \left(\int_A d\xi\, \Phi v, \int_B d\xi\, \Psi u \right)_{\mathfrak{H}}, \qquad u, v \in \mathfrak{U}. \tag{4.5}$$

This is applied to (4.4) as well.

Assume that F is a $B^+(\mathfrak{U},\mathfrak{U}^*)$-valued finitely additive measure on $\mathfrak{A}$, i.e., $F \in fa(\mathfrak{A}, B^+(\mathfrak{U},\mathfrak{U}^*))$. Define $M(A,B) = F(A\cap B)$ for $A, B \in \mathfrak{A}$. Then, M is a positive definite operator bimeasure and by Corollary 2.6 there is an $X = B(\mathfrak{U},\mathfrak{H})$-valued finitely additive gramian orthogonally scattered measure $\xi \in fagos(\mathfrak{A}, X)$ such that $M(A,A) = F(A) = [\xi(A), \xi(A)] = F_\xi(A)$ for $A \in \mathfrak{A}$ and for some Hilbert space $\mathfrak{H}$. If F is of bounded $\mathfrak{U}$-operator semivariation, i.e., $\|F\|_{\mathfrak{U},o}(\Theta) < \infty$, then the integral of $\Phi \in L^\infty(\xi\,; B(\mathfrak{U}))$ with respect to F

$$\int_A dF\,\Phi \in B(\mathfrak{U},\mathfrak{U}^*), \qquad A \in \mathfrak{A}$$

is defined first for a simple function $\Phi \in L^0(\Theta\,; B(\mathfrak{U}))$ and then by an approximation for a general $\Phi \in L^\infty(\Theta\,; B(\mathfrak{U}))$. More fully, if $\Phi = \sum_{i=1}^n a_i 1_{A_i}$, then

$$\int_A dF\,\Phi = \sum_{i=1}^n F(A\cap A_i)a_i.$$

For a general $\Phi \in L^\infty(\xi\,; B(\mathfrak{U}))$ choose a sequence $\{\Phi_n\}_{n=1}^\infty \subset L^0(\Theta\,; B(\mathfrak{U}))$ such that $\|\Phi_n - \Phi\|_{\xi,\infty} \to 0$ as $n \to \infty$ and define

$$\int_A dF\,\Phi = \lim_{n\to\infty} \int_A dF\,\Phi_n. \tag{4.6}$$

Well definedness of the above integral is clear. Note that

$$\int_A \Psi^*\,dF\,\Phi \in B(\mathfrak{U},\mathfrak{U}^*), \qquad A \in \mathfrak{A} \tag{4.7}$$

can be defined for $\Phi, \Psi \in L^\infty(\xi\,; B(\mathfrak{U}))$. This implies that in $L^\infty(\xi\,; B(\mathfrak{U}))$ we can define a gramian by

$$[\Phi,\Psi]_F = \int_\Theta \Psi^*\,dF\,\Phi, \qquad \Phi,\,\Psi \in L^\infty(\xi\,; B(\mathfrak{U})). \tag{4.8}$$

Assume that $F \in bw^*ca(\mathfrak{A}, B^+(\mathfrak{U},\mathfrak{U}^*))$, so that $\|F\|_{\mathfrak{U},o}(\Theta) < \infty$. Then the corresponding ξ is weakly countably additive, gramian orthogonally scattered, and of bounded $\mathfrak{U}$-operator semivariation, denoted $\xi \in bwcagos(\mathfrak{A}, X)$. Then, the integrals in (4.6), (4.7) and (4.8) are taken in the sense of (4.5).

5. $B(\mathfrak{U},\mathfrak{H})$-valued processes

In this section, let G be a locally compact abelian group and consider $B(\mathfrak{U},\mathfrak{H})$-valued processes on G, where $\mathfrak{U}$ is a Banach space and $\mathfrak{H}$ is a Hilbert space. As was mentioned in Section 1, these are good models for Banach space valued weak second order stochastic processes and our theory in the previous sections can be applied naturally. If $\{x(t)\}$ is a $B(\mathfrak{U},\mathfrak{H})$-valued process on G, then, for each $u \in \mathfrak{U}$, $\{x(t)u\}$ is an $\mathfrak{H}$-valued process on G. Hence, the classical theory for second order stochastic processes is readily applied to these processes. Let $\widehat{G}$ be the dual group of G and $\mathfrak{B}_{\widehat{G}}$ be its Borel σ-algebra. The duality pair of G and $\widehat{G}$ is denoted by $\chi(t) = \langle t, \chi\rangle$ for $t \in G$ and $\chi \in \widehat{G}$. As before let us denote $X = B(\mathfrak{U},\mathfrak{H})$.

DEFINITION 5.1. Let $\{x(t)\}$ be an $X = B(\mathfrak{U},\mathfrak{H})$-valued process on G.
(1) The **operator covariance function** Γ of $\{x(t)\}$ is defined by

$$\Gamma(s,t) = \big[x(s), x(t)\big] = x(t)^*x(s) \in B(\mathfrak{U},\mathfrak{U}^*), \qquad s,t \in G.$$

Let $u \in \mathfrak{U}$. For the $\mathfrak{H}$-valued process $\{x(t)u\}$ on G the **scalar covariance function** γ_u is defined by

$$\gamma_u(s,t) = \big(x(s)u, x(t)u\big)_{\mathfrak{H}} = \big\langle u, \Gamma(s,t)u\big\rangle, \qquad s,t \in G. \tag{5.1}$$

(2) The process $\{x(t)\}$ is said to be **operator stationary** if $\Gamma(s,t) = \tilde{\Gamma}(st^{-1})$, $s,t \in G$ and if $\tilde{\Gamma}(\cdot)$ is weakly continuous, i.e., $\langle u, \tilde{\Gamma}(\cdot)u\rangle$ is continuous for every $u \in \mathfrak{U}$. $\{x(t)\}$ is said to be **scalarly stationary** if the $\mathfrak{H}$-valued process $\{x(t)u\}$ is stationary for every $u \in \mathfrak{U}$, i.e., the scalar covariance function γ_u defined by (5.1) satisfies that $\gamma_u(s,t) = \tilde{\gamma}_u(st^{-1})$, $s,t \in G$ and that $\tilde{\gamma}_u(\cdot)$ is continuous on G.

(3) The process $\{x(t)\}$ is said to be **operator harmonizable** if its operator covariance function Γ has an integral representation

$$\Gamma(s,t) = \iint_{\widehat{G}^2} \langle s,\chi\rangle\overline{\langle t,\chi'\rangle}\, M(d\chi, d\chi'), \qquad s,\, t \in G,$$

where $M \in \mathfrak{M}_{rb}^{w^*}$ is a regular $B(\mathfrak{U},\mathfrak{U}^*)$-valued weak* countably additive positive definite operator bimeasure on $\mathfrak{B}_{\widehat{G}} \times \mathfrak{B}_{\widehat{G}}$ of bounded $\mathfrak{U}$-operator semivariation. Here, the regularity of a scalar or operator bimeasure is defined similarly as in the case of usual measures using the semivariation. Hence, the integral above is in the sense that

$$\big\langle u, \Gamma(s,t)v\big\rangle = \iint_{\widehat{G}^2} \langle s,\chi\rangle\overline{\langle t,\chi'\rangle}\, \big\langle u, M(d\chi, d\chi')v\big\rangle$$

for $s,\, t \in G$ and $u,\, v \in \mathfrak{U}$. $\{x(t)\}$ is said to be **scalarly harmonizable** if the $\mathfrak{H}$-valued process $\{x(t)u\}$ is weakly harmonizable for every $u \in \mathfrak{U}$ in the sense of [**7**, Definition IV.3.1], i.e., there exists a regular positive definite scalar bimeasure $m_u \in \mathbf{M}_r = \mathfrak{M}_r(\mathfrak{B}_{\widehat{G}} \times \mathfrak{B}_{\widehat{G}}\,; \mathbb{C})$ such that

$$\gamma_u(s,t) = \iint_{\widehat{G}^2} \langle s,\chi\rangle\overline{\langle t,\chi'\rangle}\, m_u(d\chi, d\chi'), \qquad s,\, t \in G.$$

(4) The process $\{x(t)\}$ is said to be **strongly continuous** if $x(\cdot)u$ is continuous on G in the norm $\|\cdot\|_{\mathfrak{H}}$ for every $u \in \mathfrak{U}$.

(5) The process $\{x(t)\}$ is said to be **operator V-bounded** if it is strongly continuous and bounded, and if there is a constant $C > 0$ such that

$$\left\| \int_G x(t)\Phi(t)u\, \varrho(dt) \right\|_X \le C\|\mathbf{F}\Phi\|_\infty \|u\|_{\mathfrak{U}}$$

for $\Phi \in L^1\big(G; B(\mathfrak{U})\big)$ and $u \in \mathfrak{U}$, where the integral is a well-defined Bochner integral and $\mathbf{F}\Phi$ is the Fourier transform of Φ given by

$$\mathbf{F}\Phi(\chi) = \int_G \langle t,\chi\rangle \Phi(t)\, \varrho(dt), \qquad \chi \in \widehat{G},$$

ϱ being the Haar measure of G. Also $L^1(G; B(\mathfrak{U}))$ is the set of all $B(\mathfrak{U})$-valued Bochner integrable functions with respect to ϱ on G. $\{x(t)\}$ is said to be **scalarly V-bounded** if the $\mathfrak{H}$-valued process $\{x(t)u\}$ is V-bounded for every $u \in \mathfrak{U}$ (cf. [**7**, Definition IV.4.1]), i.e., it is norm continuous and bounded, and there exists a constant $C_u > 0$ such that

$$\left\| \int_G \varphi(t)x(t)u\, \varrho(dt) \right\|_{\mathfrak{H}} \le C_u \|\mathbf{F}\varphi\|_\infty, \qquad \varphi \in L^1(G),$$

where $L^1(G)$ is the L^1-group algebra of G.

We shall consider the integral representation of various types of X-valued processes mentioned above. Since we are considering measures and bimeasures on a topological group G, all the measures are supposed to be **regular** and the set of such measures is denoted by $rca(\mathfrak{B}_{\widehat{G}}, X)$ or $\mathfrak{M}_r$ etc (see Section 3).

Since stationary processes are always our starting points we first collect some results on stationary processes. (2), (3) and (4) of the following proposition are given by Chobanyan and Weron [**4**] and (1) is well-known.

PROPOSITION 5.2. *Let $\{x(t)\}$ be an X-valued process on G.*

(1) *$\{x(t)\}$ is scalarly stationary if and only if, for $u \in \mathfrak{U}$, there exists a unique $\mathfrak{H}$-valued regular countably additive orthogonally scattered measure $\xi_u \in rcaos(\mathfrak{B}_{\widehat{G}}, \mathfrak{H})$ such that*

$$x(t)u = \int_{\widehat{G}} \langle t, \chi \rangle \, \xi_u(d\chi), \qquad t \in G.$$

(2) *$\{x(t)\}$ is operator stationary if and only if there exists a unique X-valued regular weakly countably additive gramian orthogonally scattered measure ξ, denoted $\xi \in rwcagos(\mathfrak{A}, X)$ and called the* **representing measure**, *such that $x(t) = \int_{\widehat{G}} \langle t, \chi \rangle \, \xi(d\chi)$ for $t \in G$, where the integral is in the weak sense, i.e.,*

$$(x(t)u, \phi)_{\mathfrak{H}} = \int_{\widehat{G}} \langle t, \chi \rangle \, (\xi(d\chi)u, \phi)_{\mathfrak{H}}, \qquad t \in G, \ u \in \mathfrak{U}, \ \phi \in \mathfrak{H}. \tag{5.2}$$

(3) *If $\{x(t)\}$ is operator stationary, then there exists a $B^+(\mathfrak{U}, \mathfrak{U}^*)$-valued regular weak* countably additive measure $F \in rw^*ca(\mathfrak{B}_{\widehat{G}}, B^+(\mathfrak{U}, \mathfrak{U}^*))$ such that*

$$\langle u, \tilde{\Gamma}(s)v \rangle = \int_{\widehat{G}} \langle s, \chi \rangle \, \langle u, F(d\chi)v \rangle, \qquad s \in G, \ u, v \in \mathfrak{U},$$

where $\tilde{\Gamma}(s) = \Gamma(s, e)$ for $s \in G$, Γ being the operator covariance function of $\{x(t)\}$.

(4) *A strongly continuous X-valued process $\{x(t)\}$ is operator stationary if and only if it is scalarly stationary.*

In the following we shall show that a harmonizable process has an integral representation.

PROPOSITION 5.3. *Let $\{x(t)\}$ be an X-valued process on G.*

(1) *$\{x(t)\}$ is scalarly harmonizable if and only if, for each $u \in \mathfrak{U}$, there exists a unique $\mathfrak{H}$-valued regular countably additive measure $\xi_u \in rca(\mathfrak{B}_{\widehat{G}}, \mathfrak{H})$ such that*

$$x(t)u = \int_{\widehat{G}} \langle t, \chi \rangle \, \xi_u(d\chi), \qquad t \in G.$$

(2) *$\{x(t)\}$ is operator harmonizable if and only if there exists a unique X-valued regular weakly countably additive measure $\xi \in rbwca(\mathfrak{B}_{\widehat{G}}, X)$ of bounded $\mathfrak{U}$-operator semivariation, called the* **representing measure**, *such that*

$$x(t) = \int_{\widehat{G}} \langle t, \chi \rangle \, \xi(d\chi), \qquad t \in G, \tag{5.3}$$

where the integral is in the weak sense (5.2).

PROOF. (1) follows from [**7**, Theorem IV.3.2].

(2) Suppose that $\{x(t)\}$ has a representation given by (5.3). Then, the operator covariance function $\Gamma(s,t)$ can be written as

$$\begin{aligned}\Gamma(s,t) &= \left[\int_{\widehat{G}} \langle s,\chi\rangle\,\xi(d\chi), \int_{\widehat{G}} \langle t,\chi'\rangle\,\xi(d\chi')\right] \\ &= \iint_{\widehat{G}^2} \langle s,\chi\rangle\overline{\langle t,\chi'\rangle}\, M_\xi(d\chi,d\chi')\end{aligned}$$

for $s,t\in G$, where the integral is in the weak sense. Clearly, $M_\xi \in \mathfrak{M}_{rb}^{w^*}$ and $\{x(t)\}$ is operator harmonizable.

Conversely, assume that $\{x(t)\}$ is operator harmonizable with an operator bimeasure $M\in\mathfrak{M}_{rb}^{w^*}$. Then, $M(\cdot,\cdot):\mathfrak{B}_{\widehat{G}}\times\mathfrak{B}_{\widehat{G}}\to B(\mathfrak{U},\mathfrak{U}^*)$ is a positive definite kernel. It follows from Proposition 2.4 that there exists a RKHS $\mathfrak{H}_M$ of M and $\eta\in rbwca(\mathfrak{B}_{\widehat{G}},Y)$ such that $M=M_\eta$, i.e., $M(A,B)=\eta(B)^*\eta(A)$ for $A,B\in\mathfrak{B}_{\widehat{G}}$, where $Y=B(\mathfrak{U},\mathfrak{H}_M)$. Define a Y-valued process $\{y(t)\}$ by $y(t)=\int_{\widehat{G}}\langle t,\chi\rangle\,\eta(d\chi)$ (a weak integral) for $t\in G$. Then we see that $\{y(t)\}$ is operator harmonizable and the operator covariance function of $\{y(t)\}$ is Γ. We can assume that

$$\begin{aligned}\mathfrak{H}_M &= \mathfrak{S}_0\{y(t)u : t\in G,\ u\in\mathfrak{U}\},\\ \mathfrak{H} &= \mathfrak{S}_0\{x(t)u : t\in G,\ u\in\mathfrak{U}\}\end{aligned}$$

(cf. Proof of (2) $\Rightarrow$ (3) of Theorem 3.12). If we define an operator $U:\mathfrak{H}_M\to\mathfrak{H}$ by $Uy(t)u=x(t)u$ for $t\in G$ and $u\in\mathfrak{U}$, then clearly U is a unitary operator and $x(t)=\int_{\widehat{G}}\langle t,\chi\rangle\,U\eta(d\chi)$ for $t\in G$, so that $\{x(t)\}$ has an integral representation with the representing measure $U\eta\in rbwca(\mathfrak{B}_{\widehat{G}},X)$. $\square$

Note that any operator harmonizable process is strongly continuous. As is expected harmonizability and V-boundedness are equivalent. To prove this we need the following lemma whose proof is similar to that of [**7**, Lemma IV.4.4, pp. 167–168] and is omitted.

LEMMA 5.4. *Let $\{x(t)\}$ be an X-valued process on G.*

(1) *If $\{x(t)\}$ is scalarly harmonizable with representing measures $\xi_u\in rca(\mathfrak{B}_{\widehat{G}},\mathfrak{H})$ for the $\mathfrak{H}$-valued processes $\{x_u(t)\}=\{x(t)u\}$, $u\in\mathfrak{U}$, then it holds that*

$$\int_G \varphi(s)x_u(s)\,\varrho(ds) = \int_{\widehat{G}} \mathbf{F}\varphi(\chi)\,\xi_u(d\chi),\qquad \varphi\in L^1(G),$$

where $\mathbf{F}\varphi$ is the Fourier transform of φ, the left side integral is a Bochner integral and the right side is a Dunford-Schwartz integral.

(2) *If $\{x(t)\}$ is operator harmonizable, then*

$$\int_G x(s)\Phi(s)u\,\varrho(ds) = \int_{\widehat{G}} \xi(d\chi)\,\mathbf{F}\Phi(\chi)u$$

for $\Phi\in L^1\big(G;B(\mathfrak{U})\big)$ and $u\in\mathfrak{U}$, where the left side integral is a Bochner integral and the right side is a Dunford-Schwartz integral.

Now we have nontrivial relations between harmonizability and V-boundedness in the following two propositions, the first of which is due to Miamee [**13**].

PROPOSITION 5.5. *Let $\{x(t)\}$ be an X-valued process on G. Then the following conditions are equivalent.*

(1) *$\{x(t)\}$ is scalarly harmonizable.*

(2) *$\{x(t)\}$ is scalarly V-bounded.*

(3) *There exists a regular weakly countably additive measure* $\xi \in rwca(\mathfrak{B}_{\widehat{G}}, X)$ *such that* $x(t) = \int_{\widehat{G}} \langle t, \chi \rangle \, \xi(d\chi)$ *for* $t \in G$, *where the integral is in the weak sense.*

PROPOSITION 5.6. *Let* $\{x(t)\}$ *be an* X*-valued process on* G. *Then, it is operator harmonizable if and only if it is operator* V*-bounded.*

PROOF. Assume that $\{x(t)\}$ is operator harmonizable with the representing measure $\xi \in rbwca(\mathfrak{B}_{\widehat{G}}, X)$ of bounded $\mathfrak{U}$-operator semivariation, so that

$$\big(x(t)\Phi(t)u, \phi\big)_{\mathfrak{H}} = \int_{\widehat{G}} \langle t, \chi \rangle \, \big(\xi(d\chi)\Phi(t)u, \phi\big)_{\mathfrak{H}} \tag{5.4}$$

for $t \in G$, $\Phi \in L^1(G; B(\mathfrak{U}))$, $u \in \mathfrak{U}$ and $\phi \in \mathfrak{H}$. Then, by Lemma 5.4(2), we see that for $\Phi \in L^1(G; B(\mathfrak{U}))$ and $u \in \mathfrak{U}$

$$\begin{aligned}\left\| \int_G x(s)\Phi(s)u \, \varrho(ds) \right\|_{\mathfrak{H}} &= \left\| \int_{\widehat{G}} \xi(d\chi) \, \mathbf{F}\Phi(\chi)u \right\|_{\mathfrak{H}} \\ &\le \|\mathbf{F}\Phi\|_\infty \|\xi\|_{\mathfrak{U},o}(\widehat{G}) \|u\|_{\mathfrak{U}}.\end{aligned}$$

Thus, we conclude that $\{x(t)\}$ is operator V-bounded with $C = \|\xi\|_{\mathfrak{U},o}(\widehat{G})$.

Conversely, suppose that $\{x(t)\}$ is operator V-bounded and define an operator $T_0 : \mathbf{F}(L^1(G; B(\mathfrak{U}))) \to X$ by

$$T_0(\mathbf{F}\Phi)u = \int_G x(s)\Phi(s)u \, \varrho(ds) \tag{5.5}$$

for $\Phi \in L^1\big(G; B(\mathfrak{U})\big)$ and $u \in \mathfrak{U}$. Then, T_0 is a bounded right $B(\mathfrak{U})$-module map since $\{x(t)\}$ is operator V-bounded with a constant $C > 0$ and hence $\|T_0(\mathbf{F}\Phi)\|_X \le C\|\mathbf{F}\Phi\|_\infty$ for $\Phi \in L^1(G; B(\mathfrak{U}))$. Now $\mathbf{F}(L^1(G; B(\mathfrak{U})))$ is dense in $C_0(\widehat{G}; B(\mathfrak{U}))$ by [**7**, Lemma IV.4.2 (1)] and T_0 can be extended uniquely to a bounded right $B(\mathfrak{U})$-module map $T : C_0(\widehat{G}; B(\mathfrak{U})) \to X$. It follows from Proposition 3.17 that there exists a unique regular weakly countably additive measure $\xi \in rwca(\mathfrak{B}_{\widehat{G}}, X)$ such that $\|\xi\|_{\mathfrak{U},o}(\widehat{G}) < \infty$ and $T(\Psi) = \int_{\widehat{G}} \xi(d\chi) \, \Psi(\chi)$ for $\Psi \in C_0(\widehat{G}; B(\mathfrak{U}))$ with $\|T\| = \|\xi\|_{\mathfrak{U},o}(\widehat{G})$, where the integral is in the sense that

$$\big(T(\Psi)u, \phi\big)_{\mathfrak{H}} = \int_{\widehat{G}} \big(\xi(d\chi)\Psi(\chi)u, \phi\big)_{\mathfrak{H}} \tag{5.6}$$

for $\Psi \in C_0(\widehat{G}; B(\mathfrak{U}))$, $u \in \mathfrak{U}$ and $\phi \in \mathfrak{H}$. For $\Phi \in L^1(G; B(\mathfrak{U}))$ it holds that by (5.6)

$$\begin{aligned}T(\mathbf{F}\Phi) &= \int_{\widehat{G}} \xi(d\chi) \, \mathbf{F}\Phi(\chi) \\ &= \int_{\widehat{G}} \xi(d\chi) \int_G \langle s, \chi \rangle \Phi(s) \, \varrho(ds) \\ &= \int_G \left(\int_{\widehat{G}} \langle s, \chi \rangle \, \xi(d\chi) \right) \Phi(s) \, \varrho(ds),\end{aligned}$$

where we have used a Fubini type theorem and the integrals are taken in the sense of (5.6). This and (5.5) imply that

$$\int_G \left(x(s) - \int_{\widehat{G}} \langle s, \chi \rangle \, \xi(d\chi) \right) \Phi(s) \, \varrho(ds) = 0, \qquad \Phi \in L^1\big(G; B(\mathfrak{U})\big),$$

which gives $x(t) = \int_{\widehat{G}} \langle t, \chi \rangle \, \xi(d\chi)$ for $t \in G$, where the integral is taken in the sense of (5.4). Thus, $\{x(t)\}$ is operator harmonizable. □

We briefly look at stationary dilation of X-valued processes.

DEFINITION 5.7. An X-valued process $\{x(t)\}$ on G is said to have an **operator stationary dilation** if there exist a Hilbert space $\mathfrak{K}$ containing $\mathfrak{H}$ as a closed subspace and a $Y = B(\mathfrak{U}, \mathfrak{K})$-valued operator stationary process $\{y(t)\}$ on G such that $x(t) = Py(t)$, $t \in G$, where P is the orthogonal projection from $\mathfrak{K}$ onto $\mathfrak{H}$. Similarly, $\{x(t)\}$ is said to have a **scalarly stationary dilation** if, for every $u \in \mathfrak{U}$, the $\mathfrak{H}$-valued process $\{x(t)u\}$ has a stationary dilation.

Obviously every scalarly stationary process is scalarly harmonizable, and every scalarly harmonizable process has a scalarly stationary dilation. As to operator stationary dilation we have the following proposition, which is a reformulation of Miamee [**13**].

PROPOSITION 5.8. *Every X-valued operator harmonizable process on G has an operator stationary dilation.*

PROOF. Assume that $\{x(t)\}$ is operator harmonizable. Then by Proposition 5.3(2) it has the representing measure $\xi \in rbwca(\mathfrak{B}_{\widehat{G}}, X)$ of bounded $\mathfrak{U}$-operator semivariation, so that $\|\xi\|_{\mathfrak{U},o}(\widehat{G}) = \|\xi\|_s(\widehat{G}) < \infty$. It follows from Theorem 3.14 that ξ has a regular weakly countably additive gramian orthogonally scattered dilation η, denoted $\eta \in rwcagos(\mathfrak{B}_{\widehat{G}}, Y)$, where $Y = B(\mathfrak{U}, \mathfrak{K})$ for some Hilbert space $\mathfrak{K}$ containing $\mathfrak{H}$ as a closed subspace and $\xi = P\eta$ with $P : \mathfrak{K} \to \mathfrak{H}$ the orthogonal projection. Let $\{y(t)\}$ be the Y-valued process defined by $y(t) = \int_{\widehat{G}} \langle t, \chi \rangle \, \eta(d\chi)$, $t \in G$, where the integral is in the weak sense, so that it is scalarly stationary. Since $\{y(t)\}$ is strongly continuous, by Proposition 5.2(4) we see that it is operator stationary and hence is an operator stationary dilation of $\{x(t)\}$ since $x(t) = Py(t)$ for $t \in G$. □

To finish this section we shall give some examples of processes based on Example 3.6 as follows.

EXAMPLE 5.9. Let $\mathfrak{U} = \ell^1$, $\mathfrak{H}$ be a Hilbert space with an orthonormal basis $\{\phi_k\}_{k=1}^{\infty}$ and $X = B(\ell^1, \mathfrak{H})$. Also let $G = \mathbb{R}$ with the Borel σ-algebra $\mathfrak{B}$.

(1) *Scalarly harmonizable but not operator harmonizable.* Let ξ be defined as in Example 3.6(6). Then, $\xi \in wca(\mathfrak{B}, X)$ with unbounded $\mathfrak{U}$-operator semivariation. Hence, the X-valued process $\{x(t)\}$ given by $x(t) = \int_{\mathbb{R}} e^{itu} \, \xi(du)$, $t \in \mathbb{R}$ is scalarly harmonizable but not operator harmonizable.

(2) *Operator harmonizable but not operator stationary.* Let ξ be defined as in Example 3.6(3). Then, $\xi \in vca(\mathfrak{B}, X)$ but not gramian orthogonally scattered. Hence, the X-valued process $\{x(t)\}$ given by $x(t) = \int_{\mathbb{R}} e^{itu} \, \xi(du)$, $t \in \mathbb{R}$ is operator harmonizable but not operator stationary.

(3) *Operator stationary but not operator harmonizable.* Let ξ be defined as in Example 3.6(5). Then, $\xi \in wcagos(\mathfrak{B}, X)$ with unbounded $\mathfrak{U}$-operator semivariation. Hence, the X-valued process $\{x(t)\}$ given by $x(t) = \int_{\mathbb{R}} e^{itu} \, \xi(du)$, $t \in \mathbb{R}$ is operator stationary but not operator harmonizable.

References

[1] S. A. Čobanjan, *The class of correlation functions of stationary stochastic processes with values in a Banach space* (Russian, with Georgian and English summaries), Sakharth. SSR Mecn. Akad. Moambe **55** (1969), 21–24. MR0272048

[2] S. A. Čobanjan, *Certain properties of positive operator measures in Banach spaces* (Russian, with Georgian and English summaries), Sakharth. SSR Mecn. Akad. Moambe **57** (1970), 273–276. MR0272049

[3] S. A. Čobanjan, *Regularity of stationary processes with values in a Banach space and factorization of operator-valued functions* (Russian, with Georgian and English summaries), Sakharth. SSR Mecn. Akad. Moambe **61** (1971), 29–32. MR0290450

[4] S. A. Chobanyan and A. Weron, *Banach-space-valued stationary processes and their linear prediction*, Dissertationes Math. (Rozprawy Mat.) **125** (1975), 45. MR451373

[5] J. Diestel and J. J. Uhl Jr., *Vector measures*, American Mathematical Society, Providence, R.I., 1977. With a foreword by B. J. Pettis; Mathematical Surveys, No. 15. MR0453964

[6] Ramesh Gangolli, *Wide-sense stationary sequences of distributions on Hilbert space and the factorization of operator valued functions*, J. Math. Mech. **12** (1963), 893–910. MR0161349

[7] Yûichirô Kakihara, *Multidimensional second order stochastic processes*, Series on Multivariate Analysis, vol. 2, World Scientific Publishing Co., Inc., River Edge, NJ, 1997, DOI 10.1142/9789812779298. MR1625379

[8] R. M. Loynes, *Linear operators in VH-spaces*, Trans. Amer. Math. Soc. **116** (1965), 167–180, DOI 10.2307/1994111. MR192359

[9] R. M. Loynes, *On generalized positive-definite functions*, Proc. London Math. Soc. (3) **15** (1965), 373–384, DOI 10.1112/plms/s3-15.1.373. MR173933

[10] R. M. Loynes, *On a generalization of second-order stationarity*, Proc. London Math. Soc. (3) **15** (1965), 385–398, DOI 10.1112/plms/s3-15.1.385. MR176531

[11] A. Makagon and H. Salehi, *Spectral dilation of operator-valued measures and its application to infinite-dimensional harmonizable processes*, Studia Math. **85** (1987), no. 3, 257–297, DOI 10.4064/sm-85-3-257-297. MR887488

[12] A. G. Miamee, *On $B(X, K)$-valued stationary stochastic processes*, Indiana Univ. Math. J. **25** (1976), no. 10, 921–932, DOI 10.1512/iumj.1976.25.25073. MR420807

[13] A. G. Miamee, *Spectral dilation of $L(B, H)$-valued measures and its application to stationary dilation for Banach space valued processes*, Indiana Univ. Math. J. **38** (1989), no. 4, 841–860, DOI 10.1512/iumj.1989.38.38040. MR1029680

[14] A. G. Miamee and H. Salehi, *On the square root of a positive $B(X, X^*)$-valued function*, J. Multivariate Anal. **7** (1977), no. 4, 535–550, DOI 10.1016/0047-259X(77)90065-3. MR467897

[15] Robert Schatten, *Norm ideals of completely continuous operators*, Ergebnisse der Mathematik und ihrer Grenzgebiete. N. F., Heft 27, Springer-Verlag, Berlin-Göttingen-Heidelberg, 1960. MR0119112

Department of Mathematics, California State University, San Bernardino, California 92407

Email address: ykakihar@csusb.edu

Contemporary Mathematics
Volume **774**, 2021
https://doi.org/10.1090/conm/774/15570

Explicit transient probabilities of various Markov models

Alan Krinik, Hubertus von Bremen, Ivan Ventura, Uyen Vietthanh Nguyen, Jeremy J. Lin, Thuy Vu Dieu Lu, Chon In (Dave) Luk, Jeffrey Yeh, Luis A. Cervantes, Samuel R. Lyche, Brittney A. Marian, Saif A. Aljashamy, Mark Dela, Ali Oudich, Pedram Ostadhassanpanjehali, Lyheng Phey, David Perez, John Joseph Kath, Malachi C. Demmin, Yoseph Dawit, Christine Carmen Marie Hoogendyk, Aaron Kim, Matthew McDonough, Adam Trevor Castillo, David Beecher, Weizhong Wong, and Heba Ayeda

Dedicated to M. M. Rao upon celebrating his 90th birthday.

Abstract. In analyzing finite-state Markov chains knowing the exact eigenvalues of the transition probability matrix P is important information for predicting the explicit transient behavior of the system. Once the eigenvalues of P are known, linear algebra and duality theory are used to find P^k where $k = 2, 3, 4, \ldots$. This article is about finding explicit eigenvalue formulas, that scale up with the dimension of P for various Markov chains. Eigenvalue formulas and expressions of P^k are first presented when P is tridiagonal and Toeplitz. These results are generalized to tridiagonal matrices with alternating birth-death probabilities. More general eigenvalue formulas and expression of P^k are obtained for non-tridiagonal transition matrices P that have both catastrophe-like and birth-death transitions. Similar results for circulant matrices are also explored. Applications include finding probabilities of sample paths restricted to a strip and generalized ballot box problems. These results generalize to Markov processes with P^k being replaced by e^{Qt} where Q is a transition rate matrix.

1. Introduction and summary

It is a pleasure to be contributing to this AMS Contemporary Mathematics Series honoring M. M. Rao's 90th birthday. We celebrated the occasion during an AMS Special Session of the Fall Western Sectional Meeting November 9-10, 2019 at the University of California, Riverside. Throughout M. M. Rao's stellar academic career, he was well known for conducting active and popular mathematical seminars. His colleagues and graduate students took turns presenting and discussing a wide range of traditional and current research topics. Generally the topics revolved

2020 *Mathematics Subject Classification.* Primary 60J10, 60J22.

Key words and phrases. Markov chains and Markov processes, eigenvalues and eigenvectors, transient probabilities, ballot box problem, dual processes, transition probability matrix, tridiagonal matrices, Toeplitz matrices, circulant matrices, spectral projectors, gambler's ruin, catastrophes.

around probability theory, integration theory, functional analysis, and random processes. Specific topics often included harmonizable processes, stochastic differential equations, Orlicz (spaces), the Radon Nikodym Theorem, vector measures, etc.

Following in the M. M. Rao tradition of an active seminar program, this article comes from our Cal Poly Pomona Research Group on Markovian models and matrix properties. This research group began in the summer of 2016 with four students and has steadily grown and remains active even during the Coronavirus pandemic. The group is almost entirely composed of students affiliated with Cal Poly Pomona. Most of its members are undergraduate students (who are doing research for the first time). Some are Cal Poly Pomona graduate students or Cal Poly Pomona alumni. The composition of the research group constantly changes as students graduate and either take jobs or enter graduate programs in mathematics or statistics.

Our Cal Poly Pomona Research Group has presented over 15 talks over the past five years. This includes presentations at the Joint Mathematics Meetings in 2017-2021 as well as local sectional meetings of the American Mathematical Society, the Mathematical Association of America, and local colloquia. Much of the group's early work is documented in the Master Theses of Uyen Nguyen [**Ngu17**] and Samuel Lyche [**Lyc18**].

This current article has 27 co-authors, consisting of 3 professors and 24 students. Our Cal Poly Pomona Research Group includes the contributions of Cal Poly Pomona Math Professors Alan Krinik, Hubertus von Bremen and Ivan Ventura.

In this article, we are interested in finding exact formulas for transient probabilities of certain finite, Markov models. Our approach, in this article, is eigenvalue centered, where we restrict ourselves to *nice* families of real $n \times n$ transition matrices, M. We assume that M has distinct and explicitly known eigenvalues (rather than numerically approximated eigenvalues). Finding transient probabilities in finite Markov models, having transition matrix M, is equivalent to finding exact expressions for:

$$(1)\ M^k \text{ when } k \in \mathbb{N}$$

$$(2)\ e^{Mt} \text{ when } t \in [0, \infty)$$

Our main goal is to find explicit formulas for M^k or e^{Mt} as a function of exact eigenvalues formulas that scale-up for large n. Most of our work assumes that M is an $n \times n$ tridiagonal matrix with distinct known eigenvalues. Our matrix applications include:

(1) Calculating the probability of going from state i to state j while being confined to a strip for discrete or continuous time birth-death models.
(2) Generalize the Classical Ballot Box problem and its solution to a birth-death chain or process setting.
(3) Duality theory allows us to generalize our results and applications to a broader class of $n \times n$ real matrices M that are neither tridiagonal nor Toeplitz

In section two, we consider $n \times n$ dimensional real matrices M that are assumed to have n-distinct eigenvalues. We highlight the following important Sylvester eigenvalue expansions that are used throughout our article:

$$M^k = A_1\omega_1^k + A_2\omega_2^k + A_3\omega_3^k + \cdots + A_n\omega_n^k \quad \text{where } k \in \mathbb{N} \tag{1.1}$$

$$e^{Mt} = A_1e^{\omega_1 t} + A_2e^{\omega_2 t} + A_3e^{\omega_3 t} + \cdots + A_ne^{\omega_n t} \quad \text{where } t \in [0,\infty) \tag{1.2}$$

and $\omega_1, \omega_2, \ldots \omega_n$ are the distinct eigenvalues of M and $A_1, A_2, \ldots A_n$ are constant $n \times n$ matrices called spectral projectors.

We initially assume that M is a tridiagonal Toeplitz matrix with a entries along the subdiagonal, b entries along the main diagonal and c entries along the superdiagonal. The eigenvalues and eigenvectors of tridiagonal Toeplitz matrices have been well-known for some time (pages 514-516 of [**Mey00**]). Later, we replace matrix M with a transition matrix, P, where entries a, b, c are replaced by transition probabilities q, r, and p respectively where $0 < q,\ p < 1$ and $0 \leq r < 1$. Section two also contains an important and useful trigonometric expression for the A_i matrices when P is a tridiagonal Toeplitz matrix. Our method for determining the A_i coefficients essentially follows the Perron outer-product of eigenvectors, see [**Ber18**]. We also explain the steady state probabilities for non-stochastic, nonnegative P matrices in section two.

Section three consists of a variety of examples demonstrating the results of section two. For example, we find the probability of all birth-death sample paths going from state i to state j in n-steps while being restricted to states within a horizontal strip. We also calculate the relative probability of sample paths taking values within nested horizontal strips. These types of problems are described in both discrete and continuous time. The elegant combinatorical solution of the classical ballot box problem is reviewed. We generalize the classical ballot box problem using birth-death chains or processes. A matrix power solution of the generalized ballot box problem in discrete time is formulated by taking the nth power of two different transition matrices. The solution is the ratio of two selected entries from each matrix. This naturally leads to an eigenvalue solution for tridiagonal Toeplitz matrices. We then calculate a variety of generalized ballot box probabilities under more general conditions.

In section four, we consider $n \times n$ transition matrices that represent Markov chains that have the general form, when $n = 4$ as shown below:

$$P_1 = \begin{bmatrix} r+q+c_0 & p+c_1 & c_2 & c_3 \\ q+c_0 & r+c_1 & p+c_2 & c_3 \\ c_0 & q+c_1 & r+c_2 & p+c_3 \\ c_0 & c_1 & q+c_2 & r+p+c_3 \end{bmatrix} \tag{1.3}$$

where $0 < p,\ q < 1$ and $0 \leq r < 1$ and $p+r+q+c_0+c_1+c_2+c_3 = 1$, where c_i is the probability of going from any state to state i, such that $0 \leq c_0,\ c_1,\ c_2,\ c_3 < 1$. We determine the transient probability functions of matrices P_1 having the preceding form when $n = 2, 3, 4, \ldots$ using the Duality Theorem. An analogous result is obtained in section four when a Markov process has a Q matrix that is similar to (1.3), as shown in Figure 28.

In section five, we now consider an $n \times n$ matrix P_2 of transition probabilities, having the following form, where n is assumed to be an odd number:

$$P_2 = \begin{bmatrix} r & p_0 & 0 & 0 & \dots & 0 & 0 \\ q_1 & r & p_1 & 0 & \dots & 0 & 0 \\ 0 & q_2 & r & p_0 & \dots & 0 & 0 \\ 0 & 0 & q_1 & r & p_1 & \dots & 0 \\ \vdots & \vdots & \vdots & \ddots & \dots & \vdots & \vdots \\ 0 & 0 & \dots & 0 & 0 & q_2 & r \end{bmatrix} \tag{1.4}$$

In Theorem 5.1, the eigenvalues and eigenvectors of the preceding matrix, P_2, are determined. This result is a consequence of Theorem 3.1 found in Kouachi's 2006 article. Theorem 5.1 is extended to include catastrophe-like transitions as shown in (1.3). Corollary 5.2.1 describes a method to determine the explicit transient probabilities of birth-death chains having the following form when H is an odd number:

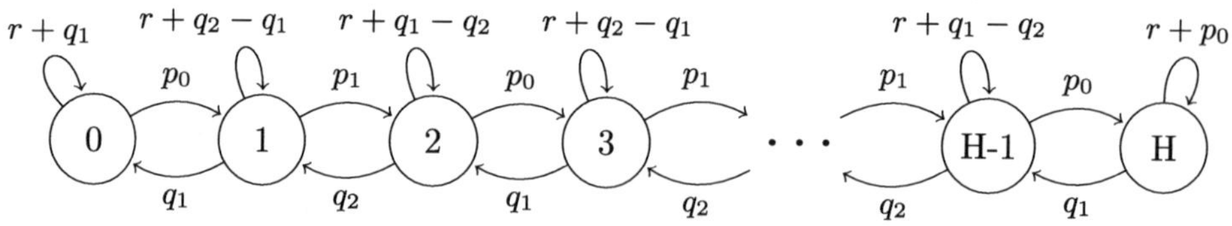

where $0 < q_1, q_2, r, p_0, p_1 < 1$, $q_1 + r + p_0 = 1$, $q_2 + r + p_1 = 1$, and $r > |q_2 - q_1|$. Our last result in section five is Theorem 5.3, which considers a birth-death chain that has the following transition probablility diagram:

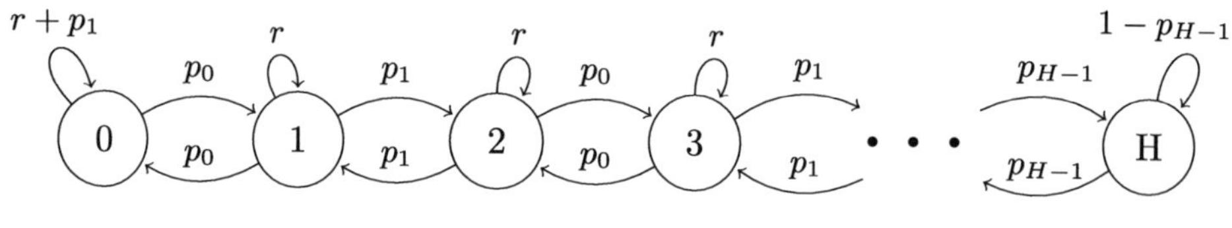

where $0 < p_0, p_1$ and $p_0 + r + p_1 = 1$. Then the eigenvalues of the matrix corresponding to this birth-death chain are explicitly known and we can find the Sylvester Eigenvalue Expansion (1.1) for this preceding birth-death matrix. We also analyze the analogous continuous time birth-death process corresponding to the preceding transition diagram.

In section six, circulant matrices are considered. Circulant matrices have distinct eigenvalues and eigenvector formulas that have been known for a long time [**Wik21**] [**Dav70**]. These formulas scale up with n where n is the dimension of the circulant matrix. Sylvester's eigenvalue expansions (1.1) and (1.2) can be used to find the transient probabilities of circulant transition matrices. In section six, we consider a three-state circular birth-death chain having a circulant matrix P as its transition probability matrix. The explicit Sylvester eigenvalue expansion of P for the three-state circular birth-death chain is determined. Section six concludes with a probability problem that connects a three-state circular birth-death chain to a three-state linear birth-death chain.

2. Matrix results

2.1. Eigenvalue expansion. Throughout this article, we take M to be a real $n \times n$ matrix having n distinct eigenvalues $\omega_1, \omega_2, \ldots, \omega_n$ and denote the u, v entry of M as $M(u,v)$. Under this setting, it can be shown that for each $k \in \mathbb{N}$, the kth power of M, that is M^k, always has an eigenvalue expansion of the form

$$M^k = A_1\omega_1^k + A_2\omega_2^k + A_3\omega_3^k + \cdots + A_n\omega_n^k \tag{2.1}$$

where the coefficients $A_1, A_2, \ldots, A_n$ are constant $n \times n$ matrices. Consequently, the u, v entry of M^k can be expressed as

$$M^k(u,v) = A_1(u,v)\omega_1^k + A_2(u,v)\omega_2^k + A_3(u,v)\omega_3^k + \cdots + A_n(u,v)\omega_n^k \tag{2.2}$$

where $1 \le u,\ v \le n$. When M is a nonnegative matrix, M has a positive eigenvalue ω_1 such that $\omega_1 > |\omega_s|$ for all $s = 2, 3, \ldots, n$ [**Ber18**].

Combining expansion (2.1) with the definition of an exponential matrix, it follows that the matrix e^{Mt} has the form

$$e^{Mt} = A_1e^{\omega_1 t} + A_2e^{\omega_2 t} + A_3e^{\omega_3 t} + \cdots + A_ne^{\omega_n t} \tag{2.3}$$

with $t \ge 0$ and $A_1, \ldots, A_n$ being the same matrices appearing in expansion (2.1).

EXAMPLE 2.1. Consider the following transition probability diagram and 2×2 P matrix:

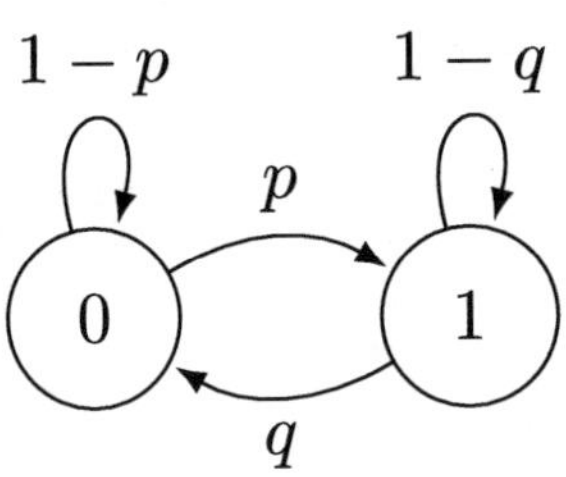

$$P = \begin{bmatrix} 1-p & p \\ q & 1-q \end{bmatrix}$$

where $0 < p, q < 1$

Matrix P has eigenvalues $\omega_1 = 1$ and $\omega_2 = 1 - p - q$. Then

$$\begin{aligned} P^k &= A_1\omega_1^k + A_2\omega_2^k \\ &= \begin{bmatrix} \frac{q}{p+q} & \frac{p}{p+q} \\ \frac{q}{p+q} & \frac{p}{p+q} \end{bmatrix} (1)^k + \begin{bmatrix} \frac{p}{p+q} & -\frac{p}{p+q} \\ -\frac{q}{p+q} & \frac{q}{p+q} \end{bmatrix} (1-p-q)^k \end{aligned}$$

EXAMPLE 2.2. Consider the following transition rate diagram and Q matrix:

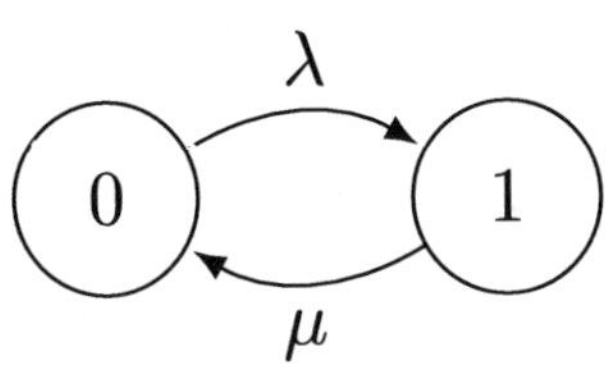

$$Q = \begin{bmatrix} -\lambda & \lambda \\ \mu & -\mu \end{bmatrix}$$

where $\lambda,\ \mu > 0$

Matrix Q has eigenvalues $\omega_1 = 0$ and $\omega_2 = -(\lambda+\mu)$. Then

$$\begin{aligned} e^{Qt} &= A_1 e^{\omega_1 t} + A_2 e^{\omega_2 t} \\ &= \begin{bmatrix} \frac{\mu}{\lambda+\mu} & \frac{\lambda}{\lambda+\mu} \\ \frac{\mu}{\lambda+\mu} & \frac{\lambda}{\lambda+\mu} \end{bmatrix} e^{(0)t} + \begin{bmatrix} \frac{\lambda}{\lambda+\mu} & -\frac{\lambda}{\lambda+\mu} \\ -\frac{\mu}{\lambda+\mu} & \frac{\mu}{\lambda+\mu} \end{bmatrix} e^{-(\lambda+\mu)t} \end{aligned}$$

Coefficient matrices $A_1, A_2, \ldots, A_n$ may be calculated several ways, and our research group had fun rediscovering two well-known methods for calculating them: Sylvester's Formula and the Perron-Frobenius technique. Sylvester's Formula characterizes these matrices as

$$A_s = \frac{(M-\omega_1 I)(M-\omega_2 I)\cdots(\widehat{M-\omega_s I})\cdots(M-\omega_n I)}{(\omega_s-\omega_1)(\omega_s-\omega_2)\cdots(\widehat{\omega_s-\omega_s})\cdots(\omega_s-\omega_n)} \tag{2.4}$$

for each $s = 1, 2, \ldots, n$ with I being the $n \times n$ identity matrix, $\widehat{M-\omega_s I} = I$, and $\widehat{\omega_s - \omega_s} = 1$ [**Wik19b**].

While Sylvester's Formula only requires knowledge of M and its eigenvalues, it may be difficult to simplify. Perron-Frobenius techniques provide a useful alternative. Provided that M is nonnegative, this formula states that the coefficient matrices can be determined by first choosing the right eigenvector, $\vec{R}_s$ of M (associated with eigenvalue ω_s) and the left eigenvector, $\vec{L}_s$ of M (associated with eigenvalue ω_s). We scale the eigenvectors so that their dot product $\vec{L}_s \cdot \vec{R}_s = 1$, for each $s = 1, 2, 3, \ldots, n$. This means multiply each eigenvector $\vec{L}_s$ and $\vec{R}_s$ by a constant c so that $c\vec{L}_s \cdot c\vec{R}_s = 1$. Obtaining A_s then follows by taking the scaled outer matrix product of $\vec{R}_s$ and $\vec{L}_s$ [**Ber18**], that is,

$$A_s = c^2 \vec{R}_s \vec{L}_s \tag{2.5}$$

To see how this works, see Appendix A when M is a tridiagonal Toeplitz matrix.

Next, associated with M, we define M_R to be the normalization of M row-wise as

$$M_R(u,v) = \frac{M(u,v)}{\sum_{v=1}^{n} M(u,v)} \tag{2.6}$$

in other words, each entry of M is divided by its corresponding row sum. Similarly, the kth power of M_R can be normalized entry-wise as $M_R^k(u,v) = M^k(u,v)/\sum_{v=1}^{n} M^k(u,v)$. We can obtain a type of steady state result for M_R as follows: scale M^k by ω_1^k for each $k \in \mathbb{N}$, then the Perron-Frobenius Theorem states

$$\lim_{k\to\infty} \frac{M^k}{\omega_1^k} = A_1 \tag{2.7}$$

Writing A_1 in terms of the scaled eigenvectors of $\vec{R}_1$ and $\vec{L}_1$ of A_1, we have

$$A_1 = \begin{bmatrix} R_1(1)L_1(1) & R_1(1)L_1(2) & \cdots & R_1(1)L_1(n) \\ \vdots & \vdots & \ddots & \vdots \\ R_1(n)L_1(1) & R_1(n)L_1(2) & \cdots & R_1(n)L_1(n) \end{bmatrix} \tag{2.8}$$

Dividing each entry of A_1 by its associated row sum, we get the normalized matrix $A_{1,R}$ with form

$$A_{1,R} = \frac{1}{\sum_{u=1}^{n} L_1(u)} \begin{bmatrix} L_1(1) & L_1(2) & \cdots & L_1(n) \\ \vdots & \vdots & \ddots & \vdots \\ L_1(1) & L_1(2) & \cdots & L_1(n) \end{bmatrix} \tag{2.9}$$

and $\lim_{k\to\infty} M_R^k = A_{1,R}$. Note that all entries on the kth column are equal.

Having closed form expressions of P^k or e^{Qt} is very helpful when working on these objects analytically, such as taking limits, etc. The closed form expressions also allow for accurately obtaining numerical values for P^k or e^{Qt}. It is well known that computing P^k directly will lead to significant loss of accuracy as k becomes large, due to the fact that the columns of P^k numerically converge to the eigenvector corresponding to the leading eigenvalue of P (the eigenvalue with largest magnitude). Methods for computing e^{Qt} using numerical schemes can present accuracy or efficiency issues as observed by [**MVL03**].

2.2. Tridiagonal Toeplitz matrices. Let constants $a, b, c \in \mathbb{R}$. We apply the above results to tridiagonal Toeplitz matrices, that is, matrix M is of the form

$$M = \begin{bmatrix} b & c & 0 & \cdots & \cdots & 0 & 0 & 0 \\ a & b & c & \cdots & \cdots & 0 & 0 & 0 \\ 0 & a & b & \cdots & \cdots & 0 & 0 & 0 \\ \vdots & \vdots & \vdots & \ddots & \ddots & \vdots & \vdots & \vdots \\ \vdots & \vdots & \vdots & \ddots & \ddots & \vdots & \vdots & \vdots \\ 0 & 0 & 0 & \cdots & \cdots & b & c & 0 \\ 0 & 0 & 0 & \cdots & \cdots & a & b & c \\ 0 & 0 & 0 & \cdots & \cdots & 0 & a & b \end{bmatrix} \tag{2.10}$$

Recall that for $s = 1, 2, \ldots, n$, the eigenvalues of M are given by

$$\omega_s = b + 2\sqrt{ac}\cos\left(\frac{s\pi}{n+1}\right) \tag{2.11}$$

with the right $\vec{R_s}$ and left $\vec{L_s}$ eigenvectors expressed component-wise as

$$\vec{R_s}(u) = (a/c)^{u/2}\sin(u\pi s/(n+1)) \qquad \vec{L_s}(u) = (c/a)^{u/2}\sin(u\pi s/(n+1))$$

as they appear in [**Mey00**, pg. 514].

Applying equation (2.5), the coefficient matrix A_s of M is

$$A_s(u,v) = \frac{2}{n+1}\left(\frac{a}{c}\right)^{\frac{u-v}{2}} \sin\left(\frac{u\pi s}{n+1}\right)\sin\left(\frac{v\pi s}{n+1}\right) \tag{2.12}$$

where $s, u, v = 1, 2, .., n$ (see Appendix A for proof.) In particular,

$$A_1(u,v) = \frac{2}{n+1}\left(\frac{a}{c}\right)^{\frac{u-v}{2}} \sin\left(\frac{u\pi}{n+1}\right)\sin\left(\frac{v\pi}{n+1}\right)$$

Using equation (2.2), M^k becomes

$$M^k(u,v) = \frac{2}{n+1}\left(\frac{a}{c}\right)^{\frac{u-v}{2}} \sum_{s=1}^{n} \sin\left(\frac{u\pi s}{n+1}\right)\sin\left(\frac{v\pi s}{n+1}\right)\omega_s^k \tag{2.13}$$

where ω_s is given by equation (2.11). This also implies

$$e^{Mt}(u,v) = \frac{2}{n+1}\left(\frac{a}{c}\right)^{\frac{u-v}{2}} \sum_{s=1}^{n} \sin\left(\frac{u\pi s}{n+1}\right) \sin\left(\frac{v\pi s}{n+1}\right) e^{\omega_s t} \tag{2.14}$$

where ω_s is again given by equation (2.11).

REMARK 2.1. In this section, we assume our matrix M is a tridiagonal Toeplitz matrix where the diagonals are the main diagonal, the sub-diagonal, and super-diagonal, i.e. the off-diagonals are adjacent to the main diagonal. The results of this section may be generalized to tridigonal Toeplitz matrices where the three diagonals are now the main diagonal and the two symmetrically placed diagonals k steps from the main diagonal. For example, when $k = 2$, the associated Markov Chain will have transition probability steps of size 2. A study of the eigenvalues and eigenvectors corresponding to this different type of tridigonal transition matrix M under different conditions appears in [**Los92**]. The generalization of Section 2 of this article to Markov chains having this new type of tridiagonal transition matrices will be addressed elsewhere.

Observe that the associated normalization of M as defined in the previous subsection is

$$M_R = \begin{bmatrix} \frac{b}{b+c} & \frac{c}{b+c} & 0 & \cdots & \cdots & 0 & 0 & 0 \\ \frac{a}{a+b+c} & \frac{b}{a+b+c} & \frac{c}{a+b+c} & \ddots & \cdots & 0 & 0 & 0 \\ 0 & \frac{a}{a+b+c} & \frac{b}{a+b+c} & \ddots & \ddots & 0 & 0 & 0 \\ \vdots & \ddots & \ddots & \ddots & \ddots & \ddots & \vdots & \vdots \\ \vdots & \vdots & \ddots & \ddots & \ddots & \ddots & \ddots & \vdots \\ 0 & 0 & 0 & \ddots & \ddots & \frac{b}{a+b+c} & \frac{c}{a+b+c} & 0 \\ 0 & 0 & 0 & \cdots & \ddots & \frac{a}{a+b+c} & \frac{b}{a+b+c} & \frac{c}{a+b+c} \\ 0 & 0 & 0 & \cdots & \cdots & 0 & \frac{a}{a+b} & \frac{b}{a+b} \end{bmatrix}$$

As $k \to \infty$, we have $M_R^k \to A_{1,R}$, where

$$A_{1,R} = \frac{1}{S}\begin{bmatrix} \left(\frac{c}{a}\right)^{1/2} \sin\left(\frac{\pi}{n+1}\right) & \cdots & \left(\frac{c}{a}\right)^{n/2} \sin\left(\frac{n\pi}{n+1}\right) \\ \vdots & & \vdots \\ \left(\frac{c}{a}\right)^{1/2} \sin\left(\frac{\pi}{n+1}\right) & \cdots & \left(\frac{c}{a}\right)^{n/2} \sin\left(\frac{n\pi}{n+1}\right) \end{bmatrix}$$

with $S = \sum_{u=1}^{n} \left(\frac{c}{a}\right)^{u/2} \sin\left(\frac{u\pi}{n+1}\right)$.

2.3. Probability applications. We would now like to apply the results above to stochastic processes. Matrix M as shown in equation (2.10) appears very similar to a transition matrix for a time-homogeneous birth-death chain with a, b, and c being probabilities such that $a + b + c = 1$. Unfortunately, the first and last rows of M would not sum to 1 under these conditions. Nevertheless, we will provide various examples in this article that M, along with its above mentioned properties, can be useful in calculating probabilities of sample paths associated with birth-death chains with state space $[0, n-1]$ given that the path stays within a strip $[L, H] \subseteq [0, n-1]$.

From here onward, we will use zero-based matrix indexing since the 0 state is commonly used in birth-death chains as the initial or first state. This means that the 1st row and 1st column of P represent transitions to or from state 0. So we modify our previous formulas appropriately by changing our previous (u, v) entry matrix notation where $1 \leq u, v \leq n$ to (i, j) where $0 \leq i, j \leq n-1$.

To handle these kinds of problems, we generalize the concept of the transition matrix. Consider matrix

$$P = \begin{bmatrix} P_{0,0} & P_{0,1} & \cdots & P_{0,n-1} \\ P_{1,0} & P_{1,1} & \cdots & P_{1,n-1} \\ \vdots & \vdots & \ddots & \vdots \\ P_{n-1,0} & P_{n-1,1} & \cdots & P_{n-1,n-1} \end{bmatrix}$$

We say P is substochastic if for all $0 \leq i, j \leq n-1$, we have $0 \leq P_{i,j} \leq 1$ and

$$0 \leq \sum_{j=0}^{n-1} P_{i,j} \leq 1$$

When P is substochastic, then for all $k \in \mathbb{N}$,

$$P^k = \begin{bmatrix} P^{(k)}_{0,0} & P^{(k)}_{0,1} & \cdots & P^{(k)}_{0,n-1} \\ P^{(k)}_{1,0} & P^{(k)}_{1,1} & \cdots & P^{(k)}_{1,n-1} \\ \vdots & \vdots & \ddots & \vdots \\ P^{(k)}_{n-1,0} & P^{(k)}_{n-1,1} & \cdots & P^{(k)}_{n-1,n-1} \end{bmatrix}$$

where

$$P^{(k)}_{i,j} = \sum_{0 \leq i_1, \ldots, i_{k-1} \leq n-1} P_{i,i_1} P_{i_1,i_2} \cdots P_{i_{k-2},i_{k-1}} P_{i_{k-1},j}$$

and for all $0 \leq i, j \leq n-1$, we have $\sum_{j=0}^{n-1} P^{(k)}_{i,j} \leq 1$. Now define

$$P^{(k)}_{i,j}([L, H]) = \sum_{L \leq i_1, \ldots, i_{k-1} \leq H} P_{i,i_1} P_{i_1,i_2} \cdots P_{i_{k-2},i_{k-1}} P_{i_{k-1},j}$$

where $L \leq i, j \leq H$. In contrast to stochastic matrices, $P^{(k)}_{i,j}([L, H])$ generally does not sum to 1, but $\sum_{j=0}^{n-1} P^{(k)}_{i,j}([L, H]) \leq 1$. However, for all $0 \leq i, j \leq n-1$,

$$\sum_{j=0}^{n-1} \frac{P^{(k)}_{i,j}([L, H])}{\sum_{j=0}^{n-1} P^{(k)}_{i,j}([L, H])} = 1$$

By restricting transitions of a birth-death chain to a solution strip, we obtain the following transition diagram and its substochastic transition matrix. Suppose p, q, and r are probabilities with $0 < p$, $q < 1$ and $0 \leq r < 1$ and $p + q + r = 1$.

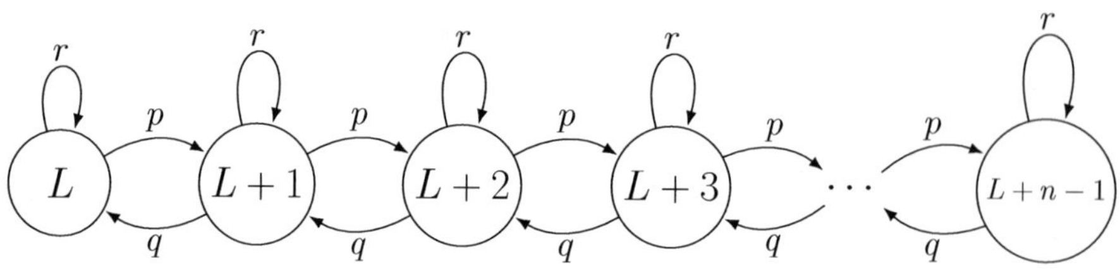

FIGURE 1

where $n, L \in \mathbb{Z}$, and $n > 1$ and $3 - n \le L \le 0$.

$$P = \begin{array}{c} \\ L \\ L+1 \\ L+2 \\ \vdots \\ L+n-1 \end{array} \begin{array}{c} \begin{array}{ccccc} L & L+1 & L+2 & \cdots & L+n-1 \end{array} \\ \begin{bmatrix} r & p & 0 & \cdots & 0 \\ q & r & p & \cdots & 0 \\ 0 & q & r & \cdots & 0 \\ \vdots & \vdots & \vdots & \ddots & \vdots \\ 0 & 0 & 0 & \cdots & r \end{bmatrix} \end{array}$$

Then:

$$\omega_s = r + 2\sqrt{pq}\cos\left(s\pi/(n+1)\right) \tag{2.15}$$

from (2.11).

$$A_s(i,j) = \frac{2}{n+1}\left(\frac{q}{p}\right)^{\frac{i-j}{2}} \sin\left(\frac{(i-j+k+2)\pi s}{2(n+1)}\right) \sin\left(\frac{(-i+j+k+2)\pi s}{2(n+1)}\right) \tag{2.16}$$

this is a suitably modified version of (2.12).

$$\begin{aligned} P^k(i,j) &= \sum_{s=1}^{n} A_s(i,j)\omega_s^k = \\ &\sum_{s=1}^{n} \frac{2}{n+1}\left(\frac{q}{p}\right)^{\frac{i-j}{2}} \sin\left(\frac{(i-j+k+2)\pi s}{2(n+1)}\right) \\ &\times \sin\left(\frac{(-i+j+k+2)\pi s}{2(n+1)}\right)\left(r + 2\sqrt{pq}\cos\left(\frac{s\pi}{n+1}\right)\right)^k \end{aligned} \tag{2.17}$$

which is (2.1). When $L = 0$, Figure 1 becomes

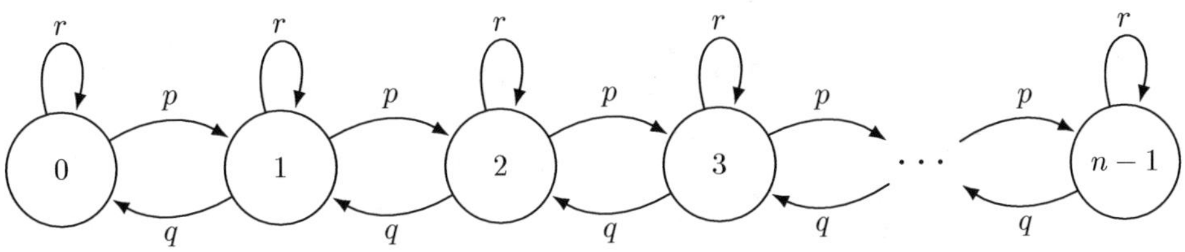

FIGURE 2

where p, r, q are probabilities such that $0 < p, q < 1$, $p + r + q = 1$, once again:

$$\omega_s = r + 2\sqrt{pq}\cos\left(s\pi/(n+1)\right) \tag{2.18}$$

$$A_s(i,j) = \frac{2}{n+1}\left(\frac{q}{p}\right)^{\frac{i-j}{2}} \sin\left(\frac{(i+1)\pi s}{n+1}\right)\sin\left(\frac{(j+1)\pi s}{n+1}\right) \tag{2.19}$$

$$P^k(i,j) = \sum_{s=1}^{n} A_s(i,j)\omega_s^k = \tag{2.20}$$

$$\sum_{s=1}^{n} \frac{2}{n+1}\left(\frac{q}{p}\right)^{\frac{i-j}{2}} \sin\left(\frac{(i+1)\pi s}{n+1}\right)\sin\left(\frac{(j+1)\pi s}{n+1}\right)\left(r + 2\sqrt{pq}\cos\left(\frac{s\pi}{n+1}\right)\right)^k$$

3. Strip probabilities and ballot box problems

3.1. Strip probabilities. Now, we apply the results of Subsection 2.3 to various probability problems.

EXAMPLE 3.1 (Probability of All Paths from $i = 2$ to $j = 5$ in a Strip). Consider the birth-death chain given in Figure 3 with $p = 1/3$, $q = 1/2$, and $r = 1/6$. We remark that for state 0 and state 7, the probabilities exiting these states do not sum to 1. What is the probability of going from state $i = 2$ to $j = 5$ in 15 steps given the process is restricted to states 0 through 7, inclusive? A sample path satisfying the given conditions as well as the restriction is shown in Figure 4. The restriction is graphically represented as a horizontal strip with lower and upper bounds $y = 0$ and $y = 7$, respectively.

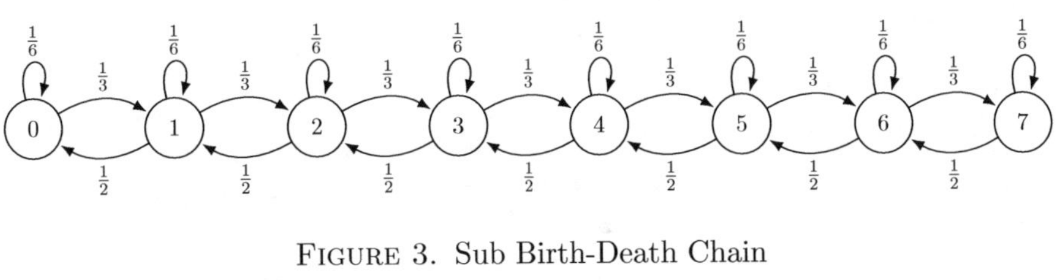

FIGURE 3. Sub Birth-Death Chain

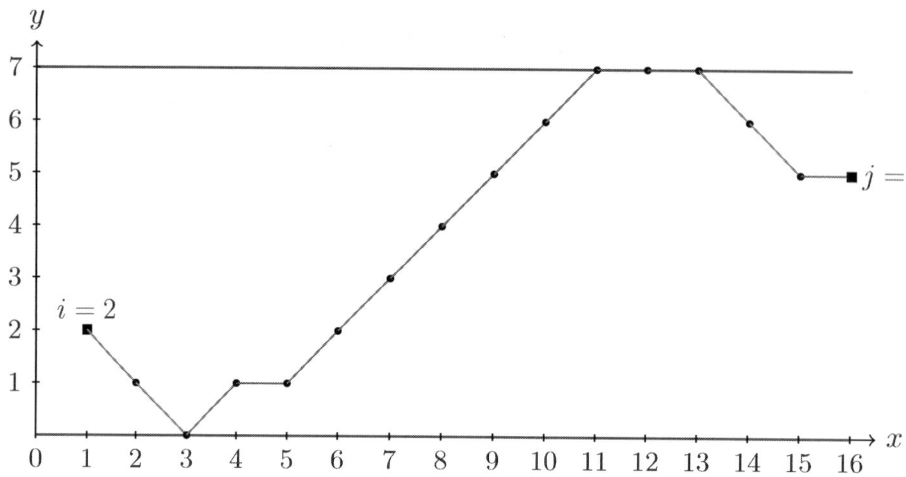

FIGURE 4. A sample path going from $i = 2$ to $j = 5$ in 15 steps staying within the strip

To solve this problem, denote P as the substochastic matrix corresponding to the sub birth-death chain of Figure 3. Note that we cannot simply take the appropriate entry of P^{15} as P is substochastic. Instead, we use the normalization P_R^{15} calculation as described below. The probability of all paths going from state 2 to state 5 in 15 steps while staying within the strip is simply $P^{15}(2,5)$ which, without resorting to matrix multiplication computations, may be computed using Equation 2.20. Substituting in the appropriate values, we obtain

$$P^{15}(2,5) = \sum_{s=1}^{8}\left[\frac{2}{9}\left(\frac{3}{2}\right)^{\frac{-3}{2}} \sin\left(\frac{3\pi s}{9}\right) \sin\left(\frac{6\pi s}{9}\right)\left(\frac{1}{6} + 2\sqrt{\frac{1}{6}} \cos\left(\frac{s\pi}{9}\right)\right)^{15}\right] \approx 0.0298$$

Next, we calculate the associated row sum, and this can be determined with the formula $\sum_{j=0}^{7} P^{15}(2,j)$. Each term as well as the sum is given on the first row of Table 1. Therefore, the solution to this problem is

$$P_R^{15}(2,5) = \frac{\sum_{s=1}^{8}\left[\frac{2}{9}\left(\frac{3}{2}\right)^{\frac{-3}{2}} \sin\left(\frac{3\pi s}{9}\right) \sin\left(\frac{6\pi s}{9}\right)\left(\frac{1}{6} + 2\sqrt{\frac{1}{6}} \cos\left(\frac{s\pi}{9}\right)\right)^{15}\right]}{\sum_{j=0}^{7}\left(\sum_{s=1}^{8}\left[\frac{2}{9}\left(\frac{3}{2}\right)^{\frac{2-j}{2}} \sin\left(\frac{3\pi s}{9}\right) \sin\left(\frac{(j+1)\pi s}{9}\right)\left(\frac{1}{6} + 2\sqrt{\frac{1}{6}} \cos\left(\frac{s\pi}{9}\right)\right)^{15}\right]\right)}$$

$$\approx \frac{0.0298}{0.3225} \approx 0.0925$$

This probability and all other probabilities of the form $P_R^{15}(2,j)$, with $j = 0, \ldots, 7$ are given on the second row of Table 1. On a related note, Table 2 provides various probabilities of the form $P_R^k(2,j)$ for different time steps k. The row labeled "steady state" refers to when $k \to \infty$, that is, $\lim_{k\to\infty} P_R^k(2,j)$, see [**Lin**].

TABLE 1. Comparison of Probabilities $P^{15}(2,j)$ and $P_R^{15}(2,j)$.

j	0	1	2	3	4	5	6	7	Total
$P^{15}(2,j)$	0.0410	0.0615	0.0647	0.0572	0.0439	0.0298	0.0171	0.0071	0.3225
$P_R^{15}(2,j)$	0.1271	0.1906	0.2008	0.1774	0.1362	0.0925	0.0531	0.0224	1

TABLE 2. Probabilities of Going from 2 to j in k Steps While Restricted to a $[0,7]$ Strip.

j	0	1	2	3	4	5	6	7	Total
$P_R^{5}(2,j)$	0.1587	0.2693	0.2118	0.1922	0.0917	0.0564	0.0141	0.0056	1
$P_R^{10}(2,j)$	0.1445	0.2066	0.2137	0.1736	0.1270	0.0768	0.0419	0.0159	1
$P_R^{15}(2,j)$	0.1271	0.1906	0.2008	0.1774	0.1362	0.0925	0.0531	0.0224	1
$P_R^{45}(2,j)$	0.1139	0.1747	0.1922	0.1784	0.1456	0.1045	0.0633	0.0275	1
Steady state	0.1138	0.1746	0.1921	0.1784	0.1456	0.1046	0.0634	0.0275	1

EXAMPLE 3.2 (Probability of All Paths from i to j in a Sub-Strip). Continuing under the same setting as Example 3.1, what is the probability of going from state $i=2$ to $j=5$ in 15 steps while being restricted to the strip $[1,6]$? Figure 5 illustrates a path satisfying these conditions. Observe that $[1,6]$ is a sub-strip of $[0,7]$, the restriction considered in the previous example.

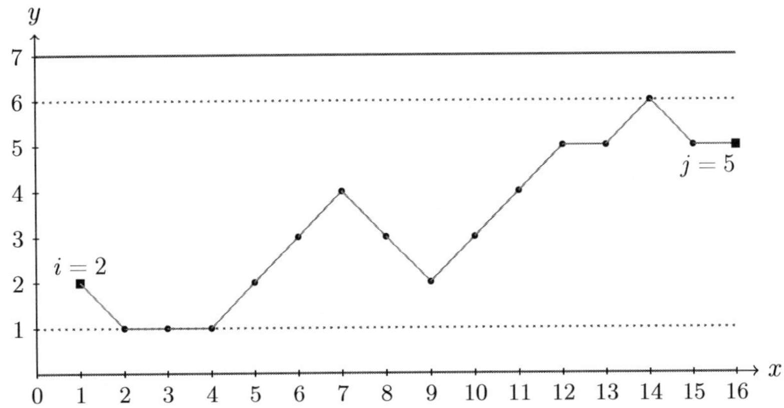

FIGURE 5. A path satisfying the conditions given in Example 3.2

Solving this is similar to the logic presented in Example 3.1. We first construct the substochastic matrix $\bar{P}$ associated with the strip $[1,6]$ which is

$$\bar{P} = \begin{bmatrix} 1/6 & 1/3 & 0 & 0 & 0 & 0 \\ 1/2 & 1/6 & 1/3 & 0 & 0 & 0 \\ 0 & 1/2 & 1/6 & 1/3 & 0 & 0 \\ 0 & 0 & 1/2 & 1/6 & 1/3 & 0 \\ 0 & 0 & 0 & 1/2 & 1/6 & 1/3 \\ 0 & 0 & 0 & 0 & 1/2 & 1/6 \end{bmatrix}$$

Next, it follows that

$$\begin{aligned} \bar{P}^{15}(2,5) &= \sum_{s=1}^{6} \left[\frac{2}{7} \left(\frac{3}{2} \right)^{\frac{-3}{2}} \sin\left(\frac{3\pi s}{7} \right) \sin\left(\frac{6\pi s}{7} \right) \left(\frac{1}{6} + 2\sqrt{\frac{1}{6}} \cos\left(\frac{s\pi}{7} \right) \right)^{15} \right] \\ &\approx 0.0199 \end{aligned}$$

Finally, we divide this value by the row sum associated with the second row of $\bar{P}^{15}$ to obtain

$$\bar{P}^{15}_{[1,6]}(2,5) \approx \frac{0.0199}{0.1623} = 0.1226$$

EXAMPLE 3.3. (Probability of all paths going from 2 to 5 in 15 steps staying in the strip [1,6] given we are restricted to be in the strip [0,7])

$$\frac{\bar{P}^{15}(2,5)}{P^{15}(2,5)} = \frac{0.0199}{0.0298} = 0.6678$$

This probability was also verified numerically using Monte Carlo simulation.

EXAMPLE 3.4. (Probability of Hitting One of the Original Boundaries) We continue under the same setting of Examples 3.1 and 3.2. What is the probability of going from state $i = 2$ to $j = 5$ in 15 steps with the requirement that the path hits states 0 or 7 at some time during its journey? To solve this, we can use the $P^{15}(2,5)$ and $\bar{P}^{15}(2,5)$ calculated in the previous examples. One way to approach this problem is to think of $\bar{P}^{15}(2,5)$ as the collection of paths that stay within the strip $[1,6]$, and we remove these paths from the set of path restricted to $[0,7]$, represented by $P^{15}(2,5)$. This difference would represent the "good paths," that is, the paths that satisfy our requirement. Our solution is then the percentage of good paths contained in $P^{15}(2,5)$, i.e.,

$$\frac{P^{15}(2,5) - \bar{P}^{15}(2,5)}{P^{15}(2,5)} = \frac{0.0298 - 0.0199}{0.0298} = \frac{0.0099}{0.0298} = 0.3313$$

3.2. Transient probabilities of a finite birth-death process restricted to a strip. We can extend our methods of Section 2.3 to calculate transient probabilities for continuous time, finite birth-death processes under similar restrictions. When S is the $n \times n$ transition rate matrix of a finite birth-death process, then the transient probability $P_{i,j}(t)$ of going from state i to state j in time t is given by $e^{St}(i,j)$. Also, recall that S has the property of being conservative, i.e., for all $i = 0, \dots, n-1$, we have $\sum_{j=0}^{n-1} S(i,j) = 0$. In other words, all rows sum to 0.

To determine transient probabilities of birth-death processes restricted to a strip, our approach involves considering a sub-block of the transition rate matrix S associated with the states contained in that strip. Denote this restricted matrix as

Q. For instance, consider the birth-death process with birth rate $\lambda > 0$ and death rate $\mu > 0$ represented by the state rate diagram in Figure 6. Here,

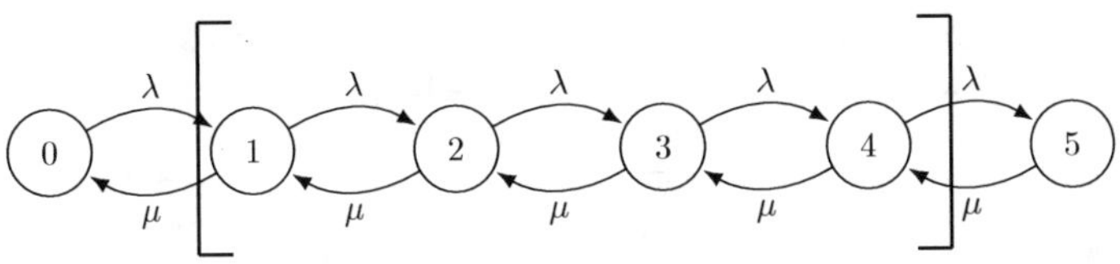

FIGURE 6. State rate diagram for a birth-death process with state space $[0, 5]$.

If S is the rate matrix corresponding to the preceding diagram, we are interested in the sub-matrix Q of S associated with states $[1, 4]$.

$$S = \begin{bmatrix} -\lambda & \lambda & 0 & 0 & 0 & 0 \\ \mu & -(\lambda+\mu) & \lambda & 0 & 0 & 0 \\ 0 & \mu & -(\lambda+\mu) & \lambda & 0 & 0 \\ 0 & 0 & \mu & -(\lambda+\mu) & \lambda & 0 \\ 0 & 0 & 0 & \mu & -(\lambda+\mu) & \lambda \\ 0 & 0 & 0 & 0 & \mu & -\mu \end{bmatrix}$$

Now, consider Q corresponding to the strip of states $[1, 4]$. Then

$$Q = \begin{bmatrix} -(\lambda+\mu) & 0 & 0 & 0 \\ \mu & -(\lambda+\mu) & \lambda & 0 \\ 0 & \mu & -(\lambda+\mu) & \lambda \\ 0 & 0 & \mu & -(\lambda+\mu) \end{bmatrix} \tag{3.1}$$

It is clear matrix Q is not conservative since the first and last row do not sum to 0. Consequently, the transient probability of paths going from state i to j in t amount of time while confined to a strip is given by the ratio

$$\frac{e^{Qt}(i,j)}{\sum_{j=0}^{n-1} e^{Qt}(i,j)} \tag{3.2}$$

and we can apply the results of Subsection 2.2 (since Q is a tridiagonal Toeplitz matrix) to obtain

$$e^{Qt}(i,j) = \sum_{s=1}^{n} \left[\frac{2}{n+1} \left(\frac{\mu}{\lambda}\right)^{\frac{i-j}{2}} \sin\left(\frac{(i+1)\pi s}{n+1}\right) \sin\left(\frac{(j+1)\pi s}{n+1}\right) e^{\omega_s t} \right] \tag{3.3}$$

where ω_s is the sth eigenvalue

$$\omega_s = -(\lambda+\mu) + 2\sqrt{\lambda\mu}\cos\left(\frac{s\pi}{n+1}\right) \tag{3.4}$$

We apply these results to various examples below. As in Examples 3.1 to 3.4, we assume the following rate diagrams represent the restricted states. In other words, we can think of each diagram as a subset of a larger rate diagram not illustrated here.

EXAMPLE 3.5 (Transient Probability Given a Strip). Consider a birth-death process with birth rate $\lambda = 1.2$ and death rate $\mu = 2.8$. We are interested in calculating the probability of the process starting at state 2 and ending at state 5 in five time units given the process is restricted to the strip $[0, 7]$. The state rate diagram and a sample path satisfying these conditions are given in Figures 7 and 8, respectively.

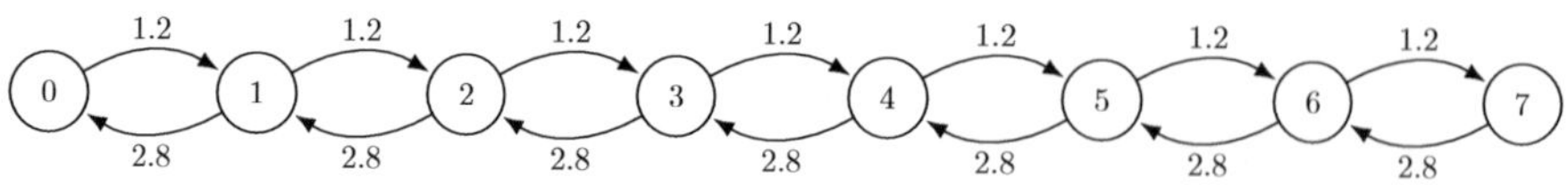

FIGURE 7. State rate diagram considered in Example 3.5

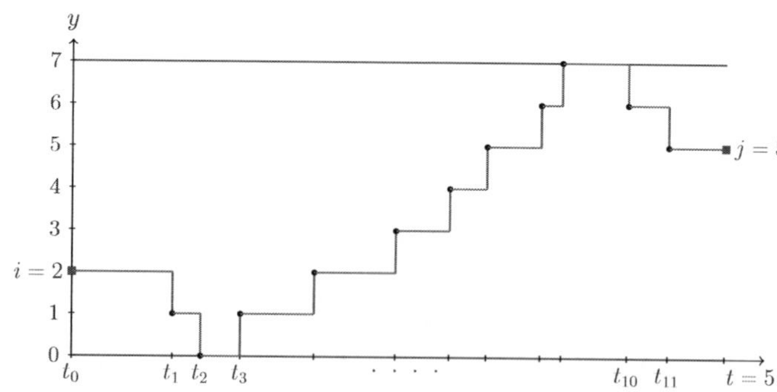

FIGURE 8. A sample path satisfying the restrictions specified in example 3.5

To calculate this probability, we simply evaluate ratio (3.2) using equation (3.3) where Q is an 8×8 matrix having the form of equation (3.1). Substituting in the given values, we obtain $e^{5Q}(2,5) \approx 0.0028$, and the transient probability assuming that the sample paths stay within the strip to be

$$\frac{e^{5Q}(2,5)}{\sum_{j=0}^{7} e^{5Q}(2,j)} \approx 0.0547$$

This transient probability was verified by running a Monte Carlo simulation for this example.

EXAMPLE 3.6 (Transient Probability of Going from i to j Given a Sub-Strip). Analogous to Example 3.5, we now calculate the probability of going from state 2 to state 5 in $t = 5$ time units given the process is restricted to the strip of states $[1, 6]$. A typical sample path that meets these conditions is illustrated in Figure 9. This means matrix Q is a 6×6 matrix of the form in equation (3.1). Calculating this probability is similar to Example 3.5 except that our restriction strip is $[1, 6]$ instead. Consequently, we relabel the beginning state from 2 to 1 and the ending state from 5 to 4 in order to use equation (3.3) properly. To be more specific, we substitute into our preceding formula $i = 1$, $j = 4$, $n = 6$, $t = 5$, $\lambda = 1.2$, and $\mu = 2.8$. This yields a conditional probability of

$$\frac{e^{5Q}(1,4)}{\sum_{j=0}^{5} e^{5Q}(1,j)} \approx \frac{0.0015}{0.0193} \approx 0.0772$$

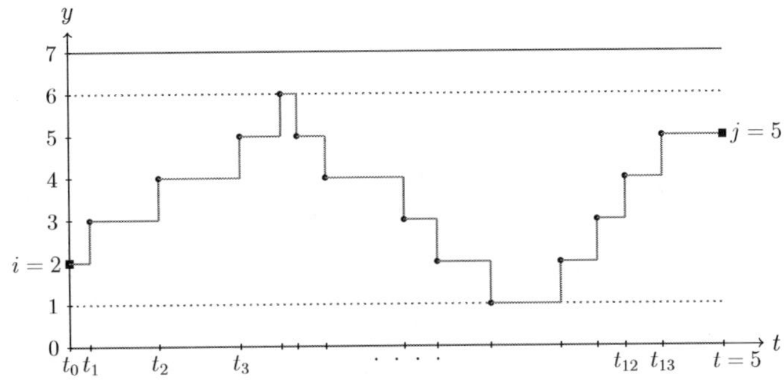

FIGURE 9. A sample path from state 2 to state 5 in 5 time units given that the process is confined to the strip $[1, 6]$.

EXAMPLE 3.7 (Probability of a Markov Process Staying Within a Small Solution Strip Given it Stays Within Large Solution Strip). Continuing with the same set up as Examples 3.5 and 3.6, suppose the process moves from state 2 to state 5 in $t = 5$ time units and its sample path never leaves the solution strip $[0, 7]$. What is the probability that the process stays within the sub-solution strip $[1, 6]$? We may calculate this probability by determining the proportion of sample paths that stay within $[1, 6]$ from the set of paths restricted to the larger solution strip $[0, 7]$, thus

$$\frac{e^{\bar{Q}\cdot 5}(1,4)}{e^{Q\cdot 5}(2,5)} = \frac{0.0015}{0.0028} \approx 0.5357$$

We also evaluated this probability numerically using Monte Carlo simulation, and simulation results are consistent with the value above.

EXAMPLE 3.8 (Transient Probability of Hitting Original Boundaries). Under the same settings of Examples 3.5 and 3.6, given the process is restricted to $[0, 7]$, we would like to calculate the probability the process hits either states 0 or 7 while traversing from states 2 to 5 in $t = 5$ time units. We can calculate this probability using the same reasoning presented in Example 3.4 to obtain

$$\frac{e^{Q\cdot 5}(2,5) - e^{\bar{Q}\cdot 5}(1,4)}{e^{Q\cdot 5}(2,5)} \approx \frac{0.0028 - 0.0015}{0.0028} \approx 0.4643$$

3.3. Combinatorial solution traditional ballot box problem. Suppose that in a two person election, candidate A receives a votes while candidate B receives b votes, where $a > b$. A classic problem in combinatorics is to compute the probability that A never falls behind B throughout the counting of the ballots. For example, suppose candidate A receives $U = 3$ up votes and candidate B receives $D = 2$ down votes. We can recast this problem in terms of lattice paths; in this context, the problem amounts to calculating the probability of obtaining a path starting at state $i = 0$ and ending at state $j = 1$ in $n = 5$ steps without going below the x-axis.

Traditionally, the solution to this problem employs the notion of good and bad paths. A **good** lattice path from i to j in n steps never goes below the x-axis while a **bad** lattice path from i to j in n steps goes below the x-axis somewhere along the way. Examples of these types of paths are illustrated in the three figures below.

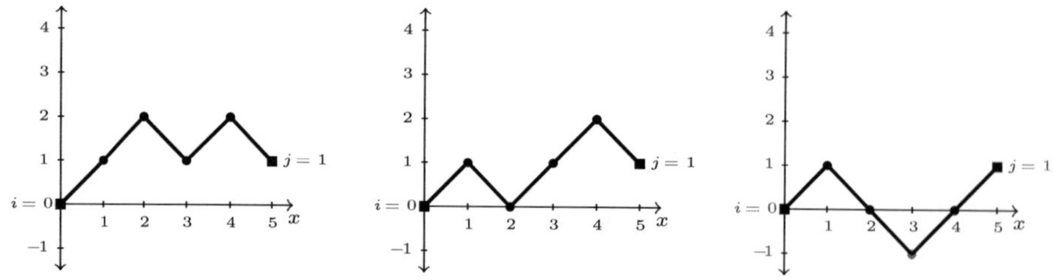

(A) Good lattice path (B) Good lattice path (C) Bad lattice path

Let $\mathcal{G}$, $\mathcal{B}$ denote the collection of **good** and **bad** lattice paths, respectively. The number of bad lattice paths $|\mathcal{B}|$ can be counted using a clever bijection method called the Reflection Principle [**Ren07, Moh14**]. An example of this bijection is shown in Figure 11. In our example, the bad paths from $i = 0$ to $j = 1$ correspond to the reflected paths from $k = -2$ to $j = 1$. Therefore, the number of bad paths is $|\mathcal{B}| = \binom{5}{4} = 5$.

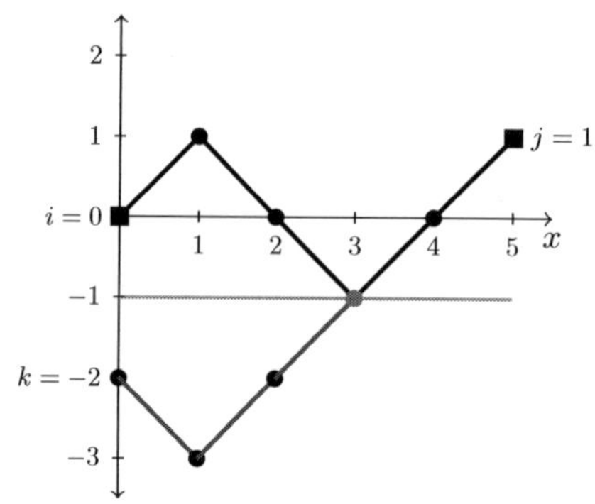

FIGURE 11. Counting bad paths using the Reflection Principle.

The Traditional Ballot Box solution

$$\left.\begin{array}{l} \mathrm{U} - \mathrm{D} = j - i = 1 \\ \mathrm{U} + \mathrm{D} = n = 5 \end{array}\right\} \implies \left\{\begin{array}{l} \mathrm{U} = 3 \\ \mathrm{D} = 2 \end{array}\right.$$

$$\begin{aligned} \frac{|\mathcal{G}|}{|\mathcal{A}|} = \frac{|\mathcal{A}| - |\mathcal{B}|}{|\mathcal{A}|} &= \frac{\text{all lattice paths} - \text{bad lattice paths}}{\text{all lattice paths}} \\ &= \frac{\binom{5}{3} - \binom{5}{4}}{\binom{5}{3}} \\ &= \frac{10 - 5}{10} = \frac{1}{2} \end{aligned} \tag{3.5}$$

3.4. Matrix-power solution of birth-death chain ballot box problem. In this section, we first will solve the traditional ballot box problem posed in Section 3.3 by using a matrix-power approach. We then will utilize this matrix-power method to solve a more general ballot box problem.

EXAMPLE 3.9 (Traditional Ballot Box Problem using Matrix-Power Solution). Similar to Section 3.3, to calculate the solution of going from state 0 to state 1 in 5 steps we need to calculate the probability of good paths and all paths. If a lattice path is good and goes from state $i = 0$ to state $j = 1$ in 5-steps, then all the steps will take place in the solution strip between 0 and 3 as shown in Figure 12.

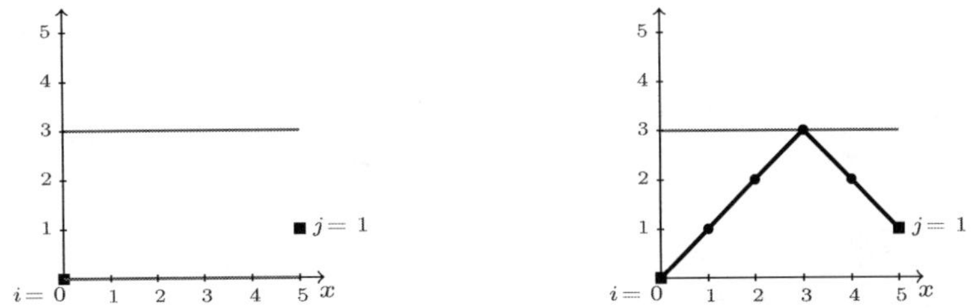

(A) Solution Strip of Good Paths

(B) Example of a Good Path

FIGURE 12

Figure 12 (A) shows the solution strip where all good paths reside. Figure 12 (B) is a good path going from 0 to 1 in 5 steps having maximal height.

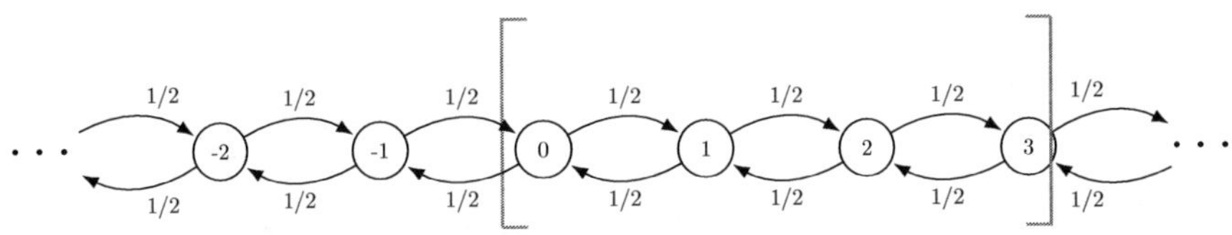

FIGURE 13. Birth-death state transition diagram

$$\mathcal{G} = \begin{array}{c} \\ 0 \\ 1 \\ 2 \\ 3 \end{array} \begin{array}{c} \begin{array}{cccc} 0 & 1 & 2 & 3 \end{array} \\ \begin{bmatrix} 0 & \frac{1}{2} & 0 & 0 \\ \frac{1}{2} & 0 & \frac{1}{2} & 0 \\ 0 & \frac{1}{2} & 0 & \frac{1}{2} \\ 0 & 0 & \frac{1}{2} & 0 \end{bmatrix} \end{array}$$

1-step transition probability matrix for good paths

The probability of all paths starting at state $i = 0$ and ending at state $j = 1$ in 5 steps and never going outside the boundaries of the strip is $\mathcal{G}^5(0,1) = \frac{5}{32}$ as shown below.

$$\mathcal{G}^5 = \begin{bmatrix} 0 & \frac{1}{2} & 0 & 0 \\ \frac{1}{2} & 0 & \frac{1}{2} & 0 \\ 0 & \frac{1}{2} & 0 & \frac{1}{2} \\ 0 & 0 & \frac{1}{2} & 0 \end{bmatrix}^5 = \begin{bmatrix} 0 & \boxed{\frac{5}{32}} & 0 & \frac{3}{32} \\ \frac{5}{32} & 0 & \frac{1}{4} & 0 \\ 0 & \frac{1}{4} & 0 & \frac{5}{32} \\ \frac{3}{32} & 0 & \frac{5}{32} & 0 \end{bmatrix}$$

Next, we will compute the probability of the all paths, starting at state $i = 0$ and ending at state $j = 1$ in $n = 5$ steps without any restrictions. Again $y = 3$ is the highest state that we can reach and still get back to $j = 1$ in $n = 5$ steps. And $y = -2$ is the lowest state that we reach and still get back to $j = 1$ in $n = 5$ steps.

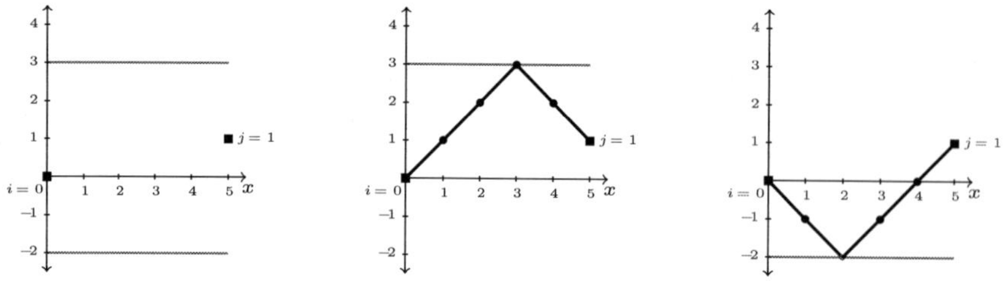

(A) Solution Strip of All Paths (B) Upper Boundary (C) Lower Boundary

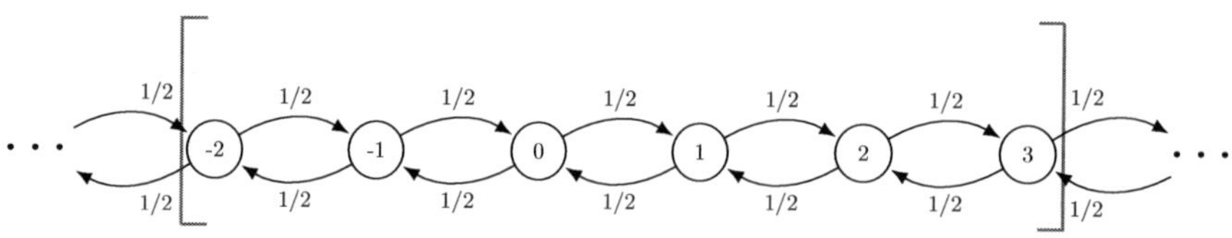

$$\mathcal{A} = \begin{array}{c} \\ -2 \\ -1 \\ 0 \\ 1 \\ 2 \\ 3 \end{array} \begin{array}{c} \begin{array}{cccccc} -2 & -1 & 0 & 1 & 2 & 3 \end{array} \\ \begin{bmatrix} 0 & \frac{1}{2} & 0 & 0 & 0 & 0 \\ \frac{1}{2} & 0 & \frac{1}{2} & 0 & 0 & 0 \\ 0 & \frac{1}{2} & 0 & \frac{1}{2} & 0 & 0 \\ 0 & 0 & \frac{1}{2} & 0 & \frac{1}{2} & 0 \\ 0 & 0 & 0 & \frac{1}{2} & 0 & \frac{1}{2} \\ 0 & 0 & 0 & 0 & \frac{1}{2} & 0 \end{bmatrix} \end{array}$$

1-step transition probability matrix for all paths

The probability of starting at state $i = 0$ and ending at state $j = 1$ in 5 steps with no restrictions is $\mathcal{A}(0,1)^5 = \frac{10}{32}$ as shown below.

$$\mathcal{A}^5 = \begin{bmatrix} 0 & \frac{1}{2} & 0 & 0 & 0 & 0 \\ \frac{1}{2} & 0 & \frac{1}{2} & 0 & 0 & 0 \\ 0 & \frac{1}{2} & 0 & \frac{1}{2} & 0 & 0 \\ 0 & 0 & \frac{1}{2} & 0 & \frac{1}{2} & 0 \\ 0 & 0 & 0 & \frac{1}{2} & 0 & \frac{1}{2} \\ 0 & 0 & 0 & 0 & \frac{1}{2} & 0 \end{bmatrix}^5 = \begin{bmatrix} 0 & \frac{5}{32} & 0 & \frac{4}{32} & 0 & \frac{1}{32} \\ \frac{5}{32} & 0 & \frac{9}{32} & 0 & \frac{5}{32} & 0 \\ 0 & \frac{9}{32} & 0 & \boxed{\frac{10}{32}} & 0 & \frac{4}{32} \\ \frac{4}{32} & 0 & \frac{10}{32} & 0 & \frac{9}{32} & 0 \\ 0 & \frac{5}{32} & 0 & \frac{9}{32} & 0 & \frac{5}{32} \\ \frac{1}{32} & 0 & \frac{4}{32} & 0 & \frac{5}{32} & 0 \end{bmatrix}$$

So of all paths starting at $i = 0$ and ending at $j = 1$ in $n = 5$ steps, the probability of never leaving the strip $0 \le y \le 3$ is

$$\frac{\mathcal{G}^5(0,1)}{\mathcal{A}^5(0,1)} = \frac{\frac{5}{32}}{\frac{10}{32}} = \frac{1}{2} \tag{3.6}$$

which agrees with the combinatorial solution (3.5).

EXAMPLE 3.10. Suppose we are now moving on the following birth-death chain:

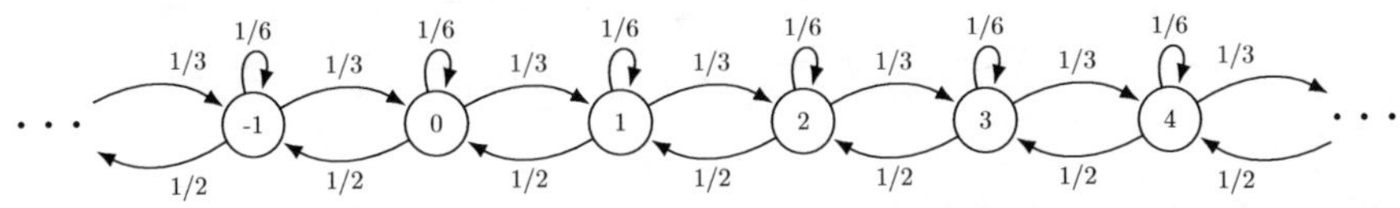

Suppose candidate A receives 3 votes and candidate B receives 2 votes. Calculate the probability of going from state $i = 0$ to state $j = 1$ in $n = 5$ steps without going below 0. To solve this problem we will use the same matrix-power method used in Example 3.9.

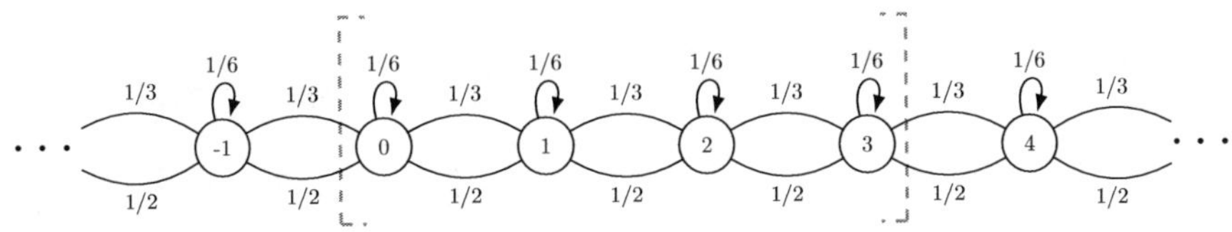

FIGURE 15. Transition Diagram for the Good Path Solution Strip

If a lattice path is good and goes from state $i = 0$ to state $j = 1$ in 5-steps, then all steps take place in the following strip.

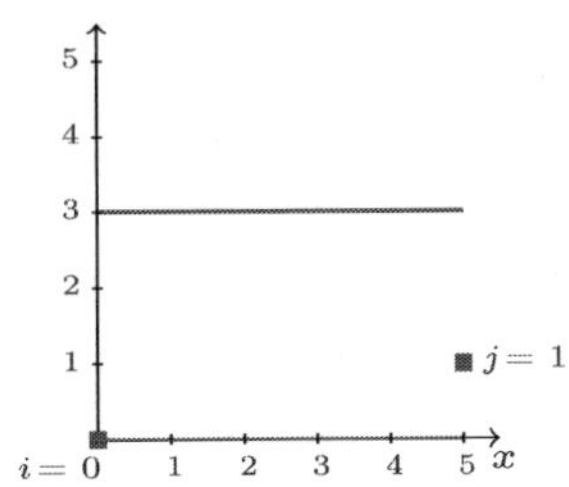

FIGURE 16. Solution Strip of Good Paths

$$\mathcal{G} = \begin{array}{c} \\ 0 \\ 1 \\ 2 \\ 3 \end{array} \begin{array}{c} \begin{array}{cccc} 0 & 1 & 2 & 3 \end{array} \\ \begin{bmatrix} \frac{1}{6} & \frac{1}{3} & 0 & 0 \\ \frac{1}{2} & \frac{1}{6} & \frac{1}{3} & 0 \\ 0 & \frac{1}{2} & \frac{1}{6} & \frac{1}{3} \\ 0 & 0 & \frac{1}{2} & \frac{1}{6} \end{bmatrix} \end{array} \tag{3.7}$$

1-step transition probability matrix for good paths.

The probability of all paths starting at $i = 0$ and ending at $j = 1$ in 5 steps and never going outside the boundaries of the strip is $\mathcal{G}^5(0,1) = \frac{305}{3888}$.

$$\mathcal{G}^5 = \begin{bmatrix} \frac{1}{6} & \frac{1}{3} & 0 & 0 \\ \frac{1}{2} & \frac{1}{6} & \frac{1}{3} & 0 \\ 0 & \frac{1}{2} & \frac{1}{6} & \frac{1}{3} \\ 0 & 0 & \frac{1}{2} & \frac{1}{6} \end{bmatrix}^5 = \begin{bmatrix} \frac{421}{7776} & \boxed{\frac{305}{3888}} & \frac{25}{486} & \frac{7}{243} \\ \frac{305}{2592} & \frac{1021}{7776} & \frac{473}{3888} & \frac{25}{486} \\ \frac{25}{216} & \frac{473}{2592} & \frac{1021}{7776} & \frac{305}{3888} \\ \frac{7}{72} & \frac{25}{216} & \frac{305}{2592} & \frac{219}{4045} \end{bmatrix}$$

Next, we calculate the probability of starting at state $i = 0$ and ending at state $j = 1$ in $n = 5$ steps without any restrictions. Again $y = 3$ is the highest state that we can reach and still get back to $j = 1$ in $n = 5$ steps. And $y = -2$ is the lowest state that we reach and still get back to $j = 1$ in $n = 5$ steps.

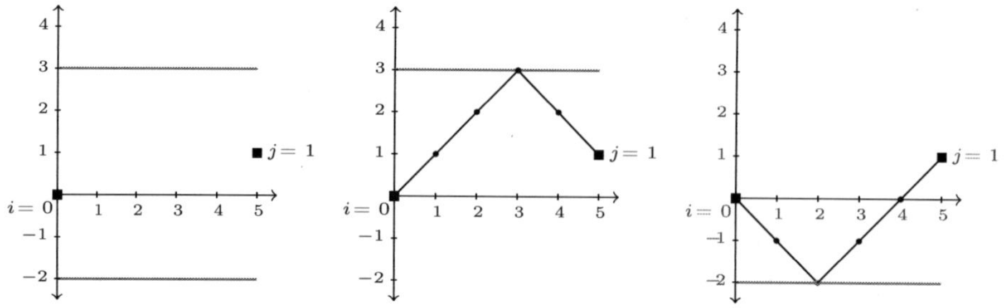

(A) Solution Strip of All Paths (B) Upper Boundary (C) Lower Boundary

FIGURE 17

Figure 17(A) represents the strip where all possible lattice paths of 5 steps starting at $i = 0$ and ending at $j = 1$ can occur. Figure 17(B) represents the path where candidate A has the largest possible lead over candidate B. Figure 17(C) represents the path where candidate B has the largest possible lead over candidate A.

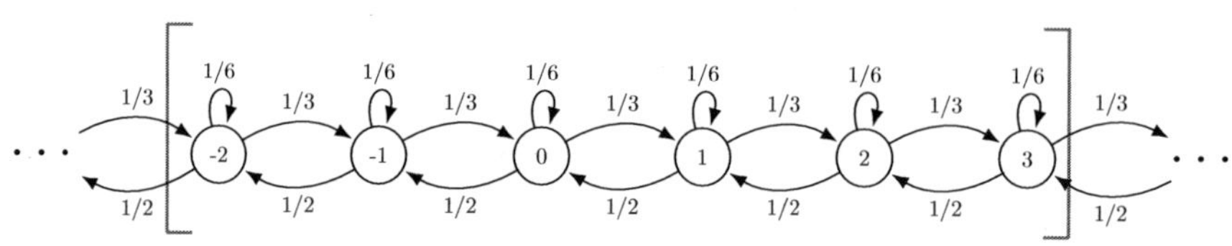

FIGURE 18. Transition Diagram of **all** paths solution strip

$$\mathcal{A} = \begin{array}{c} \\ -2 \\ -1 \\ 0 \\ 1 \\ 2 \\ 3 \end{array} \begin{array}{c} \begin{array}{cccccc} -2 & -1 & 0 & 1 & 2 & 3 \end{array} \\ \begin{bmatrix} \frac{1}{6} & \frac{1}{3} & 0 & 0 & 0 & 0 \\ \frac{1}{2} & \frac{1}{6} & \frac{1}{3} & 0 & 0 & 0 \\ 0 & \frac{1}{2} & \frac{1}{6} & \frac{1}{3} & 0 & 0 \\ 0 & 0 & \frac{1}{2} & \frac{1}{6} & \frac{1}{3} & 0 \\ 0 & 0 & 0 & \frac{1}{2} & \frac{1}{6} & \frac{1}{3} \\ 0 & 0 & 0 & 0 & \frac{1}{2} & \frac{1}{6} \end{bmatrix} \end{array} \tag{3.8}$$

Matrix 3.8 is the 1-step transition probability matrix for all paths.

The probability of starting at state $i = 0$ and ending at state $j = 1$ in 5 steps with no restrictions is $A^5(0,1) = \frac{545}{3888}$ as shown below.

$$\mathcal{A}^5 = \begin{bmatrix} \frac{1}{6} & \frac{1}{3} & 0 & 0 & 0 & 0 \\ \frac{1}{2} & \frac{1}{6} & \frac{1}{3} & 0 & 0 & 0 \\ 0 & \frac{1}{2} & \frac{1}{6} & \frac{1}{3} & 0 & 0 \\ 0 & 0 & \frac{1}{2} & \frac{1}{6} & \frac{1}{3} & 0 \\ 0 & 0 & 0 & \frac{1}{2} & \frac{1}{6} & \frac{1}{3} \\ 0 & 0 & 0 & 0 & \frac{1}{2} & \frac{1}{6} \end{bmatrix}^5$$

$$= \begin{bmatrix} \frac{421}{7776} & \frac{305}{3888} & \frac{25}{486} & \frac{17}{486} & \frac{5}{486} & \frac{1}{26} \\ \frac{305}{2592} & \frac{1021}{7776} & \frac{509}{3888} & \frac{65}{972} & \frac{10}{243} & \frac{5}{486} \\ \frac{25}{216} & \frac{509}{2592} & \frac{1201}{7776} & \boxed{\frac{545}{3888}} & \frac{65}{972} & \frac{17}{486} \\ \frac{17}{144} & \frac{65}{432} & \frac{545}{2592} & \frac{1201}{7776} & \frac{509}{3888} & \frac{25}{486} \\ \frac{5}{96} & \frac{5}{36} & \frac{65}{432} & \frac{509}{2592} & \frac{1021}{7776} & \frac{305}{3888} \\ \frac{1}{32} & \frac{5}{96} & \frac{17}{144} & \frac{25}{216} & \frac{305}{2592} & \frac{421}{7776} \end{bmatrix}$$

So of all paths starting at $i = 0$ and ending at $j = 1$ in $n = 5$ steps, the probability of never leaving the strip $0 \le y \le 3$ is

$$\frac{\mathcal{G}^5(0,1)}{\mathcal{A}^5(0,1)} = \frac{\left(\frac{305}{3888}\right)}{\left(\frac{545}{3888}\right)} = \frac{305}{545} \approx 0.560 \tag{3.9}$$

Remarks:

Question #1: How do we check our work, that is, how do we verify that (3.5) and (3.9) are correct?

Answer #1: We could enumerate and count all of the sample paths that take us from $i = 0$ to $j = 1$ in five steps.

Question #2: Why is the probability of (3.9) larger than the probability of (3.5)?

Answer #2: Intuitively, because i and j are close to each other, there are more good paths having multiple abstentions than there are bad paths having multiple abstentions.

3.5. Birth-death chain ballot box problem solution in terms of eigenvalues. Recall equation (2.20):

$$P^k(i,j)=\sum_{s=1}^{n} A_s(i,j)\omega_s^k$$
$$=\sum_{s=1}^{n}\frac{2}{n+1}\left(\frac{q}{p}\right)^{\frac{i-j}{2}}\sin\left(\frac{(i+1)\pi s}{n+1}\right)\sin\left(\frac{(j+1)\pi s}{n+1}\right)\left(r+2\sqrt{pq}\cos\left(\frac{s\pi}{n+1}\right)\right)^k$$

For **good** lattice paths in the previous example, $i=0$, $j=1$, $k=5$, $n=4$, $q=\frac{1}{2}$, $p=\frac{1}{3}$, $r=\frac{1}{6}$. Then,

$$\mathcal{G}^5(0,1)=\sum_{s=1}^{4} A_s(0,1)\omega_s^5 \approx \frac{179}{2531}+\frac{30}{12727}+\frac{1}{1188665}+\frac{66}{12301}\approx\frac{305}{3888}$$

which agrees with the appropriate entry of the fifth power of the G matrix.

Recall equation (2.17):

$$P^k(i,j)=\sum_{s=1}^{n} A_s(i,j)\omega_s^k$$
$$=\sum_{s=1}^{n}\frac{2}{n+1}\left(\frac{q}{p}\right)^{\frac{i-j}{2}}\sin\left(\frac{(i-j+k+2)\pi s}{2(n+1)}\right)$$
$$\times\sin\left(\frac{(-i+j+k+2)\pi s}{2(n+1)}\right)\left(r+2\sqrt{pq}\cos\left(\frac{s\pi}{n+1}\right)\right)^k$$

For **all** lattice paths with $i=0$, $j=1$, $k=5$, $n=6$, $q=\frac{1}{2}$, $p=\frac{1}{3}$, $r=\frac{1}{6}$,

$$\mathcal{A}^5(0,1)=\sum_{s=1}^{6}\tilde{A}_s(0,1)\omega_s^5\approx\frac{37}{279}+\frac{-109}{17615}+\frac{24}{32809}+0+\frac{-23}{111265}+\frac{112}{8471}\approx\frac{545}{3888}$$

which agrees with the appropriate entry of the fifth power of the A matrix. Thus, taking the ratio between those values found, we obtain the probability of going from state $i=0$ to $j=1$ in 5 steps without going out of the solution strip:

$$\frac{\mathcal{G}^5(0,1)}{\mathcal{A}^5(0,1)}=\frac{\frac{305}{3888}}{\frac{545}{3888}}=\frac{305}{545}\approx 0.560 \tag{3.10}$$

Note that our answer found using eigenvalues (3.10) agrees with our answer found using the matrix-power method (3.9).

3.6. Exponential Matrix solution of birth-death process ballot box problem. Consider going from $i = 2$ to $j = 1$ in time $t = 5$ on a birth-death process restricted to the following transition diagram. These paths will visit states in the following solution strip. An example of such a path is pictured below and is called a good path.

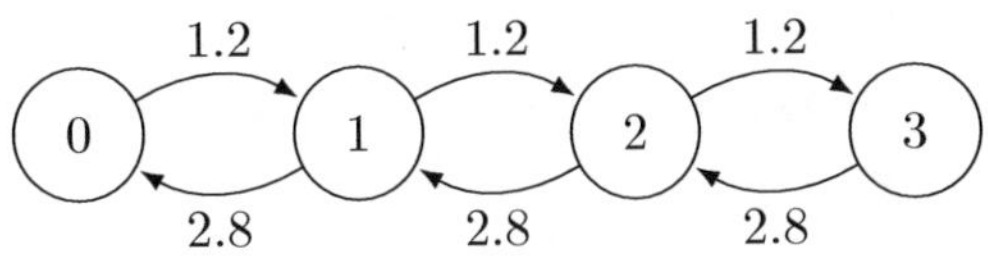

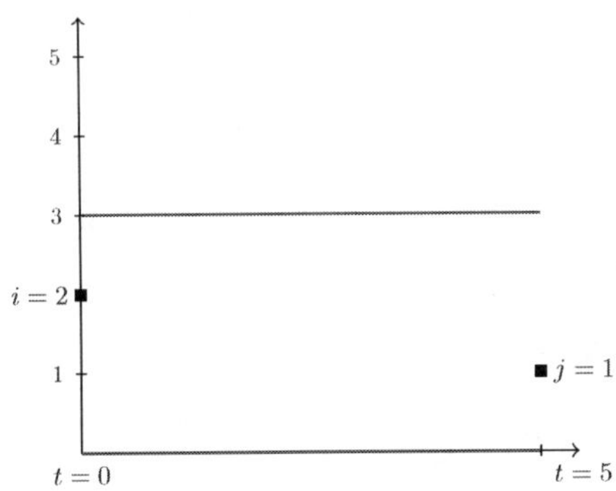

FIGURE 19. Solution Strip of Good Paths

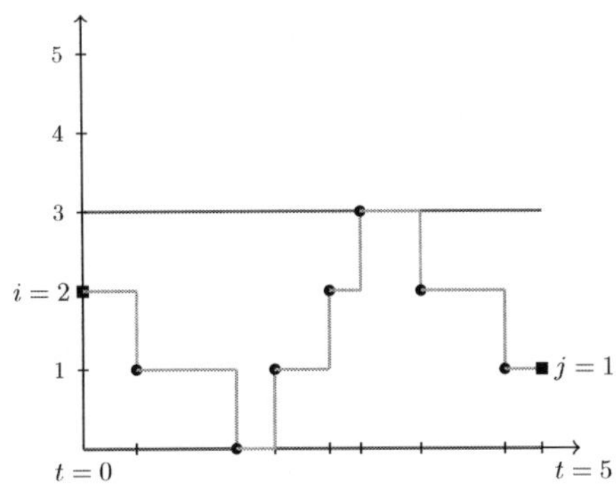

FIGURE 20. A Good Path

$$\mathcal{G} = \begin{array}{c} \\ 0 \\ 1 \\ 2 \\ 3 \end{array} \begin{array}{c} \begin{array}{cccc} 0 & 1 & 2 & 3 \end{array} \\ \begin{bmatrix} -4 & 1.2 & 0 & 0 \\ 2.8 & -4 & 1.2 & 0 \\ 0 & 2.8 & -4 & 1.2 \\ 0 & 0 & 2.8 & -4 \end{bmatrix} \end{array}$$

1-step rate transition matrix for good paths

$$e^{\mathcal{G}5} = \begin{array}{c} \\ 0 \\ 1 \\ 2 \\ 3 \end{array} \begin{array}{c} \begin{array}{cccc} 0 & 1 & 2 & 3 \end{array} \\ \begin{bmatrix} 0.000785431 & 0.000831829 & 0.000544446 & 0.000220245 \\ 0.00194093 & 0.0020558 & 0.00134573 & 0.000544446 \\ 0.0029642 & \boxed{0.00314004} & 0.0020558 & 0.00831829 \\ 0.00279792 & 0.0029642 & 0.00194093 & 0.000785431 \end{bmatrix} \end{array}$$

Exponential matrix of good paths computed using Wolfram Alpha

Transition Diagram, Solution Strip of All Paths

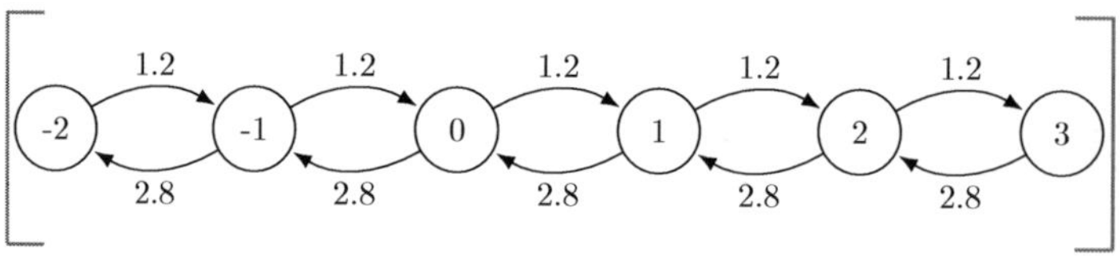

FIGURE 21. Birth-death state rate transition diagram

$$\mathcal{A} = \begin{array}{c} \\ -2 \\ -1 \\ 0 \\ 1 \\ 2 \\ 3 \end{array} \begin{array}{c} \begin{array}{cccccc} -2 & -1 & 0 & 1 & 2 & 3 \end{array} \\ \left[\begin{array}{cccccc} -4 & 1.2 & 0 & 0 & 0 & 0 \\ 2.8 & -4 & 1.2 & 0 & 0 & 0 \\ 0 & 2.8 & -4 & 1.2 & 0 & 0 \\ 0 & 0 & 2.8 & -4 & 1.2 & 0 \\ 0 & 0 & 0 & 2.8 & -4 & 1.2 \\ 0 & 0 & 0 & 0 & 2.8 & -4 \end{array}\right] \end{array}$$

1-step transition rate matrix for all paths

Consider **all** paths going from $i = 2$ to $j = 1$ in time $t = 5$ having states in the preceding solution strip, Figure 21, of all paths. When a path goes below the x-axis as shown below, it is considered to be a **bad** path.

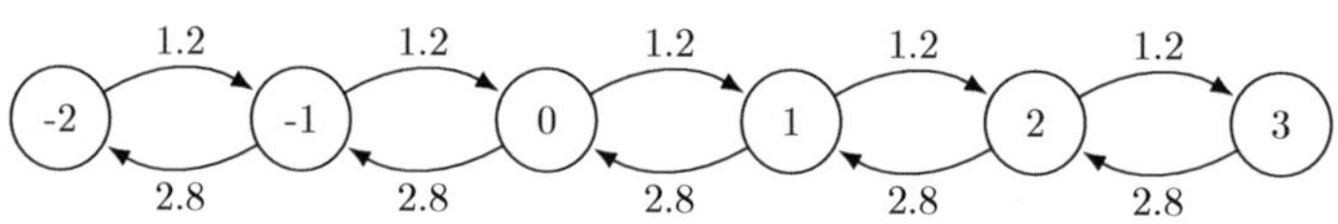

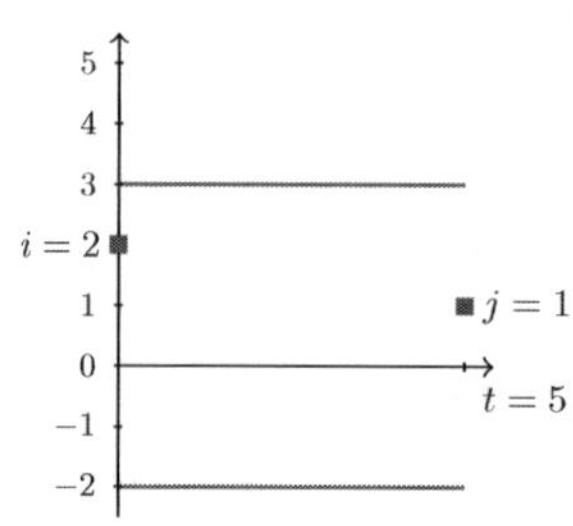

FIGURE 22. Solution Strip of All Paths

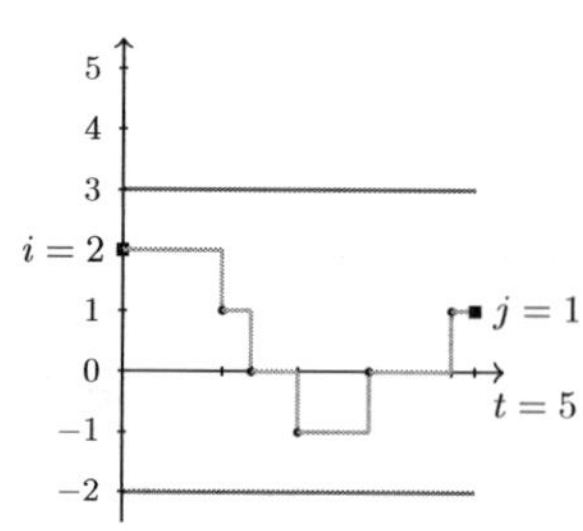

FIGURE 23. A Bad Path

$e^{\mathcal{A}5} =$

	−2	−1	0	1	2	3
−2	0.00168196	0.00197206	0.00159567	0.00103431	0.000538134	0.000194285
−1	0.00460147	0.00540519	0.00438544	0.00285132	0.00148764	0.000538134
0	0.00868754	0.0102327	0.00833503	0.00544321	0.00285132	0.00103431
1	0.0131395	0.0155238	0.0127008	0.00833503	0.00438544	0.00159567
2	0.0159514	0.0188985	0.0155238	0.0102327	0.00540519	0.00197206
3	0.0134376	0.0159514	0.0131395	0.00868754	0.00460147	0.0016819

Exponential matrix of all paths computed using Wolfram Alpha

So of **all** paths of a birth-death process going from $i = 2$ to $j = 1$ in time $t = 5$, and taking values in the [-2, 3] solution strip, where $\lambda = 1.2$ and $\mu = 2.8$, the probability of never leaving the solution strip [0, 3] is

$$\frac{e^{\mathcal{G}5}(2,1)}{e^{\mathcal{A}5}(2,1)} \approx \frac{0.00314004}{0.0102327} \approx 0.306$$

This probability was also confirmed by running Monte Carlo simulations.

3.7. Birth-death process ballot box problem solution in terms of eigenvalues. The proceeding probability can also be found by using the following eigenvalues expansion:

$$e^{Qt}(i,j) = A_1(i,j)e^{\omega_1 t} + A_2(i,j)e^{\omega_2 t} + A_3(i,j)e^{\omega_3 t} + \cdots + A_n(i,j)e^{\omega_n t}$$

$$\omega_s = -(\lambda+\mu) + 2\sqrt{\lambda\mu}\cos\left(\frac{s\pi}{n+1}\right)$$

If $Q = \mathcal{G}$, $t = 5$, $i = 2$, $j = 1$, $\lambda = 1.2$, $\mu = 2.8$, $n = 4$ then $e^{\mathcal{G}5}(2,1) \approx 0.00314004$

If $Q = \mathcal{A}$, $t = 5$, $i = 2$, $j = 1$, $\lambda = 1.2$, $\mu = 2.8$, $n = 6$ then $e^{\mathcal{A}5}(2,1) \approx 0.0102327$

REMARK. In the birth-death chain ballot box problems of Examples 3.9 and 3.10 the solution strips of ***good*** and ***all*** paths are determined by the states i, j, and the number of voters, k. However, in the birth-death process ballot box model of Section 3.6, the good and bad path strips are arbitrarily defined.

4. Birth-death models with catastrophes

4.1. Dual chains and processes and the duality theorem.

Formal definition of the dual matrix. Assume $P = [P(i,j)]$ for i,j going from $0, 1, \ldots, H$ is an $(H+1) \times (H+1)$ stochastic matrix. Suppose that the $(H+2) \times (H+2)$ matrix $P^* = [P^*(i,j)]$ where:

$$P^*(i,j) = \sum_{s=i+1}^{H} [P(j+1,s) - P(j,s)] \tag{4.1}$$

is also a stochastic matrix under the following conventions:

- $P(-1,s) = 0$ if $-1 < s \le H$
- $P^*(H,H) = 1$
- $P^*(i,H) = 1 - \sum_{s=0}^{H-1} P^*(i,s)$

Then P^* is called the dual matrix of P. Note that even though P^* may not be a stochastic matrix, P and P^* will still have the same set of eigenvalues, see [**Lyc18**]. A very similar definition holds for the dual of a Markov process having transition rate matrix Q, see either [**KMR04**] or [**KRM05**].

Duality Theorem Suppose P is a stochastic $(H+1)\times(H+1)$ matrix and its dual matrix P^* exists and is also a stochastic matrix, then the transient probabilities of the original process and its dual process are related as follows:

$$P^{*(n)}(i,j) = \sum_{s=i+1}^{H} \left[P^{(n)}(j+1,s) - P^{(n)}(j,s)\right] \text{ for } i,j = -1,0,1,\dots,H \tag{4.2}$$

$$P^{(n)}(i,j) = \sum_{s=i}^{H} \left[P^{*(n)}(j,k) - P^{*(n)}(j-1,s)\right] \text{ for } i,j = 0,1,2,\dots,H \tag{4.3}$$

for $n = 1,2,3,\dots$ with the conventions:

- $P^{(n)}(-1,s) = 0$ if $-1 < s \le H$
- $P^{*(n)}(H,H) = 1$
- $P^{*(n)}(i,H) = 1 - \sum_{s=0}^{H-1} P^{*(n)}(i,s)$

Below is an example of a Birth-Death Chain Matrix and its corresponding dual matrix. Suppose

$$\begin{array}{cccc} 0 & 0 & 0 & 0 \end{array}$$
$$P = \begin{bmatrix} 0.7 & 0.3 & 0 \\ 0.2 & 0.6 & 0.2 \\ 0 & 0.7 & 0.3 \end{bmatrix} \begin{array}{c} 0 \\ 0 \\ 0 \end{array}$$
$$\begin{array}{cccc} 0 & 0 & 0 & 1 \end{array}$$

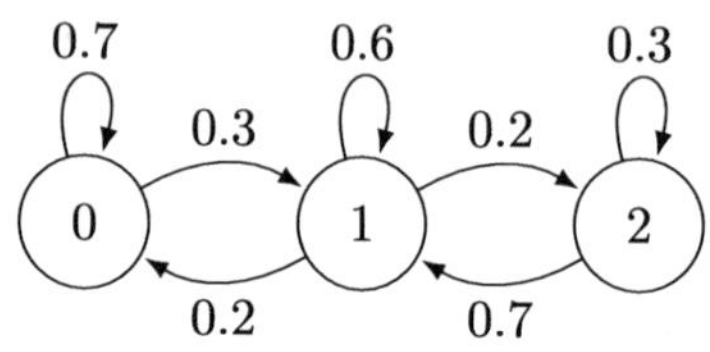

The dual matrix P^* is given by

$$P^* = \begin{bmatrix} 1 & 0 & 0 & 0 \\ 0.3 & 0.5 & 0.2 & 0 \\ 0 & 0.2 & 0.1 & 0.7 \\ 0 & 0 & 0 & 1 \end{bmatrix}$$

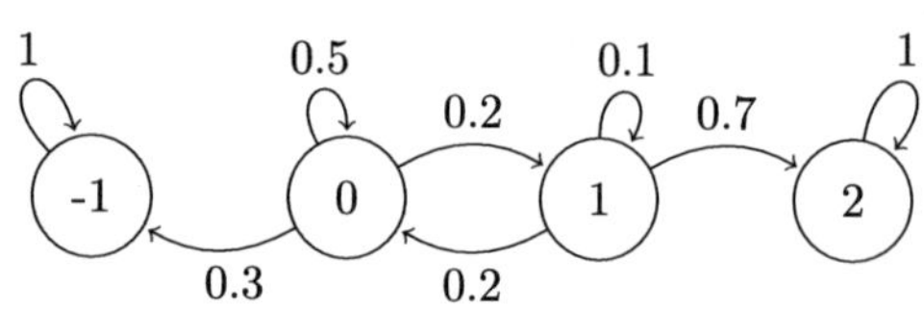

REMARK 4.1. The definition of the dual of a finite Markov process having infinitesimal transition rate matrix Q parallels the preceding approach. For Markov processes, the dual may be obtained by surrounding the Q matrix with a border entirely of 0's and algorithmically follow the same procedure. Having a 0 in the lower right hand boundary corner reflects that the rows of Q matrices usually sum to 0. The Duality Theorem still holds. The reader is referred to [**And91**], [**KMR04**], [**KM10**], and [**KRM05**] for more details. For a more recent approach generalizing stochastic duality to linear algebraic duality see [**RK21**].

4.2. Transient probabilities of more general Markov chains. Consider a four state Markov chain, having the form as shown in Figure 24. States 0, 1, 2 allow a 1-step upward transition having a *birth* probability p; states 1, 2, 3 allow 1-step downward transitions with *death* probability q; and all states 0, 1, 2, 3 have 1-step *return* probability r. This chain also has catastrophe-like probabilities c_0, c_1, c_2, c_3 of transitioning to the states 0, 1, 2, 3 respectively from anywhere in the state space. All of these conditions generalize and scale up naturally to an analogous Markov chain on state space $\mathcal{S} = \{0, 1, 2, \ldots, n-1\}$.

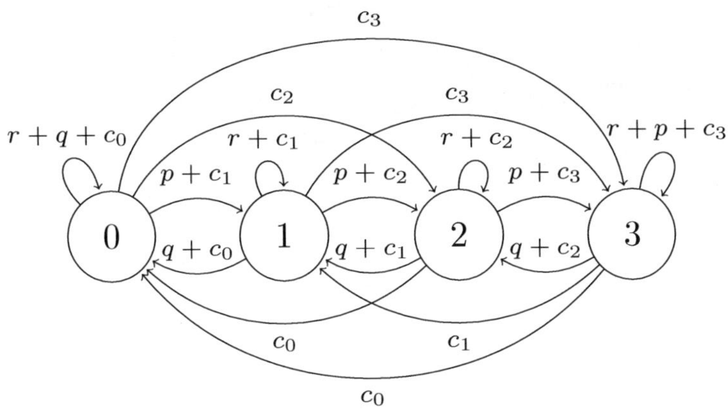

FIGURE 24

where $0 \leq r, c_0, c_1, c_2, c_3 < 1$ and $0 < p, q < 1$ and $p+q+r+c_0+c_1+c_2+c_3 = 1$

$$P = \begin{array}{c} \\ 0 \\ 1 \\ 2 \\ 3 \end{array} \begin{array}{c} \begin{array}{cccc} 0 & 1 & 2 & 3 \end{array} \\ \begin{bmatrix} r+q+c_0 & p+c_1 & c_2 & c_3 \\ q+c_0 & r+c_1 & p+c_2 & c_3 \\ c_0 & q+c_1 & r+c_2 & p+c_3 \\ c_0 & c_1 & q+c_2 & r+p+c_3 \end{bmatrix} \end{array} \tag{4.4}$$

Using (4.1), we calculate the dual of P as given below:

$$P^* = \begin{array}{c} \\ -1 \\ 0 \\ 1 \\ 2 \\ 3 \end{array} \begin{array}{c} \begin{array}{ccccc} -1 & 0 & 1 & 2 & 3 \end{array} \\ \begin{bmatrix} 1 & 0 & 0 & 0 & 0 \\ p+c_1+c_2+c_3 & r & q & 0 & c_0 \\ c_2+c_3 & p & r & q & c_0+c_1 \\ c_3 & 0 & p & r & c_0+c_1+c_2+q \\ 0 & 0 & 0 & 0 & 1 \end{bmatrix} \end{array} \tag{4.5}$$

We will explicitly determine P^k, where $k \in \mathbb{N}$, using two different methods.

Method 1. Notice that the center 3×3 matrix within P^* is $T_{P^*} = \begin{bmatrix} r & q & 0 \\ p & r & q \\ 0 & p & r \end{bmatrix}$, which is a tridiagonal Toeplitz matrix. By (2.11), T_{P^*} has eigenvalues

$$\omega_s = r + 2\sqrt{pq}\cos\left(\frac{s\pi}{n+1}\right) \tag{4.6}$$

where $n = 3$, and $s = 1, 2, 3$. So P^* has the following eigenvalues:

$$1 \quad 1 \quad r + 2\sqrt{pq}\cos\left(\frac{\pi}{4}\right) \quad r + 2\sqrt{pq}\cos\left(\frac{2\pi}{4}\right) \quad r + 2\sqrt{pq}\cos\left(\frac{3\pi}{4}\right)$$

Recall that P and P^* have the same set of eigenvalues, see [**Lyc18**]. The eigenvalues of P are distinct and explicitly known to be:

$$\omega_1 = 1 \quad \omega_2 = r + 2\sqrt{pq}\cos\left(\frac{\pi}{4}\right) \quad \omega_3 = r + 2\sqrt{pq}\cos\left(\frac{2\pi}{4}\right) \quad \omega_4 = r + 2\sqrt{pq}\cos\left(\frac{3\pi}{4}\right)$$

From these known eigenvalues, we can precisely calculate right and left eigenvectors, $\vec{R}_s$ and $\vec{L}_s$. Recall the spectral projectors, A_s, are given by (2.5) as shown below:

$$A_s = c^2 \vec{R}_s \vec{L}_s$$

By (2.1), the explicit transient probabilities of this Markov chain are:

$$P^k = A_1\omega_1^k + A_2\omega_2^k + A_3\omega_3^k + A_4\omega_4^k \tag{4.7}$$

The preceding argument works for any $n \in \mathbb{N}$ because forming P^* will always have a central matrix, which is a tridiagonal Toeplitz matrix. This means P has an eigenvalue of 1 and those eigenvalues given by (4.6) therefore P^k is known from (4.7). Finally, notice that by the Duality Theorem (4.2) that the transient probabilities of the dual process, $(P^*)^k$, of Figure 25 are explicitly known.

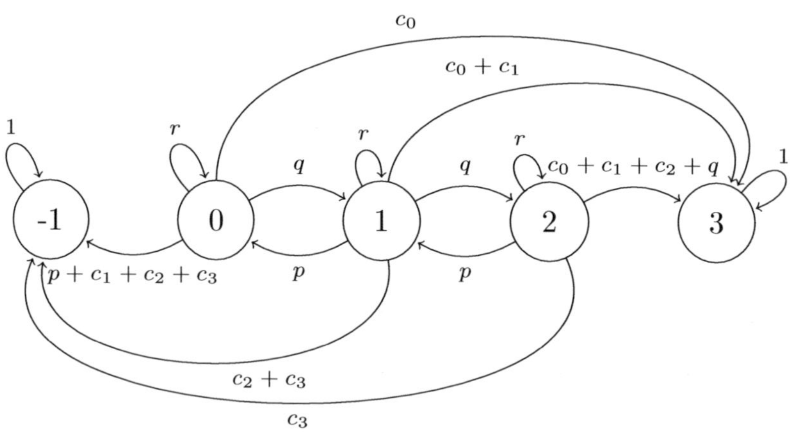

FIGURE 25

Method 2. Notice from Figure 25 that states -1 and 3 are absorbing states, which means that once we visit these states, we will be unable to leave them. We can use this information to help us calculate the entries of $P^{*(k)}$ by considering four different cases.

CASE 1 ($P^{*(k)}(i,\, j)$, where $i, j = 0, 1, 2$).

$$P^{*(k)} = \begin{bmatrix} 1 & 0 & 0 & 0 & 0 \\ - & \begin{bmatrix} P^{*(k)}(0,0) & P^{*(k)}(0,1) & P^{*(k)}(0,2) \\ P^{*(k)}(1,0) & P^{*(k)}(1,1) & P^{*(k)}(1,2) \\ P^{*(k)}(2,0) & P^{*(k)}(2,1) & P^{*(k)}(2,2) \end{bmatrix} & & & - \\ - & & & & - \\ - & & & & - \\ 0 & 0 & 0 & 0 & 1 \end{bmatrix}$$

Since we want to calculate the probability from state i to j, where $i = 0, 1, 2$ and $j = 0, 1, 2$, Notice that the center of the dual matrix P^* is a tridiagonal Toeplitz matrix, hence we can use formula (2.20) to calculate $P^{*(k)}_{i,j}$,$i, j = 0, 1, 2$ of the k-th power of the dual matrix $P^{*(k)}$:

$$P^{*(k)}(i,\, j) = \frac{2}{4}\left(\frac{p}{q}\right)^{\frac{i-j}{2}} \sum_{s=1}^{3} \sin\left(\frac{(i+1)\pi s}{4}\right) \sin\left(\frac{(j+1)\pi s}{4}\right)\left(r + 2\sqrt{pq}\cos\left(\frac{s\pi}{4}\right)\right)^k \tag{4.8}$$

where $i, j = 0, 1, 2$ and $k \in \mathbb{N}$.

CASE 2 ($P^{*(k)}(i, -1)$ where $i = 0, 1, 2$).

$$P^{*(k)} = \begin{bmatrix} 1 & 0 & 0 & 0 & 0 \\ P^{*(k)}(0,-1) & \begin{bmatrix} - & - & - \\ - & - & - \\ - & - & - \end{bmatrix} & & & - \\ P^{*(k)}(1,-1) & & & & - \\ P^{*(k)}(2,-1) & & & & - \\ 0 & 0 & 0 & 0 & 1 \end{bmatrix}$$

To calculate the probability going from state $i = 0, 1, 2$ to state -1, we use the fact that the -1 state is an absorbing state. This means that the probability of going from state i to state -1 is obtained by conditioning upon l, the number of steps taken before moving to state -1, and which state we are at just before we transition to state -1.

$$\begin{aligned} P^{*(k)}(i,\, -1) &= (p + c_1 + c_2 + c_3) \sum_{l=0}^{k-1} P^{*(l)}(i, 0) \\ &\quad + (c_2 + c_3) \sum_{l=0}^{k-1} P^{*(l)}(i, 1) + (c_3) \sum_{l=0}^{k-1} P^{*(l)}(i, 2) \end{aligned} \tag{4.9}$$

where $i = 0, 1, 2$

CASE 3 ($P^{*(k)}(i, 3)$ where $i = 0, 1, 2$).

$$P^{*(k)} = \begin{bmatrix} 1 & 0 & 0 & 0 & 0 \\ - & \begin{bmatrix} - & - & - \\ - & - & - \\ - & - & - \end{bmatrix} & & & P^{*(k)}(0,3) \\ - & & & & P^{*(k)}(1,3) \\ - & & & & P^{*(k)}(2,3) \\ 0 & 0 & 0 & 0 & 1 \end{bmatrix}$$

We can then calculate the probability of going from state i to state j in k steps, where $i = 0$ or 1 or 2, by taking the complement of being at the following states: -1, 0, 1, 2 after k steps. These probabilities were already calculated in Case 1 and Case 2. Thus:

$$P^{*(k)}(i,\, 3) = 1 - \sum_{s=-1}^{2} P^{*(k)}(i,\, s)$$

CASE 4 ($P^{*(k)}(i, j)$, where $i = -1, 3$ and $j = -1, 0, 1, 2, 3$). Let i be state -1 or 3. Because these are absorbing states, $P^{*(k)}(i, i) = 1$ and $P^{*(k)}(i, j) = 0$, when $i \neq j$.

Since in Method 2, we have solved for $P^{*(k)}$, we can use the Duality Theorem (4.3) to calculate each element of our $P^{(k)}$ matrix as shown below:

$$P^{(k)}(i, j) = \sum_{s=i}^{H} \left[P^{*(k)}(j, s) - P^{*(k)}(j-1, s) \right]$$

for $n \geq 0$ and for all states $i, j = 0, 1, 2, \ldots, H$ with the conventions:

- $P^{*(k)}(i, H) = 1 - \sum_{s=0}^{H-1} P^{*(k)}(i, s)$
- $P^{(k)}(-1, s) = 0$ if $s > -1$
- $P^{*(k)}(H, H) = 1$

REMARK 4.2. There are two special cases that deserve special mention:

CASE 1 ($c_0 = c_1 = c_2 = c_3 = 0$ and $0 < p, q < 1$ and $0 \leq r < 1 \quad p + q + r = 1$). In this case, Figure 24 becomes Figure 26

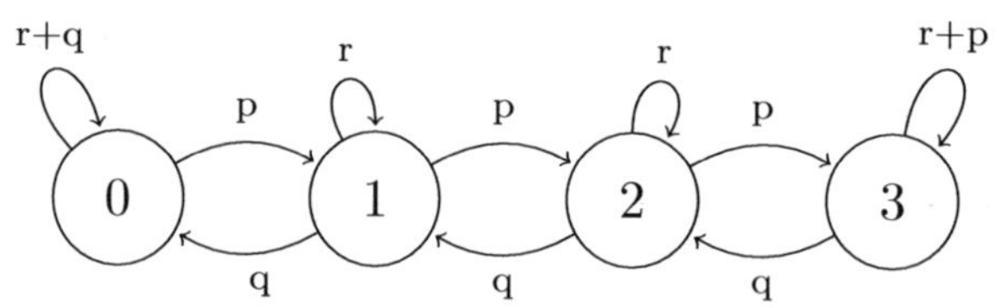

FIGURE 26

CASE 2 ($c_1 = c_2 = c_3 = 0$ and $c_0 = c$ and $0 < p, q, c < 1$ and $0 \leq r < 1$ and $p + q + r + c = 1$). In this case, Figure 24 becomes Figure 27:

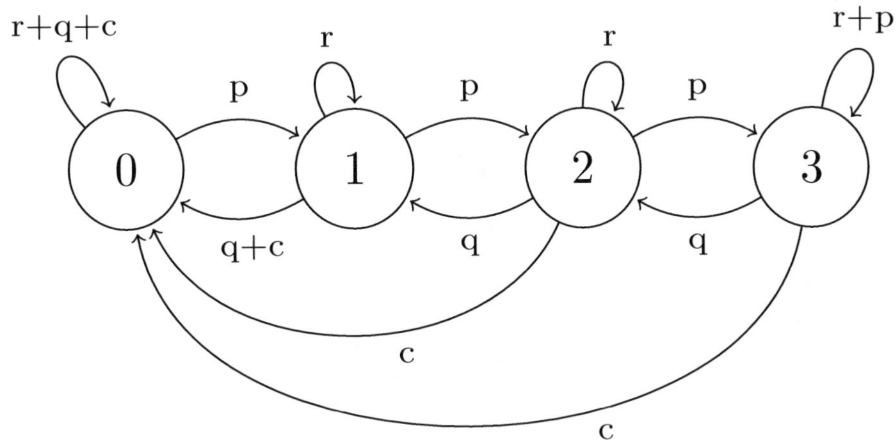

FIGURE 27

So the transient probabilities of Markov chains having transitions as shown in Figures 26 and 27 are explicitly known according to Sylvester's eigenvalue expansion (2.1). That is, the eigenvalues of P corresponding to Figures 26 and 27 are given by (4.6) along with the eigenvalue of 1, and the spectral projectors A_s can be explicitly calculated according to (2.5).

REMARK 4.3. An alternative problem of interest is captured by the dual Markov chain pictured in Figure 25. Notice that the transition diagram of Figure 25 has transition probability matrix (4.5). In particular, states -1 and 3 are absorbing states. We can assume that we start at a state $i = 0, 1$ or 2 and ask: what is the probability of being absorbed at state -1 after k units of time? The solution of this problem may be thought of as a finite-time gambler's ruin probability. By Method 1, we know the eigenvalues of the dual chain P^*, and therefore the eigenvalues of the original chain P. By (4.7) we know the entries of P^k, so by the Duality Theorem (4.2), we know the entries of $(P^*)^k$. In particular, we know the finite-time gambler's ruin probability $P^{*(k)}(i, -1)$, see [**HKN08**] and [**Lor17**].

REMARK 4.4. Even though the preceding explanations and remarks were presented for a Markov chain having 4 states, the preceding methods, results and remarks all hold for Markov chains having transition probability diagram looking like Figure 24 but having n states where $n = 3, 4, 5, \ldots$.

4.3. Transient probabilities of more general Markov processes. Consider a four state Markov process, having the transitions as shown in Figure 28. States 0, 1, 2 allow a 1-step upward *birth* rate λ; states 1, 2, 3 allow 1-step downward *death* rate μ. This process also has catastrophe-like rates γ_0, γ_1, γ_2, γ_3 that transition to the states 0, 1, 2, 3 respectively from anywhere in the state space. All of these conditions generalize and scale up naturally to an analogous Markov process on state space $\mathcal{S} = \{0, 1, 2, \ldots, n-1\}$.

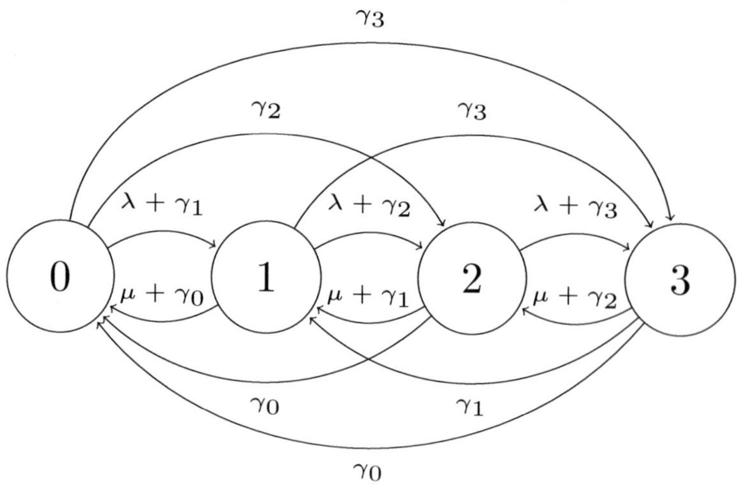

FIGURE 28. State Rate Transition Diagram I

where $0 \leq \gamma_0, \gamma_1, \gamma_2, \gamma_3$ and $0 < \lambda, \mu$ and $k = \lambda + \mu + \gamma_0 + \gamma_1 + \gamma_2 + \gamma_3$

$$Q = \begin{array}{c} \\ 0 \\ 1 \\ 2 \\ 3 \end{array} \begin{array}{c} \begin{array}{cccc} 0 & 1 & 2 & 3 \end{array} \\ \begin{bmatrix} \gamma_0 + \mu - k & \lambda + \gamma_1 & \gamma_2 & \gamma_3 \\ \mu + \gamma_0 & \gamma_1 - k & \lambda + \gamma_2 & \gamma_3 \\ \gamma_0 & \mu + \gamma_1 & \gamma_2 - k & \lambda + \gamma_3 \\ \gamma_0 & \gamma_1 & \mu + \gamma_2 & \lambda + \gamma_3 - k \end{bmatrix} \end{array} \tag{4.10}$$

Using (4.1) for Markov processes, we calculate the dual of Q as given below:

$$Q^* = \begin{array}{c} \\ -1 \\ 0 \\ 1 \\ 2 \\ 3 \end{array} \begin{array}{c} \begin{array}{ccccc} -1 & 0 & 1 & 2 & 3 \end{array} \\ \begin{bmatrix} 0 & 0 & 0 & 0 & 0 \\ \lambda + \gamma_1 + \gamma_2 + \gamma_3 & -k & \mu & 0 & \gamma_0 \\ \gamma_2 + \gamma_3 & \lambda & -k & \mu & \gamma_0 + \gamma_1 \\ \gamma_3 & 0 & \lambda & -k & \gamma_0 + \gamma_1 + \gamma_2 + \mu \\ 0 & 0 & 0 & 0 & 0 \end{bmatrix} \end{array} \tag{4.11}$$

There are different methods to determine e^{Qt}, where $t \geq 0$. For simplicity, we use a modification of Method 1 in Section 4.2. Notice that the center 3×3 matrix within Q^* is $T_{Q^*} = \begin{bmatrix} -k & \mu & 0 \\ \lambda & -k & \mu \\ 0 & \lambda & -k \end{bmatrix}$, which is a tridiagonal Toeplitz matrix. By (2.11), T_{Q^*} has eigenvalues:

$$\omega_s = -k + 2\sqrt{\lambda\mu}\cos\left(\frac{s\pi}{n+1}\right) \tag{4.12}$$

where $n = 3$, $s = 1, 2, 3$, and $k = \lambda + \mu + \gamma_0 + \gamma_1 + \gamma_2 + \gamma_3$. So Q^* has eigenvalues:

$$0 \quad 0 \quad -k + 2\sqrt{\lambda\mu}\cos\left(\frac{\pi}{4}\right) \quad -k + 2\sqrt{\lambda\mu}\cos\left(\frac{2\pi}{4}\right) \quad -k + 2\sqrt{\lambda\mu}\cos\left(\frac{3\pi}{4}\right)$$

As before, Q and Q^* have the same set of eigenvalues, see [**Lyc18**]. The eigenvalues of Q are distinct and explicitly known to be:

$$\omega_1 = 0 \quad \omega_2 = -k + 2\sqrt{\lambda\mu}\cos\left(\frac{\pi}{4}\right)$$

$$\omega_3 = -k + 2\sqrt{\lambda\mu}\cos\left(\frac{2\pi}{4}\right) \quad \omega_4 = -k + 2\sqrt{\lambda\mu}\cos\left(\frac{3\pi}{4}\right)$$

From these known eigenvalues, we can precisely calculate right and left eigenvectors, $\vec{R}_s$ and $\vec{L}_s$. Recall the spectral projectors, A_s, are given by (2.5) as shown below:

$$A_s = c^2\vec{R}_s\vec{L}_s \tag{4.13}$$

By (2.3), the explicit transient probabilities of this Markov process are:

$$e^{Qt} = A_1e^{\omega_1 t} + A_2e^{\omega_2 t} + A_3e^{\omega_3 t} + A_4e^{\omega_4 t} \tag{4.14}$$

The preceding argument works for any $n = 3, 4, 5, \ldots$ because forming Q^* will always have a central matrix, which is a tridiagonal Toeplitz matrix. This means Q has 0 as an eigenvalue and the other eigenvalues given by (4.12).

REMARK 4.5. Once again, consider the two following special cases:

CASE 1 ($\gamma_0 = \gamma_1 = \gamma_2 = \gamma_3 = 0$ and $0 < \lambda$, μ and $k = \lambda + \mu$). In this case, Figure 28 becomes Figure 29:

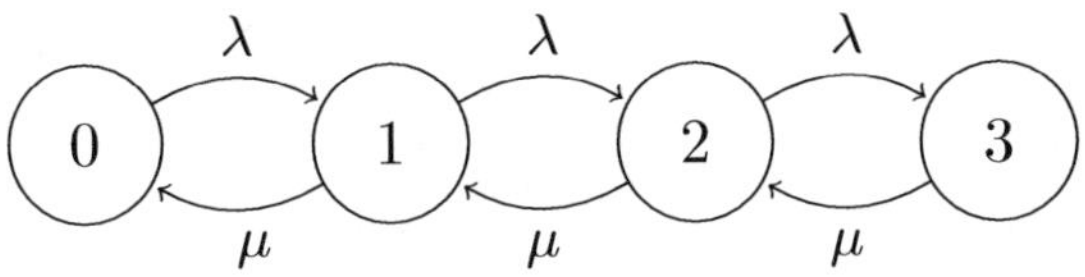

FIGURE 29

The birth-death process depicted in Figure 29 is also known as the single server queueing system with capacity 3. In the queueing literature, it is denoted as the M/M/1/3 queueing system, see [**STGH18**].

CASE 2 ($\gamma_1 = \gamma_2 = \gamma_3 = 0$ and $\gamma_0 = \gamma$ and $0 < \lambda, \mu, \gamma$ and $k = \lambda + \mu + \gamma$). In this case, Figure 28 becomes Figure 30:

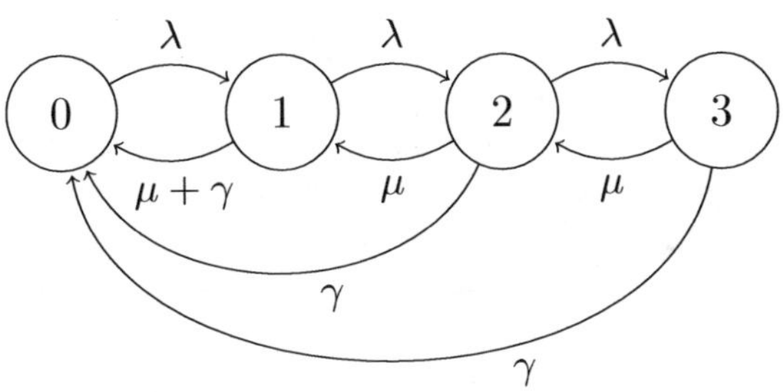

FIGURE 30

The Markov process shown in Figure 30 is often called the single server queueing system having capacity 3 with constant catastrophes, see [**KMR04**] or [**KRM05**]. So the transient probabilities of the Markov processes having transition rates as shown in Figure 29 and Figure 30 are explicitly known according to Sylvester's eigenvalue expansion (4.14). That is, the eigenvalues of Q corresponding to Figure 29 and Figure 30 are given by (4.12) along with the eigenvalue of 0, and the spectral projectors A_s can be explicitly calculated according to (4.13). Note that the eigenvalues of the Q matrix of Figure 30 are just the eigenvalues of the Q matrix of Figure 29 translated by the catastrophe rate γ. It would be fun to explore if the spectral projectors of each Markov process shown in Figures 29 and 30 are also related to each other as a simple function of γ.

REMARK 4.6. Since the dual matrix of the process shown in Figure 28 is given in (4.11), we can picture the transition rates of this dual process as shown in Figure 31. Assume we start at state $i = 0, 1$, or 2, what is the probability of being absorbed at state -1 after time t? This problem may be considered a continuous-time, gambler's ruin problem.

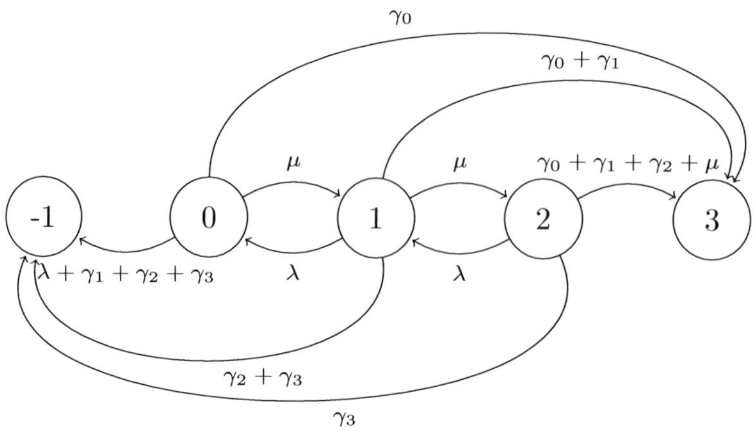

FIGURE 31

By our preceding discussion we can identify the eigenvalues of Q^* and thereby identify the eigenvalues of Q. This means we can determine the entries of e^{Qt}, and thereby the entries of e^{Q^*t} using the Duality Theorem. So, we can find $e^{Q^*t}(i, -1)$, which is the ruin probability of going from state i to state -1 in Figure 31 in time t.

REMARK 4.7. The preceding methods and remarks work in general for Markov processes having n states. In Sections 4.2 and 4.3, we have shown a method to compute the explicit n-step transient probabilities of Markov chains having the structure of Figure 24 and the transient probability functions of Markov processes having the structure of Figure 28. Our explicit eigenvalue solution forms are (1.1) and (1.2) where the eigenvalues are the closed formulas for eigenvalues of certain tridiagonal, Toeplitz matrices.

The results in Section 4.3 apply directly to computing the explicit transient probability functions of the M/M/1/K queueing system and for the M/M/1/K system that also has constant catastrophe rates γ to 0. These results appear in Sections

4 and 5 of [**KRM05**]. In that 2005 article, the eigenvalues and spectral projectors were found by using lattice path combinatorics to count sample paths. The transient probability functions of the M/M/1/K queueing system continue to stimulate research interest, see, for example, [**EHMP21**] and the thorough references within this interesting article.

The material in Section 4.2 of our current article complements the Markov process results of [**KRM05**]. Duality theory plays a key role in finding the eigenvalues. Once the eigenvalues are known, there are linear algebraic and probabilistic ways to calculate the spectral projectors see Methods 1 and 2 in Section 4.2. For Markov processes having an infinite number of states, see [**KMR04**] and [**KRM05**].

5. Odd tridiagonal matrices having constant main diagonal entries and alternating entries on the remaining diagonals

We now consider tridiagonal matrices, P, having dimension $n = 2m+1$, where $m = 1, 2, 3, \ldots$. Assume that the subdiagonal and the superdiagonal entries are alternating as shown in (5.1).

$$P = \begin{bmatrix} r & p_0 & 0 & 0 & \ldots & 0 & 0 \\ q_1 & r & p_1 & 0 & \ldots & 0 & 0 \\ 0 & q_2 & r & p_0 & \ldots & 0 & 0 \\ 0 & 0 & q_1 & r & p_1 & \ldots & 0 \\ \vdots & \vdots & \vdots & \ddots & \ldots & \vdots & \vdots \\ 0 & 0 & \ldots & 0 & 0 & q_2 & r \end{bmatrix} \tag{5.1}$$

and suppose $q_1\, p_0 = {d_1}^2$, $q_2\, p_1 = {d_2}^2$, where $d_1 \neq 0 \neq d_2$.

THEOREM 5.1. [**Kou06**, pg. 124] *For $n = 2m+1$, where $m = 1, 2, 3, \ldots$, suppose matrix P given by (5.1). Then P has distinct eigenvalues given by*

$$\omega_k = \begin{cases} r + \sqrt{d_1^2 + d_2^2 + 2d_1 d_2 \cos(\theta_k)}, & \text{if } k = 1, 2, \ldots, m \\ r - \sqrt{d_1^2 + d_2^2 + 2d_1 d_2 \cos(\theta_k)}, & \text{if } k = m+1, m+2, \ldots, 2m \\ r, & \text{if } k = n \end{cases} \tag{5.2}$$

and the corresponding eigenvectors $\vec{R}_1, \vec{R}_2, \ldots, \vec{R}_n$ are given below:

CASE 1. *The kth eigenvector, where $k = 1, 2, 3, \ldots, n-1$, is given by*

$$\vec{R}_k(j) = \begin{cases} F_j d_1 d_2 \sin\left[\left(\frac{n-j}{2} + 1\right)\theta_k\right] + d_1^2 \sin\left[\left(\frac{n-j}{2}\right)\theta_k\right], & \text{if } j = 1, 3, 5, \ldots, n \\ F_j \sqrt{d_1 d_2}(r - \lambda_k)\ \sin\left[\left(\frac{n-j+1}{2}\right)\theta_k\right], & \text{if } j = 2, 4, 6, \ldots, n-1 \end{cases}$$

where F_j and θ_k can be expressed as

$$F_j = \begin{cases} (-\sqrt{d_1 d_2})^{(n-j)} (q_1\, q_2)^{\frac{j-1}{2}}, & \text{for } j = 1, 3, 5, \ldots \\ (-\sqrt{d_1 d_2})^{(n-j)} q_1^{\frac{j}{2}}\, q_2^{\frac{j}{2}-1}, & \text{for } j = 2, 4, 6, \ldots \end{cases} \tag{5.3}$$

$$\theta_k = \begin{cases} \frac{2k\pi}{(n+1)} & \textit{for } k = 1, 2, \ldots, m \\ \\ \frac{2(k-m)\pi}{(n+1)} & \textit{for } k = m+1, m+2, \ldots, 2m \end{cases} \tag{5.4}$$

whenever $k = 1, 2, \ldots, 2m$.

CASE 2. *The nth eigenvector is given by*

$$\vec{R}_n(j) = \begin{cases} (q_1\, q_2)^{\frac{j-1}{2}}\, (-d_2^2)^{\frac{n-j}{2}} & \textit{for } j = 1, 3, 5, \ldots, n \\ \\ 0 & \textit{for } j = 2, 4, 6, \ldots, n-1 \end{cases} \tag{5.5}$$

REMARK 5.1. Theorem 5.1 is a special case of Theorem 3.1 in [**Kou06**], proved on page 124. Kouachi's Theorem 3.1 applies more generally for real number entries of p, q and r. However, since our paper is interested in transition probability matrices, we often set p, q, and r to be probabilities.

REMARK 5.2. To calculate A_k in (2.1). Let $\vec{R}_k$ be the right eigenvectors of P as given in Theorem 5.1, and let $\vec{L}_k$ be the left hand eigenvectors of P. We assume that $\vec{R}_k$ and $\vec{L}_k$ are normalized, which means $c^2\vec{L}_k \cdot \vec{R}_k = 1$ for all $k = 1, 2, \ldots, n$. Then from (2.5), we know that $A_k = c^2\vec{R}_k\vec{L}_k$ for $k = 1, 2, \ldots, n$.

Although Theorem 5.1 calculates the right eigenvectors of matrix P, we can use a slight modification of this theorem to find the left eigenvectors of P. Using the definition of the left eigenvector, we know for the kth eigenvalue

$$\vec{L}_k P = \lambda_k \vec{L}_k$$

Transposing both sides of this equation

$$(\vec{L}_k P)^T = (\lambda_k \vec{L}_k)^T$$

simplifying the preceding equation, we obtain:

$$P^T(\vec{L}_k)^T = \lambda_k \vec{L}_k^T$$

This equation shows that the kth right eigenvector of matrix P^T is equal to the transpose of the kth left eigenvector of matrix P.

P^T is also a tridiagonal matrix with the q's and p's switching places. Hence, Theorem 5.1 can be applied to calculate the eigenvalues and the eigenvectors of matrix P^T. It's known that the eigenvalues of P and P^T are the same. $\vec{R}_k$ is given in Theorem (5.1). $\vec{L}_k$ can be determined using Theorem 5.1 for P^T theorem with the following minor differences:

(1) (5.3) becomes

$$F_j = \begin{cases} (-\sqrt{d_1 d_2})^{(n-j)}(p_0\, p_1)^{\frac{j-1}{2}}, & \text{for } j = 1, 3, 5, \ldots \\ \\ (-\sqrt{d_1 d_2})^{(n-j)} p_0^{\frac{j}{2}}\, p_1^{\frac{j}{2}-1}, & \text{for } j = 2, 4, 6, \ldots \end{cases}$$

(2) (5.5) becomes

$$\vec{L}_n^T(j) = \begin{cases} (p_0\, p_1)^{\frac{j-1}{2}}\, (-d_2^2)^{\frac{n-j}{2}} & \text{for } j = 1, 3, 5, \ldots, n \\ \\ 0 & \text{for } j = 2, 4, 6, \ldots, n-1 \end{cases}$$

REMARK 5.3. Our Cal Poly Pomona research group wrote a program that calculates the eigenvalues and eigenvectors based upon the algorithm of Theorem 5.1. In the next few examples, we will apply our program and (2.5) using MATLAB to solve for various numerical and symbolic matrix expressions.

EXAMPLE 5.2 (Generalized Ballot Box Problem). Suppose candidate A receives 2 votes and candidate B receives 2 votes. Calculate the probability of going from state $i = 0$ to state $j = 0$ in $n = 4$ steps so that A never falls behind B throughout the counting of the ballots.

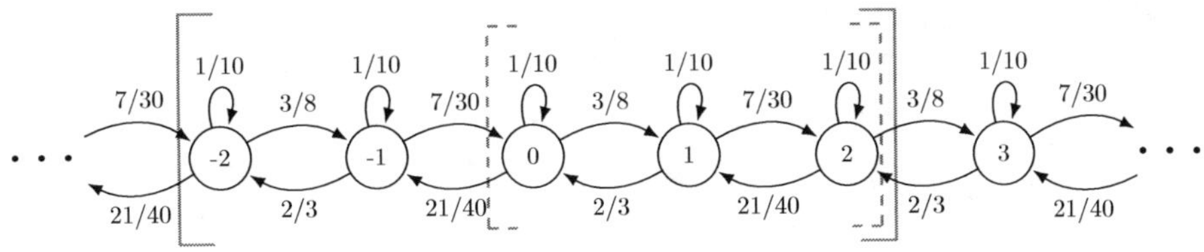

$$\mathcal{G} = \begin{array}{c} \\ 0 \\ 1 \\ 2 \end{array}\begin{array}{c} \begin{array}{ccc} 0 & 1 & 2 \end{array} \\ \begin{bmatrix} \frac{1}{10} & \frac{3}{8} & 0 \\ \frac{2}{3} & \frac{1}{10} & \frac{7}{30} \\ 0 & \frac{21}{40} & \frac{1}{10} \end{bmatrix} \end{array} \qquad \mathcal{A} = \begin{array}{c} \\ -2 \\ -1 \\ 0 \\ 1 \\ 2 \end{array}\begin{array}{c} \begin{array}{ccccc} -2 & -1 & 0 & 1 & 2 \end{array} \\ \begin{bmatrix} \frac{1}{10} & \frac{3}{8} & 0 & 0 & 0 \\ \frac{2}{3} & \frac{1}{10} & \frac{7}{30} & 0 & 0 \\ 0 & \frac{21}{40} & \frac{1}{10} & \frac{3}{8} & 0 \\ 0 & 0 & \frac{2}{3} & \frac{1}{10} & \frac{7}{30} \\ 0 & 0 & 0 & \frac{21}{40} & \frac{1}{10} \end{bmatrix} \end{array}$$

By Theorem 5.1, the eigenvalues and eigenvectors of $\mathcal{G}$ are known. The spectral projectors are found using (2.5). The Sylvester eigenvalue expansion of $\mathcal{G}^4$ produces:

$$\mathcal{G}^4(0,0) = \sum_{s=1}^{3} \tilde{A}_s(0,0)\tilde{\omega}_s^4 = \tilde{A}_1(0,0)\tilde{\omega}_1^4 + \tilde{A}_2(0,0)\tilde{\omega}_2^4 + \tilde{A}_3(0,0)\tilde{\omega}_3^4$$

$$= \frac{50}{149}\left(\frac{\sqrt{149}}{20} + \frac{1}{10}\right)^4 + \frac{50}{149}\left(\frac{1}{10} - \frac{\sqrt{149}}{20}\right)^4 + \frac{49}{149}\left(\frac{1}{10}\right)^4 \approx 0.1082$$

By Theorem 5.1, the eigenvalues and eigenvectors of $\mathcal{A}$ are known. The spectral projectors are found using (2.5). The Sylvester eigenvalue expansion of $\mathcal{A}^4$ produces:

$$\mathcal{A}^4(0,0) = \sum_{s=1}^{5} A_s(0,0)\omega_s^4$$

$$= A_1(0,0)\omega_1^4 + A_2(0,0)\omega_2^4 + A_3(0,0)\omega_3^4 + A_4(0,0)\omega_4^4 + A_5(0,0)\omega_5^4$$

$$= \frac{289}{876}\left(\frac{\sqrt{219}}{20} + \frac{1}{10}\right)^4 + \frac{9}{316}\left(\frac{\sqrt{79}}{20} + \frac{1}{10}\right)^4$$

$$+ \frac{289}{876}\left(\frac{1}{10} - \frac{\sqrt{219}}{20}\right)^4 + \frac{9}{316}\left(\frac{1}{10} - \frac{\sqrt{79}}{20}\right)^4 + \frac{4900}{17301}\left(\frac{1}{10}\right)^4 \approx 0.2225$$

So the solution of the ballot box problem is $\frac{\mathcal{G}^4(0,0)}{\mathcal{A}^4(0,0)} \approx 0.4865$

Corollary 5.2.1. *(A) Consider the birth-death chain having the following state diagram and transition probability matrix P where H is an odd number:*

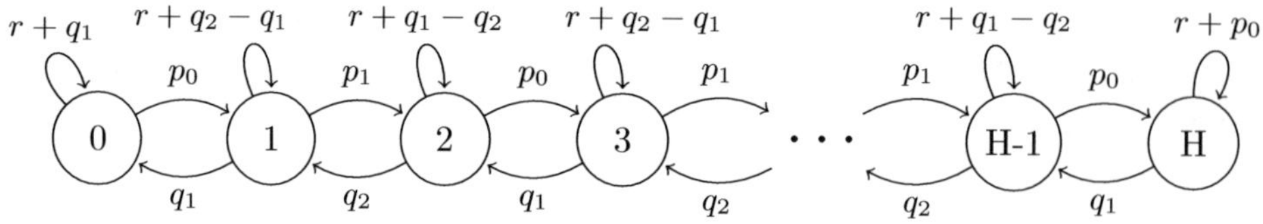

Figure 32. Alternating p's and q's, H is odd.

$$(5.6)\quad P=\begin{bmatrix} r+q_1 & p_0 & 0 & \cdots & 0 & 0 & 0\\ q_1 & r+q_2-q_1 & p_1 & \cdots & 0 & 0 & 0\\ 0 & q_2 & r+q_1-q_2 & \cdots & 0 & 0 & 0\\ \vdots & \vdots & \vdots & \ddots & \vdots & \vdots & \vdots\\ 0 & 0 & 0 & \cdots & r+q_2-q_1 & p_1 & 0\\ 0 & 0 & 0 & \cdots & q_2 & r+q_1-q_2 & p_0\\ 0 & 0 & 0 & \cdots & 0 & q_1 & r+p_0 \end{bmatrix}$$

where $0 < q_1,\, q_2,\, r,\, p_0,\, p_1 < 1,\ \ q_1+r+p_0 = 1,\ \ q_2+r+p_1 = 1,$ *and* $r > |q_2 - q_1|$. *Then* $P^k = A_1\omega_1^k + A_2\omega_2^k + A_3\omega_3^k + \cdots + A_n\omega_n^k$ *where the eigenvalues,* ω_i*, which come from Theorem* 5.1 *and the spectral projectors,* A_i*, which can be found from* (4.13).

(B) Suppose a Markov process has state rate diagram Figure 33 *and transition rate matrix Q as shown below. H, once again, is assumed to be an odd number.*

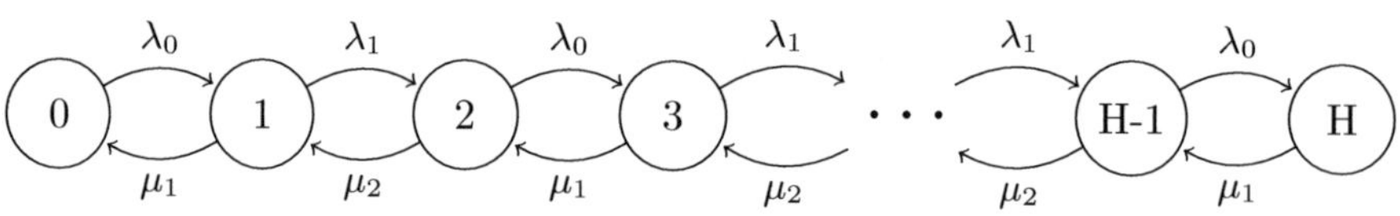

Figure 33. Alternating λ's and μ's, H is odd.

$$(5.7)\qquad Q=\begin{bmatrix} -\lambda_0 & \lambda_0 & 0 & \cdots & 0 & 0 & 0\\ \mu_1 & -\lambda_1-\mu_1 & \lambda_1 & \cdots & 0 & 0 & 0\\ 0 & \mu_2 & -\lambda_0-\mu_2 & \cdots & 0 & 0 & 0\\ \vdots & \vdots & \vdots & \ddots & \vdots & \vdots & \vdots\\ 0 & 0 & 0 & \cdots & -\lambda_1-\mu_1 & \lambda_1 & 0\\ 0 & 0 & 0 & \cdots & \mu_2 & -\lambda_0-\mu_2 & \lambda_0\\ 0 & 0 & 0 & \cdots & 0 & \mu_1 & -\mu_1 \end{bmatrix}$$

where $0 < \mu_1,\, \mu_2,\, \lambda_0,\, \lambda_1$ *and* $\lambda_0 + \mu_1 = \lambda_1 + \mu_2$.

Then $e^{Qt} = A_1e^{\omega_1 t} + A_2e^{\omega_2 t} + A_3e^{\omega_3 t} + \cdots + A_ne^{\omega_n t}$ *where the eigenvalues,* ω_i, *come from Theorem* 5.1 *and the spectral projectors,* A_i, *which can be found from* (4.13).

PROOF. We present the idea of the proof for $H = 3$, the general proof follows similarly for H being any odd number.

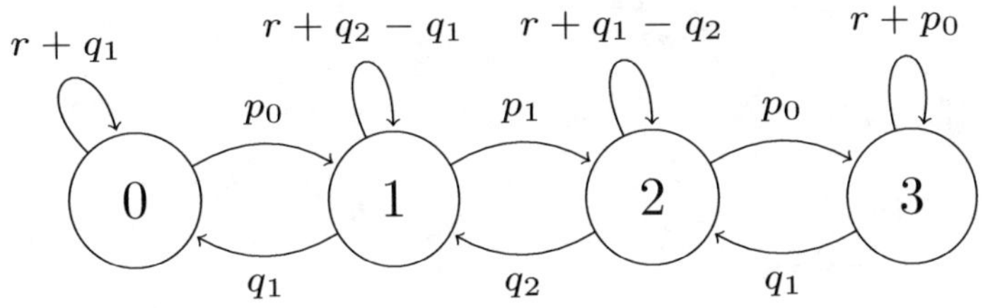

FIGURE 34

The following transition matrix, P, corresponds to Figure 34, with $0 < q_1, q_2, r, p_0, p_1 < 1, \quad q_1 + r + p_0 = 1, \quad q_2 + r + p_1 = 1, \quad r > |q_2 - q_1|$:

$$P = \begin{bmatrix} r+q_1 & p_0 & 0 & 0 \\ q_1 & r+q_2-q_1 & p_1 & 0 \\ 0 & q_2 & r+q_1-q_2 & p_0 \\ 0 & 0 & q_1 & r+p_0 \end{bmatrix}$$

The dual of matrix P^* is as follows:

$$P^* = \begin{bmatrix} 1 & 0 & 0 & 0 & 0 \\ p_0 & r & q_1 & 0 & 0 \\ 0 & p_1 & r & q_2 & 0 \\ 0 & 0 & p_0 & r & q_1 \\ 0 & 0 & 0 & 0 & 1 \end{bmatrix}$$

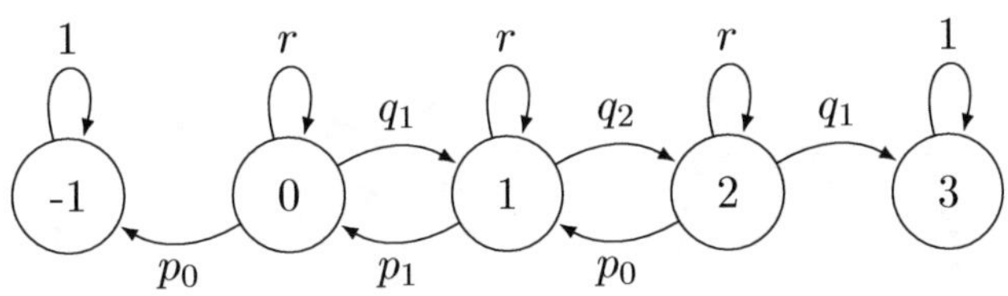

FIGURE 35

The center matrix of P^* satisfies Theorem 5.1, and therefore has eigenvalues given by (5.2). Since P and P^* have the same set of eigenvalues, see [**Lyc18**], we know the eigenvalues of P. Using (4.13) we can determine the spectral projectors of P. And (4.7) gives us the desired formula for P^k. The proof of part (B) follows in a similar way. □

REMARK 5.4. Along the lines of Figure 24 in Section 4.2, we can generalize Corollary 5.2.1 to a family of Markov chains having catastrophe-like transitions with H being odd, shown in Figure 36 when $H = 3$.

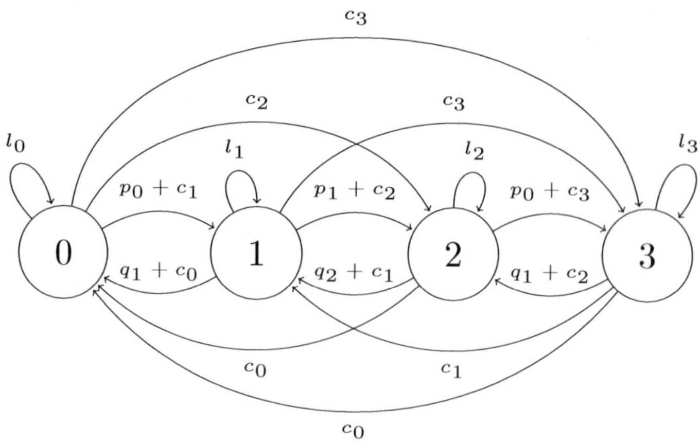

FIGURE 36. Alternating Transition Probabilities Diagram

where: $l_0 = r + q_1 + c_0$ and $l_1 = r + q_2 - q_1 + c_1$ and $l_2 = r + q_1 - q_2 + c_2$

$l_3 = r + p_0 + c_3$ and $0 < q_1,\, q_2,\, r,\, p_0,\, p_1 < 1$

$q_1 + r + p_0 + c_0 + c_1 + c_2 + c_3 = 1$ and $q_2 + r + p_1 + c_0 + c_1 + c_2 + c_3 = 1$

Assume $q_1 \le q_2 + r + c_1$ and $q_2 \le q_1 + r + c_2$.

Let P be the transition probability matrix of the Markov chain shown in Figure 36, then

$$P = \begin{bmatrix} r + q_1 + c_0 & p_0 + c_1 & c_2 & c_3 \\ q_1 + c_0 & r + q_2 - q_1 + c_1 & p_1 + c_2 & c_3 \\ c_0 & q_2 + c_1 & r + q_1 - q_2 + c_2 & p_0 + c_3 \\ c_0 & c_1 & q_1 + c_2 & r + p_0 + c_3 \end{bmatrix}$$

The dual of P is then

$$P^* = \begin{bmatrix} 1 & \begin{matrix} 0 & 0 & 0 \end{matrix} & 0 \\ p_0 + c_1 + c_2 + c_3 & \multirow{3}{*}{$\begin{bmatrix} r & q_1 & 0 \\ p_1 & r & q_2 \\ 0 & p_0 & r \end{bmatrix}$} & c_0 \\ c_2 + c_3 & & c_0 + c_1 \\ c_3 & & q_1 + c_0 + c_1 + c_2 \\ 0 & \begin{matrix} 0 & 0 & 0 \end{matrix} & 1 \end{bmatrix}$$

Once again, the center matrix of P^* satisfies Theorem 5.1, and therefore has eigenvalues given by (5.2). As before, since P and P^* have the same set of eigenvalues, see [**Lyc18**], we know the eigenvalues of P. Using (4.13) we can determine the spectral projectors of P. And (4.7) gives us the desired formula for P^k. Once again, Theorem 5.1 still applies for the analogous Markov process with H being odd and having alternating transition rates and catastrophe-like rates.

THEOREM 5.3.

CASE 1. *Suppose a birth-death chain has the following state transition diagram and transition matrix as given below:*

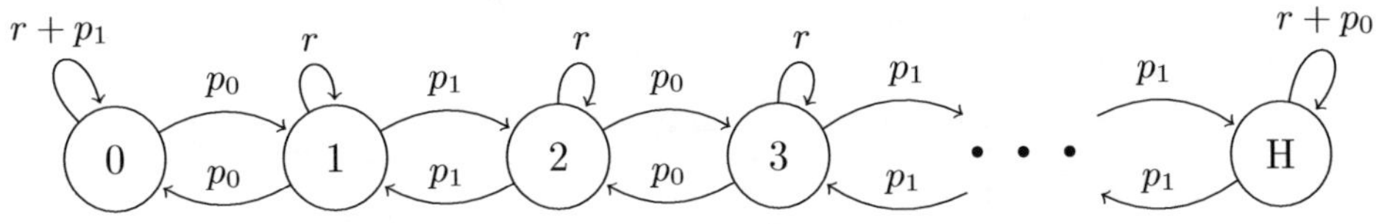

FIGURE 37. Transition probability diagram with alternating p's.

where H is even, $0 < p_0$, p_1 and $p_0 + r + p_1 = 1$.

$$P_1 = \begin{bmatrix} r+p_1 & p_0 & 0 & \dots & 0 & 0 & 0 \\ p_0 & r & p_1 & \dots & 0 & 0 & 0 \\ 0 & p_1 & r & \dots & 0 & 0 & 0 \\ \vdots & \vdots & \vdots & \ddots & \vdots & \vdots & \vdots \\ 0 & 0 & 0 & \dots & r & p_0 & 0 \\ 0 & 0 & 0 & \dots & p_0 & r & p_1 \\ 0 & 0 & 0 & \dots & 0 & p_1 & r+p_0 \end{bmatrix} \tag{5.8}$$

CASE 2. *Suppose a birth-death chain has the following state transition diagram and transition matrix as given below:*

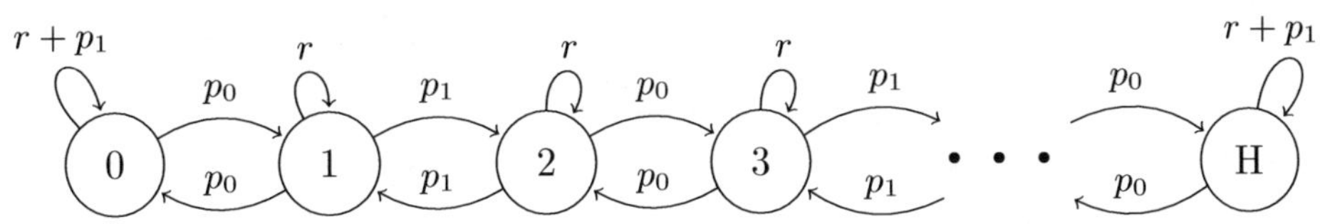

FIGURE 38. Transition probability diagram with alternating p's.

where H is odd, $0 < p_0$, p_1 and $p_0 + r + p_1 = 1$.

$$
(5.9)\qquad P_2 = \begin{bmatrix} r+p_1 & p_0 & 0 & \dots & 0 & 0 & 0 \\ p_0 & r & p_1 & \dots & 0 & 0 & 0 \\ 0 & p_1 & r & \dots & 0 & 0 & 0 \\ \vdots & \vdots & \vdots & \ddots & \vdots & \vdots & \vdots \\ 0 & 0 & 0 & \dots & r & p_1 & 0 \\ 0 & 0 & 0 & \dots & p_1 & r & p_0 \\ 0 & 0 & 0 & \dots & 0 & p_0 & r+p_1 \end{bmatrix}
$$

Then in either case of H being even or odd the eigenvalues of P_1 and P_2 are explicitly known and we can find the Sylvester Eigenvalue Expansion (1.1) *for P_1^k and P_2^k.*

PROOF. The dual matrix of P_1 is

$$
P_1^* = \begin{bmatrix} 1 & 0 & 0 & \dots & 0 & 0 & 0 \\ p_0 & 1-2p_0 & p_0 & \dots & 0 & 0 & 0 \\ 0 & p_1 & 1-2p_1 & \dots & 0 & 0 & 0 \\ \vdots & \vdots & \vdots & \ddots & \vdots & \vdots & \vdots \\ 0 & 0 & 0 & \dots & 1-2p_0 & p_0 & 0 \\ 0 & 0 & 0 & \dots & p_1 & 1-2p_1 & p_1 \\ 0 & 0 & 0 & \dots & 0 & 0 & 1 \end{bmatrix}
$$

Excluding the known eigenvalue of $\omega_1 = 1$, the remaining eigenvalues of P_1 are identical to the eigenvalues of the central matrix of P_1^*. The main, sub, and superdiagonals of the central matrix of P_1^* have alternating entries and satisfy Theorem 2 of Kouachi's 2008 article, which is reproduced below as Theorem 5.4 for the reader's convenience. Therefore the eigenvalues of P_1 are explicitly known, and the spectral projectors can then be determined by (2.5) and therefore the Sylvester Eigenvalue Expansion (1.1) of P_1 is known. The same argument produces the Sylvester Eigenvalue Expansion of P_2. □

REMARK 5.5. Note if $p_0 = 0.6$ and $p_1 = 0.3$, then P_1^* has some negative entries on the main diagonal. So P_1^* is not stochastic however by Theorem 5.4, we are still able to find the eigenvalues of P_1^*, and therefore the eigenvalues of P_1.

REMARK 5.6. In Theorem 5.3, the transition diagram of P_1^* is shown below when $0 \le p_0, p_1 \le .5$:

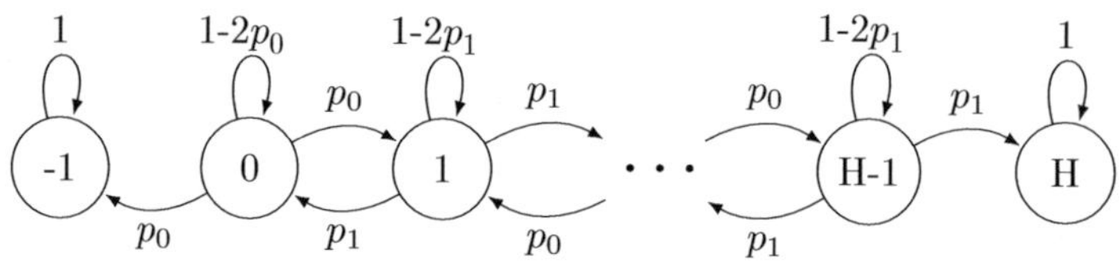

FIGURE 39

By the Duality Theorem we can find the k-step transient probabilities corresponding to Figure 39. Thus the k-step transient probabilities $P_1^{*(k)}(i,-1)$, for $i = 0, 1, 2, \cdots, H-1$, which are also known as finite-time gambler's ruin probabilities, are explicitly determined in terms of the eigenvalues of P_1, see related problems Remark 4.2 in [**HKN08**] and [**Lor17**].

REMARK 5.7. Figure 40 is the natural generalization of Theorem 5.3 to include catastrophe-like transition probabilities.

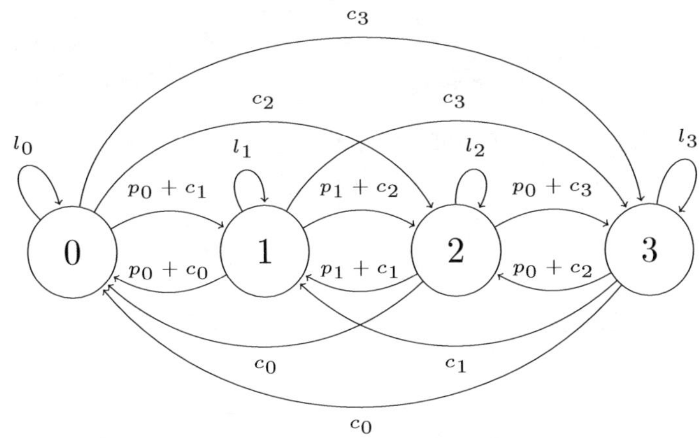

FIGURE 40. Alternating Birth-Death Probabilities including Catastrophe-like Transitions

where: $l_0 = r + p_1 + c_0 \qquad l_1 = r + c_1 \qquad l_2 = r + c_2$

$l_3 = r + p_1 + c_3 \qquad 0 < p_0,\, p_1 < 1$

$p_0 + p_1 + r + c_0 + c_1 + c_2 + c_3 = 1$

Let P be the transition probability matrix of the Markov chain shown in Figure 40, then

$$P = \begin{bmatrix} r+p_1+c_0 & p_0+c_1 & c_2 & c_3 \\ p_0+c_0 & r+c_1 & p_1+c_2 & c_3 \\ c_0 & p_1+c_1 & r+c_2 & p_0+c_3 \\ c_0 & c_1 & p_0+c_2 & r+p_1+c_3 \end{bmatrix}$$

The dual of P is then

$$P^* = \begin{bmatrix} 1 & 0 & 0 & 0 & 0 \\ p_0+c_1+c_2+c_3 & \begin{bmatrix} r+p_1-p_0 \end{bmatrix.} & p_0 & 0 & 0 \\ c_2+c_3 & p_1 & r+p_0-p_1 & p_1 & c_0+c_1 \\ c_3 & 0 & p_0 & \left. r+p_1-p_0 \right] & p_0+c_0+c_1+c_2 \\ 0 & 0 & 0 & 0 & 1 \end{bmatrix}$$

The eigenvalues of the central matrix of P^* are given by Theorem 5.4 and therefore the eigenvalues of P are also known. This argument scales up for $n > 3$ states.

REMARK 5.8. A similar version of Theorem 5.3 holds for birth-death processes having the following form whether H is even or odd: Theorem 5.3 extends to birth-death processes of the following form:

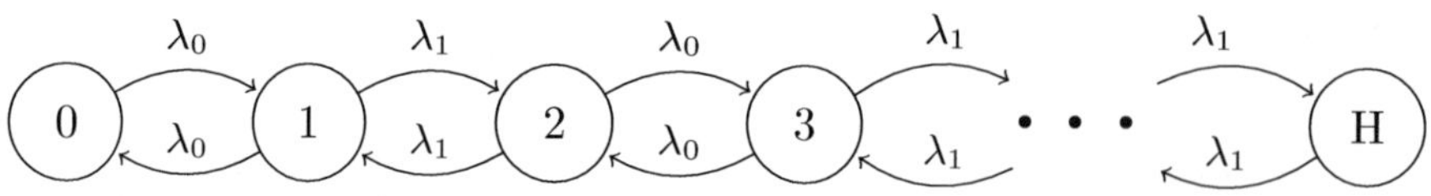

FIGURE 41(A). H is even

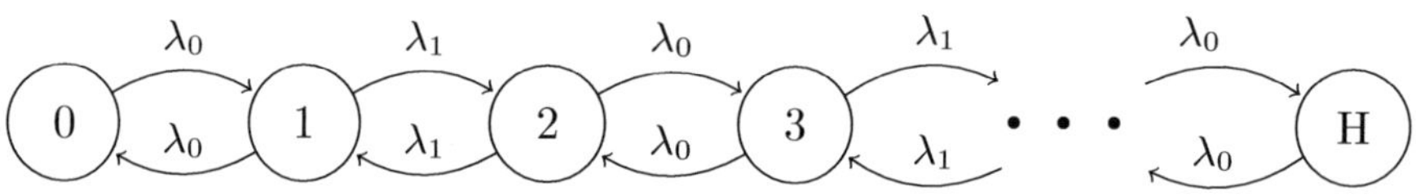

FIGURE 41(B). H is odd

Then the eigenvalues of the Q_1 and Q_2 matrices corresponding to Figures 41(A): H even and 41(B): H odd are explicitly known and the Sylvester eigenvalue expansions of (1.2) for $e^{Q_1 t}$ and $e^{Q_2 t}$ can be explicitly determined. In fact, catastrophe-like transition rates can also be added to Figures 41(A) and 41(B) to yield results along the lines mentioned in Remark 5.7.

The following Theorem appears in Kouachi's 2008 article [**Kou08**].

THEOREM 5.4. *Consider tridiagonal matrices of the form:*

$$A_N = \begin{bmatrix} b_1 & c_1 & 0 & 0 & \cdots & 0 \\ a_1 & b_2 & c_2 & 0 & \cdots & \vdots \\ 0 & a_2 & b_1 & \ddots & \ddots & \vdots \\ 0 & 0 & \ddots & \ddots & \ddots & 0 \\ \vdots & \vdots & \ddots & \ddots & \ddots & c_{N-1} \\ 0 & \cdots & \cdots & 0 & a_{N-1} & b_N \end{bmatrix} \tag{5.10}$$

where a_j and c_j, $j = 1, \ldots, N-1$, are complex numbers and d and b_j, $j = 1, \ldots, N$, are also complex numbers. assume that

$$a_j c_j = d^2, \qquad \text{for } j = 1, \ldots, N-1 \text{ where } d \neq 0$$

and

$$b_j = \begin{cases} b_1 & \text{if } j \text{ is odd,} \\ b_2 & \text{if } j \text{ is even,} \end{cases} \qquad j = 1, \ldots, N$$

Note that diagonal entries of A_N are assumed to alternate. Suppose A_N satisfies the preceding conditions. Then the eigenvalues ω_k of A_N are:

When $N = 2m$ where $m \in \mathbb{N}$.

$$\omega_k = \begin{cases} \dfrac{(b_1+b_2) - \sqrt{(b_1-b_2)^2 + 16d^2\cos^2\theta_k}}{2}, & k = 1, 2, \ldots, m \\[2ex] \dfrac{(b_1+b_2) + \sqrt{(b_1-b_2)^2 + 16d^2\cos^2\theta_k}}{2}, & k = m+1, \ldots, 2m \end{cases}$$

$$\textit{where } \theta_k = \begin{cases} \dfrac{k\pi}{2m+1}, & k = 1, 2, \ldots, m \\[2ex] \dfrac{(k-m)\pi}{2m+1}, & k = m+1, \ldots, 2m \end{cases}$$

When $N = 2m+1$ where $m \in \mathbb{N}$.

$$\omega_k = \begin{cases} \dfrac{(b_1+b_2) - \sqrt{(b_1-b_2)^2 + 16d^2\cos^2\theta_k}}{2}, & k = 1, \ldots, m \\[2ex] \dfrac{(b_1+b_2) + \sqrt{(b_1-b_2)^2 + 16d^2\cos^2\theta_k}}{2}, & k = m+1, \ldots, 2m \\[2ex] b_1, & k = N \end{cases}$$

$$\textit{where } \theta_k = \begin{cases} \dfrac{k\pi}{2m+2}, & k = 1, 2, \ldots, m \\[2ex] \dfrac{(k-m)\pi}{2m+2}, & k = m+1, \ldots, 2m \end{cases}$$

6. Circulant matrices

Consider the constant vector $\vec{c} = [c_0,\ c_{n-1},\ \cdots,\ c_2,\ c_1]$ then the $n \times n$ matrix C having the form:

$$C = \begin{bmatrix} c_0 & c_{n-1} & \cdots & c_2 & c_1 \\ c_1 & c_0 & c_{n-1} & \cdots & c_2 \\ \vdots & c_1 & c_0 & \ddots & \vdots \\ c_{n-2} & \cdots & \ddots & \ddots & c_{n-1} \\ c_{n-1} & c_{n-2} & \cdots & c_1 & c_0 \end{bmatrix} \tag{6.1}$$

is called a circulant matrix. The theory of circulant matrices is described in [**Wik21**] and [**Dav70**]. The eigenvalues of C are known to be:

$$\omega_s = c_0 + c_{n-1}\rho_s + c_{n-2}\rho_s^2 + \cdots + c_1\rho_s^{n-1} \tag{6.2}$$

where $\rho_s = \exp\left(\frac{2\pi s i}{n}\right)$, $s = 1, \ldots, n-1, n$, and i is the imaginary unit. The normalized eigenvectors of C are known to have the following form:

$$\vec{v_s} = \frac{1}{\sqrt{n}}\left[1, \rho_s, \rho_s^2, \cdots, \rho_s^{(n-1)}\right] \tag{6.3}$$

EXAMPLE 6.1. For illustration, we explore the three state circular birth-death chain:

$$P = \begin{bmatrix} r & p & q \\ q & r & p \\ p & q & r \end{bmatrix}$$

$p, q > 0 \quad r \geq 0 \quad p + q + r = 1$

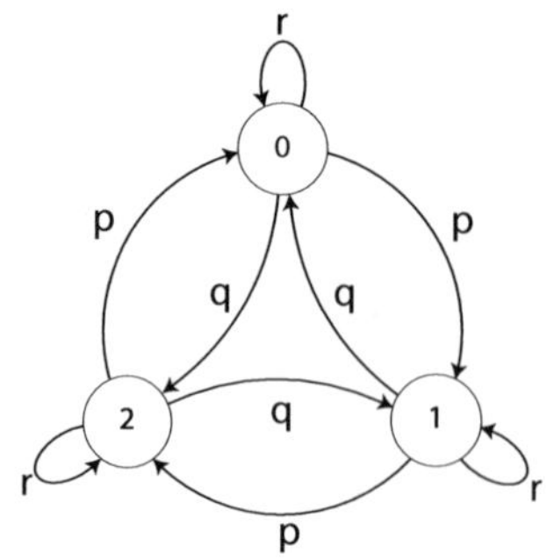

FIGURE 42
Circular 1-step transition probability diagram

where the eigenvalues are given by (6.2)

$$\begin{aligned} \omega_1 &= -\frac{q}{2} + r - \frac{p}{2} + i\left(-\frac{\sqrt{3}q}{2} + \frac{\sqrt{3}p}{2}\right) \\ \omega_2 &= -\frac{q}{2} + r - \frac{p}{2} + i\left(\frac{\sqrt{3}q}{2} - \frac{\sqrt{3}p}{2}\right) \\ \omega_3 &= 1 \end{aligned} \tag{6.4}$$

In order to obtain the Sylvester eigenvalue expansion of P:

$$P^k = A_1\omega_1^k + A_2\omega_2^k + A_3 \tag{6.5}$$

we obtained the following expression for A_s:

$$A_s(u, v) = \frac{1}{n}\rho^{s(u-v)} = \frac{1}{n}\exp\left(\frac{2\pi s(u-v)i}{n}\right) \tag{6.6}$$

when $p \neq q$. Note that the A_s are independent of p, q and r. We consider the following numerical example

$$P = \begin{bmatrix} \frac{1}{6} & \frac{1}{3} & \frac{1}{2} \\ \frac{1}{2} & \frac{1}{6} & \frac{1}{3} \\ \frac{1}{3} & \frac{1}{2} & \frac{1}{6} \end{bmatrix}$$

P is a 3×3 **circulant** matrix.

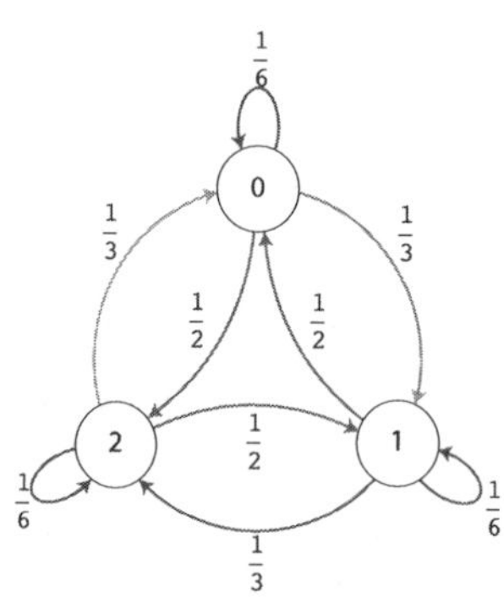

FIGURE 43
Circular birth-death 1-step tranition probability diagram

Eigenvalues are given by (6.4)

$$\omega_1 = -\frac{1}{4} - \frac{\sqrt{3}i}{12} \qquad \text{and} \qquad \omega_2 = -\frac{1}{4} + \frac{\sqrt{3}i}{12} \qquad \text{and} \qquad \omega_3 = 1$$

Given these eigenvalues, then by (6.5) this holds

$$\begin{bmatrix} \frac{1}{6} & \frac{1}{3} & \frac{1}{2} \\ \frac{1}{2} & \frac{1}{6} & \frac{1}{3} \\ \frac{1}{3} & \frac{1}{2} & \frac{1}{6} \end{bmatrix}^k = \begin{bmatrix} \frac{1}{3} & -\frac{1}{6} - \frac{\sqrt{3}i}{6} & -\frac{1}{6} + \frac{\sqrt{3}i}{6} \\ -\frac{1}{6} + \frac{\sqrt{3}i}{6} & \frac{1}{3} & -\frac{1}{6} - \frac{\sqrt{3}i}{6} \\ -\frac{1}{6} - \frac{\sqrt{3}i}{6} & -\frac{1}{6} + \frac{\sqrt{3}i}{6} & \frac{1}{3} \end{bmatrix} \left(-\frac{1}{4} - \frac{\sqrt{3}i}{12}\right)^k$$

$$+ \begin{bmatrix} \frac{1}{3} & -\frac{1}{6} + \frac{\sqrt{3}i}{6} & -\frac{1}{6} - \frac{\sqrt{3}i}{6} \\ -\frac{1}{6} - \frac{\sqrt{3}i}{6} & \frac{1}{3} & -\frac{1}{6} + \frac{\sqrt{3}i}{6} \\ -\frac{1}{6} + \frac{\sqrt{3}i}{6} & -\frac{1}{6} - \frac{\sqrt{3}i}{6} & \frac{1}{3} \end{bmatrix} \left(-\frac{1}{4} + \frac{\sqrt{3}i}{12}\right)^k$$

$$+ \begin{bmatrix} \frac{1}{3} & \frac{1}{3} & \frac{1}{3} \\ \frac{1}{3} & \frac{1}{3} & \frac{1}{3} \\ \frac{1}{3} & \frac{1}{3} & \frac{1}{3} \end{bmatrix}$$

We now consider a probability problem related to this example. Of all paths going from $i = 1$ to $j = 1$ in 3 steps in Figure 43, what is the probability of those paths that do not transition between 0 and 2?

This problem will be solved by our eigenvalue expansion method and checked by a lattice path counting method. We consider the linear sub birth-death chain shown below along with its 1-step probability matrix P_R:

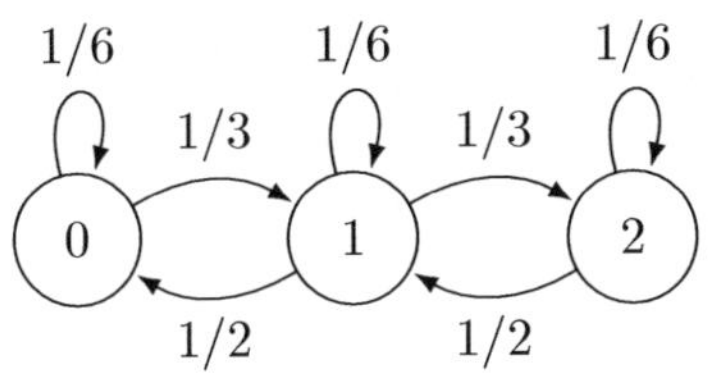

$$P_R = \begin{bmatrix} \frac{1}{6} & \frac{1}{3} & 0 \\ \frac{1}{2} & \frac{1}{6} & \frac{1}{3} \\ 0 & \frac{1}{2} & \frac{1}{6} \end{bmatrix}$$

and having eigenvalues:

$$\omega_1 = \frac{1}{6}\left(1 + 2\sqrt{3}\right) \qquad \omega_2 = \frac{1}{6}\left(1 - 2\sqrt{3}\right) \qquad \omega_3 = \frac{1}{6}$$

The probability of going from $i = 1$ to $j = 1$ in 3 steps without going around the circle equals the strip probability $P_R^3(1,1)$, and the probability of going from $i = 1$ to $j = 1$ on Figure 43 is $P^3(1,1)$. So the answer to our probability problem using eigenvalues is

$$\frac{P_R^3(1,1)}{P^3(1,1)} = \frac{\frac{1}{2}\left(\frac{1}{6}(1+2\sqrt{3})\right)^3 + \frac{1}{2}\left(\frac{1}{6}(1-2\sqrt{3})\right)^3 + 0\left(\frac{1}{6}\right)^3}{\frac{1}{3}\left(-\frac{1}{4} - \frac{\sqrt{3}i}{12}\right)^3 + \frac{1}{3}\left(-\frac{1}{4} + \frac{\sqrt{3}i}{12}\right)^3 + \frac{1}{3}} = 0.513\overline{8}$$

This answer is confirmed by path counting, where the L, U, and D represent loop, up, and down steps respectively with $P(L) = \frac{1}{6}$, $P(U) = \frac{1}{3}$, $P(D) = \frac{1}{2}$. There are nine possible paths to consider:

a) LLL	b) LUD	c) LDU
d) ULD	e) DLU	f) UDL
g) DUL	h) UUU	i) DDD

$$\frac{P_R^3(1,1)}{P^3(1,1)} = \frac{\frac{1}{2}\times\frac{1}{3}+\left(\frac{1}{6}\right)^3}{\left(\frac{1}{2}\right)^3+\left(\frac{1}{3}\right)^3+\frac{1}{2}\times\frac{1}{3}+\left(\frac{1}{6}\right)^3} = 0.513\overline{8}$$

which confirms our previous answer. Note here that $\frac{1}{2}\times\frac{1}{3}$ is the contribution from paths b) through g).

For general n,

$$\frac{P_R^n(1,1)}{P^n(1,1)} = \frac{\frac{1}{2}\left(\frac{1}{6}(1+2\sqrt{3})\right)^n+\frac{1}{2}\left(\frac{1}{6}(1-2\sqrt{3})\right)^n+0\left(\frac{1}{6}\right)^n}{\frac{1}{3}\left(-\frac{1}{4}-\frac{\sqrt{3}i}{12}\right)^n+\frac{1}{3}\left(-\frac{1}{4}+\frac{\sqrt{3}i}{12}\right)^n+\frac{1}{3}}$$

which is significantly simpler to calculate than path counting.

REMARK 6.1. The preceding example of a circular birth-death chain having a circulant transition matrix scales up to n states. This is true because circulant matrices have nice, compact formulas for their distinct eigenvalues and their eigenvectors. These P matrices are almost tridiagonal Toeplitz matrices with the addition of extra nonzero entries in the $(0,\, n-1)$ and $(n-1,\, 0)$ places.

REMARK 6.2. Similar results hold for circular birth-death processes in continuous time t having infinitesimal rate transition matrix Q as shown and diagrammed below. We can explicitly determine the transient probability functions of this system using the Sylvester eigenvalue expansion 2.3. This system may be referred to as the circular M/M/1/3. Explicit transient solutions of the circular M/M/1/K queueing system model follow in a similar manner from having explicit eigenvalue and eigenvector formulas corresponding to the circulant Q transition rate matrix. It would be interesting to explore whether one can still find explicit solutions if catastrophe-like transition rates are included in Q.

$$Q = \begin{bmatrix} -(\lambda+\mu) & \lambda & \mu \\ \mu & -(\lambda+\mu) & \lambda \\ \lambda & \mu & -(\lambda+\mu) \end{bmatrix}$$

$$\lambda,\, \mu > 0$$

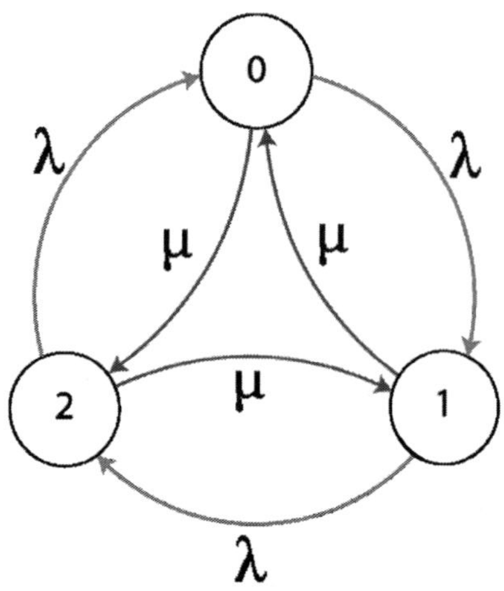

Appendix A. Appendix

Suppose

$$M = \begin{pmatrix} b & c & 0 & \cdots & \cdots & 0 & 0 & 0 \\ a & b & c & \ddots & \cdots & 0 & 0 & 0 \\ 0 & a & b & \ddots & \ddots & 0 & 0 & 0 \\ \vdots & \ddots & \ddots & \ddots & \ddots & \ddots & \vdots & \vdots \\ \vdots & \vdots & \ddots & \ddots & \ddots & \ddots & \ddots & \vdots \\ 0 & 0 & 0 & \ddots & \ddots & b & c & 0 \\ 0 & 0 & 0 & \cdots & \ddots & a & b & c \\ 0 & 0 & 0 & \cdots & \cdots & 0 & a & b \end{pmatrix}$$

It's known that a real $n \times n$ tridiagonal, Toeplitz matrix, M, has distinct eigenvalues $\omega_1, \omega_2, \ldots, \omega_n$ as follows:

$$\omega_s = b + 2\sqrt{ac}\cos\left(\frac{s\pi}{n+1}\right) \quad \text{where } s = 1, 2, 3, \ldots, n$$

with the corresponding right eigenvectors:

$$\vec{R_s} = k \cdot \begin{bmatrix} \left(\frac{a}{c}\right)^{\frac{1}{2}} \sin\left(\frac{\pi s}{n+1}\right) \\ \left(\frac{a}{c}\right)^{\frac{2}{2}} \sin\left(\frac{2\pi s}{n+1}\right) \\ \vdots \\ \left(\frac{a}{c}\right)^{\frac{n}{2}} \sin\left(\frac{n\pi s}{n+1}\right) \end{bmatrix}$$

To calculate the left eigenvector of the corresponding ω_s, use this following equation:

$$\begin{aligned} (\vec{L_s} \cdot M)^T &= (\omega_s \vec{L_s})^T \\ M^T \cdot \vec{L_s}^T &= \omega_s (\vec{L_s})^T \end{aligned}$$

This means that $\vec{L_s}$ is the transpose of the right eigenvector of M^T hence

$$\vec{L_s} = k \cdot \left[\left(\frac{c}{a}\right)^{\frac{1}{2}} \sin\left(\frac{\pi s}{n+1}\right) \quad \left(\frac{c}{a}\right)^{\frac{2}{2}} \sin\left(\frac{2\pi s}{n+1}\right) \quad \cdots \quad \left(\frac{c}{a}\right)^{\frac{n}{2}} \sin\left(\frac{n\pi s}{n+1}\right)\right]$$

To find the spectral projectors we want to find k such that

$$A_s = k^2 \vec{L_s} \cdot \vec{R_s} = 1 =$$

$$k^2 \begin{bmatrix} \left(\frac{a}{c}\right)^{\frac{1}{2}} \sin\left(\frac{\pi s}{n+1}\right) \\ \left(\frac{a}{c}\right)^{\frac{2}{2}} \sin\left(\frac{2\pi s}{n+1}\right) \\ \vdots \\ \left(\frac{a}{c}\right)^{\frac{n}{2}} \sin\left(\frac{n\pi s}{n+1}\right) \end{bmatrix} \left[\left(\frac{c}{a}\right)^{\frac{1}{2}} \sin\left(\frac{\pi s}{n+1}\right) \quad \left(\frac{c}{a}\right)^{\frac{2}{2}} \sin\left(\frac{2\pi s}{n+1}\right) \quad \cdots \quad \left(\frac{c}{a}\right)^{\frac{n}{2}} \sin\left(\frac{n\pi s}{n+1}\right)\right]$$

$$k^2 \left[\sin^2\left(\frac{\pi s}{n+1}\right) + \sin^2\left(\frac{2\pi s}{n+1}\right) + \cdots + \sin^2\left(\frac{n\pi s}{n+1}\right)\right]$$

Writing this in summation form and simplifying by using the half angle trigonometric formula gives:

$$1 = k^2 \left[\sum_{i=1}^{n} \sin^2 \left(\frac{i\pi s}{n+1} \right) \right] = k^2 \left[\frac{n}{2} - \frac{1}{2} \sum_{i=1}^{n} \cos \left(\frac{2i\pi s}{n+1} \right) \right]$$

Further simplifying the summation of the cosine terms using Euler's relation and applying the product-to-sum trigonometric formula produces:

$$1 = k^2 \left[\frac{n}{2} - \frac{1}{4} \frac{\cos\left(\frac{(n+1)\pi s}{n+1}\right) \sin\left(\frac{(n)\pi s}{n+1}\right)}{\sin\left(\frac{\pi s}{n+1}\right)} \right]$$

$$1 = k^2 \left[\frac{n}{2} - \frac{1}{2\sin\left(\frac{\pi s}{n+1}\right)} \left(\sin\left(\frac{(2n+1)\pi s}{n+1}\right) - \sin\left(\frac{\pi s}{n+1}\right) \right) \right]$$

$$1 = k^2 \left[\frac{n}{2} - \frac{1}{4} \left(\frac{\sin\left(\frac{(2n+1)\pi s}{n+1}\right)}{\sin\left(\frac{\pi s}{n+1}\right)} - 1 \right) \right] \tag{A.1}$$

Using some trigonometric manipulations and the fact that our variables s and n are integers leads to:

$$\begin{aligned}
\sin\left(\frac{(2n+1)\pi s}{n+1}\right) &= \sin\left(\pi s \left(\frac{2n+2-1}{n+1}\right)\right) \\
&= \sin\left(\pi s \left(\frac{2n+2}{n+1} - \frac{1}{n+1}\right)\right) \\
&= \sin\left(2\pi s - \frac{\pi s}{n+1}\right) \\
&= \sin(2\pi s)\cos\left(-\frac{\pi s}{n+1}\right) + \cos(2\pi s)\sin\left(-\frac{\pi s}{n+1}\right) \\
&= 0 + (1)\sin\left(-\frac{\pi s}{n+1}\right)
\end{aligned}$$

$$\text{So, } \sin\left(\frac{(2n+1)\pi s}{n+1}\right) = -\sin\left(\frac{\pi s}{n+1}\right)$$

substituting this equality into (A.1) produces:

$$k^2 \left[\frac{n}{2} - \frac{1}{4} \left(-\frac{\sin\left(\frac{\pi s}{n+1}\right)}{\sin\left(\frac{\pi s}{n+1}\right)} - 1 \right) \right] = 1$$

$$k^2 \left[\frac{n+1}{2} \right] = 1$$

$$k^2 = \frac{2}{n+1} \tag{A.2}$$

To obtain the matrix coefficient A_s use

$$A_s = k^2 \vec{R_s} \cdot \vec{L_s}$$

$$= k^2 \left[\left(\tfrac{c}{a}\right)^{\frac{1}{2}} \sin\left(\tfrac{\pi s}{n+1}\right) \quad \left(\tfrac{c}{a}\right)^{\frac{2}{2}} \sin\left(\tfrac{2\pi s}{n+1}\right) \quad \cdots \quad \left(\tfrac{c}{a}\right)^{\frac{n}{2}} \sin\left(\tfrac{n\pi s}{n+1}\right) \right] \begin{bmatrix} \left(\tfrac{a}{c}\right)^{\frac{1}{2}} \sin\left(\tfrac{\pi s}{n+1}\right) \\ \left(\tfrac{a}{c}\right)^{\frac{2}{2}} \sin\left(\tfrac{2\pi s}{n+1}\right) \\ \vdots \\ \left(\tfrac{a}{c}\right)^{\frac{n}{2}} \sin\left(\tfrac{n\pi s}{n+1}\right) \end{bmatrix}$$

where $k^2 = \frac{2}{n+1}$. Therefore the (u,v) entries of matrix A_s is:

$$\frac{2}{n+1} \left(\frac{a}{c}\right)^{\frac{u}{2}} \left(\frac{a}{c}\right)^{\frac{-v}{2}} \sin\left(\frac{u\pi s}{n+1}\right) \sin\left(\frac{v\pi s}{n+1}\right)$$

REMARK A.1. A special case of the preceding result when $b = 0$ was proved using lattice path combinatorics, see Proposition 1, page 134 of [**KRM05**].

REMARK A.2. This formula can be used as a R function "VB.R" that can be found in markovcpp Github page [**Lin**].

Acknowledgments

We are pleased to acknowledge the assistance of Professors Gerardo Rubino and Ryan Szypowski and the contributions of other student members of our Cal Poly Pomona Research Group as listed below:

Ryan Kmet, Vivian P. Hernandez, Hung (Erik) T. Doan, Jianfeng (Tony) Sun, Tanner J. Thomas, Connor L. Adams, Thomas A. Sargent, Zhang, Yu, Jonathan L. Cohen, David Nguyen, Shane J. Hernandez, Stephen J. Shu, Stephen Olsen, Kwok Wai (Kobe) Cheung, Noah J. Chung, Christian Ibarra, Oscar G. Rivera, Hakeem T. Frank, Steven L. Marquez, Ruifan Wu, Anthony J. Torres, Mac Elroyd Fernandez, Jiheng Nie, Joshua C. Johnson, Diana L. Morales, Godwin Liang, Lorenzo R. Soriano, Jorge A. Flores, Noha Abdulhadi, Evelyn J. Guerra.

References

[And91] William J. Anderson, *Continuous-time Markov chains*, Springer Series in Statistics: Probability and its Applications, Springer-Verlag, New York, 1991. An applications-oriented approach, DOI 10.1007/978-1-4612-3038-0. MR1118840

[Arr89] Kenneth J. Arrow, *A "dynamic" proof of the Frobenius-Perron theorem for Metzler matrices*, Probability, statistics, and mathematics, Academic Press, Boston, MA, 1989, pp. 17–26. MR1031275

[Ber18] Chris Bernhardt, *Powers of positive matrices*, Math. Mag. **91** (2018), no. 3, 218–227, DOI 10.1080/0025570X.2018.1446615. MR3808784

[Dav70] Philip J. Davis, *Circulant matrices*, John Wiley & Sons, New York-Chichester-Brisbane, 1979. A Wiley-Interscience Publication; Pure and Applied Mathematics. MR543191

[EHMP21] Emmanuel Ekwedike, Robert C. Hampshire, William A. Massey, and Jamol J. Pender, *Group Symmetries and Bike Sharing for M/M/1/k Queueing Transcience*, August 2021, preprint.

[FH15] Stefan Felsner and Daniel Heldt, *Lattice path enumeration and Toeplitz matrices*, J. Integer Seq. **18** (2015), no. 1, Article 15.1.3, 16. MR3303764

[HJ92] Roger A. Horn and Charles R. Johnson, *Topics in matrix analysis*, Cambridge University Press, Cambridge, 1991, DOI 10.1017/CBO9780511840371. MR1091716

[HKN08] B. Hunter, A. C. Krinik, C. Nguyen, J. M. Switkes, and H. F. von Bremen, *Gambler's ruin with catastrophes and windfalls*, J. Stat. Theory Pract. **2** (2008), no. 2, 199–219, DOI 10.1080/15598608.2008.10411871. MR2524462

[KM10] Alan Krinik and Gopal Mohanty, *On batch queueing systems: a combinatorial approach, J. Statist. Plann. Inference* **140** (2010), no. 8, 2271-2284, DOI 10.1016/j.jspi.2010.01.023.

[KMR04] Alan Krinik, Carrie Mortensen, and Gerardo Rubino, *Connections between birth-death processes*, Stochastic processes and functional analysis, Lecture Notes in Pure and Appl. Math., vol. 238, Dekker, New York, 2004, pp. 219–240. MR2059909

[Kou06] Said Kouachi, *Eigenvalues and eigenvectors of tridiagonal matrices*, Electron. J. Linear Algebra **15** (2006), 115–133, DOI 10.13001/1081-3810.1223. MR2223768

[Kou08] S. Kouachi, *Eigenvalues and eigenvectors of some tridiagonal matrices with non-constant diagonal entries*, Appl. Math. (Warsaw) **35** (2008), no. 1, 107–120, DOI 10.4064/am35-1-7. MR2407056

[KRM05] Alan Krinik, Gerardo Rubino, Daniel Marcus, Randall J. Swift, Hassan Kasfy, and Holly Lam, *Dual processes to solve single server systems*, J. Statist. Plann. Inference **135** (2005), no. 1, 121–147, DOI 10.1016/j.jspi.2005.02.010. MR2202343

[KS] Alan Krinik and Jennifer Switkes, *An element in the kth power of an $n \times n$ matrix*, Preprint.

[Lin] Jeremy Lin, *Program finding the a's matrices.*

[Lor17] Paweł Lorek, *Generalized gambler's ruin problem: explicit formulas via Siegmund duality*, Methodol. Comput. Appl. Probab. **19** (2017), no. 2, 603–613, DOI 10.1007/s11009-016-9507-6. MR3649560

[Los92] L. Losonczi, *Eigenvalues and eigenvectors of some tridiagonal matrices*, Acta Math. Hungar. **60** (1992), no. 3-4, 309–322, DOI 10.1007/BF00051649. MR1177259

[Lyc18] Samuel Lyche, *On deep learning and neural networks*, Master's thesis, California State Polytechnic University, Pomona, 2017.

[Mey00] Carl Meyer, *Matrix analysis and applied linear algebra*, Society for Industrial and Applied Mathematics (SIAM), Philadelphia, PA, 2000. With 1 CD-ROM (Windows, Macintosh and UNIX) and a solutions manual (iv+171 pp.), DOI 10.1137/1.9780898719512. MR1777382

[Moh14] Sri Gopal Mohanty, *Lattice path counting and applications*, Academic Press [Harcourt Brace Jovanovich, Publishers], New York-London-Toronto, Ont., 1979. Probability and Mathematical Statistics. MR554084

[MVL03] Cleve Moler and Charles Van Loan, *Nineteen dubious ways to compute the exponential of a matrix, twenty-five years later*, SIAM Rev. **45** (2003), no. 1, 3–49, DOI 10.1137/S00361445024180. MR1981253

[ND88] Ben Noble and James W. Daniel, *Applied linear algebra*, 2nd ed., Prentice-Hall, Inc., Englewood Cliffs, N.J., 1977. MR0572995

[Ngu17] Uyen Nguyen, *Tridiagonal stochastic matrices*, Master's thesis, California State Polytechnic University, Pomona, 2017.

[Ren07] Marc Renault, *Four proofs of the ballot theorem*, Math. Mag. **80** (2007), no. 5, 345–352, DOI 10.1080/0025570x.2007.11953509. MR2362634

[RK21] Gerardo Rubino and Alan Krinik, *The exponential-dual matrix method: Applications to Markov chain analysis*,in Stochastic Processes and Functional Analysis, New Perspectives, AMS Contemporary Mathematics Series, Volume 774, edited by Randall Swift, Alan Krinik, Jennifer Switkes and Jason Park (2021), pp. 217–235.

[Sen79] E. Seneta, *Coefficients of ergodicity: structure and applications*, Adv. in Appl. Probab. **11** (1979), no. 3, 576–590, DOI 10.2307/1426955. MR533060

[STGH18] John F. Shortle, James M. Thompson, Donald Gross, and Carl M. Harris, *Fundamentals of Queueing Theory*, 5th edition, Wiley 2018, ISBN 978-1-118-94352-6, 576 pages.

[Wik19a] Wikipedia Contributers, *Frobenius covariant–Wikipedia, the free encyclopedia*, 2019, [Online; accessed 27-May-2020].

[Wik19b] Wikipedia Contributers, *Sylvester's formula–Wikipedia, the free encyclopedia*, 2019, [Online; accessed 27-May-2020].

[Wik20] Wikipedia Contributers, *Perron-Frobenius theorem–Wikipedia, the free encyclopedia*, 2020, [Online; accessed 31-May-2020].

[Wik21] Wikipedia Contributers, *Circulant matrices–Wikipedia, the free encyclopedia*, 2021, [Online; accessed 11-Feb-2021].

Alan Krinik, California State Polytechnic University, Pomona

Hubertus von Bremen, California State Polytechnic University, Pomona

Ivan Ventura, California State Polytechnic University, Pomona

Uyen Vietthanh Nguyen, California State Polytechnic University, Pomona

Jeremy J. Lin, University of California, Irvine

Thuy Vu Dieu Lu, University of California, Irvine

Chon In (Dave) Luk, California State Polytechnic University, Pomona

Jeffrey Yeh, California State Polytechnic University, Pomona

Luis A. Cervantes, Pacific Life

Samuel R. Lyche, Booz Allen Hamilton

Brittney A. Marian, University of Southern California

Saif A. Aljashamy, California State Polytechnic University, Pomona

Mark Dela, California State Polytechnic University, Pomona

Ali Oudich, Pitzer College

Pedram Ostadhassanpanjehali, UPS

Lyheng Phey, U.S. Navy

David Perez, California State Polytechnic University, Pomona

John Joseph Kath, Claremont Graduate University

Malachi C. Demmin, California State Polytechnic University, Pomona

Yoseph Dawit, California State Polytechnic University, Pomona

Christine Carmen Marie Hoogendyk, Oregon State University

Aaron Kim, Raytheon Technologies

Matthew McDonough, University of California, Santa Barbara

Adam Trevor Castillo, California State Polytechnic University, Pomona

David Beecher, California State Polytechnic University, Pomona

Weizhong Wong, California State Polytechnic University, Pomona

Heba Ayeda, California State Polytechnic University, Pomona

Contemporary Mathematics
Volume **774**, 2021
https://doi.org/10.1090/conm/774/15571

On the use of Markovian stick-breaking priors

William Lippitt and Sunder Sethuraman

Dedicated to Professor M.M. Rao on his 90th birthday

ABSTRACT. Recently, a 'Markovian stick-breaking' process which generalizes the Dirichlet process (μ, θ) with respect to a discrete base space $\mathfrak{X}$ was introduced. In particular, a sample from from the 'Markovian stick-breaking' processs may be represented in stick-breaking form $\sum_{i\geq 1} P_i \delta_{T_i}$ where $\{T_i\}$ is a stationary, irreducible Markov chain on $\mathfrak{X}$ with stationary distribution μ, instead of i.i.d. $\{T_i\}$ each distributed as μ as in the Dirichlet case, and $\{P_i\}$ is a GEM(θ) residual allocation sequence. Although the previous motivation was to relate these Markovian stick-breaking processes to empirical distributional limits of types of simulated annealing chains, these processes may also be thought of as a class of priors in statistical problems. The aim of this work in this context is to identify the posterior distribution and to explore the role of the Markovian structure of $\{T_i\}$ in some inference test cases.

1. Introduction

Let $\mathfrak{X} \subseteq \mathbb{N}$ be a discrete space, either finite or countable. Let also μ be a measure on $\mathfrak{X}$, and $\theta > 0$ be a parameter. The Dirichlet process on $\mathfrak{X}$ with respect to pair (θ, μ) is an object with fundamental applications to Bayesian nonparametric statistics (cf. books [**17**], [**24**]). Formally, the Dirichlet process is a probability measure on the space of probability measures on $\mathfrak{X}$ such that a sample $\mathcal{P}$, with respect to any finite partition $(A_1, \ldots, A_k)$ of $\mathfrak{X}$, has the property that the distribution of $(\mathcal{P}(A_1), \ldots, \mathcal{P}(A_k))$ is Dirichlet with parameters $(\theta\mu(A_1), \ldots, \theta\mu(A_k))$ (cf. [**14**], [**6**]). Importantly, the Dirichlet process has a 'stick-breaking' representation: A sample $\mathcal{P}$ can be represented in form $\sum_{j=1}^{\infty} P_j \delta_{T_j}$ where $\mathbf{P} = \{P_j\}_{j\geq 1}$ is a GEM(θ) residual allocation sequence, and $\{T_j\}_{j\geq 1}$ is an independent sequence of independent and identically distributed (i.i.d.) random variables on $\mathfrak{X}$ with common distribution μ (cf. [**27**], [**26**]). Here, a GEM sequence is one where $P_1 = X_1$ and $P_j = X_j\big(1 - \sum_{i=1}^{j-1} P_i\big)$ for $j \geq 2$, and $\{X_j\}_{j\geq 1}$ are i.i.d. Beta$(1, \theta)$ random variables.

There are several types of generalizations of the Dirichlet process in the literature such as Polya tree, and species sampling processes [**17**][Ch. 14], [**21**], among others. In [**11**], a different generalization where $\{T_j\}_{j\geq 1}$ is a Markov chain was introduced: Let $G = \{G_{i,j} : i, j \in \mathfrak{X}\}$ be a generator matrix, that is $G_{i,j} \geq 0$ for

2020 *Mathematics Subject Classification.* Primary 60E99, 60G57, 62G20, 62G05.

This research was partly supported by ARO-W911NF-18-1-0311 and a Simons Foundations Sabbatical grant.

$i \neq j$ and $G_{i,i} = -\sum_{j \neq i} G_{i,j}$, which is irreducible with suitably bounded entries, and has μ as its stationary distribution. Let also $Q = I + G/\theta$ be a Markov transition kernel on $\mathfrak{X}$ with stationary distribution μ. Now, define $\mathbf{T} = \{T_j\}_{j \geq 1}$ as the stationary Markov chain with transition kernel Q. The 'Markovian stick-breaking' process is then represented as $\sum_{j \geq 1} P_j \delta_{T_j}$ where again $\mathbf{P}$ is an independent GEM(θ) sequence.

Although it was shown in [**11**], [**12**] that such Markovian stick-breaking processes connect to the limiting empirical distribution of certain simulated annealing chains, it is natural to consider their use as priors in statistical problems, the aim of this article. We first give a formula for the moments of the Markovian stick-breaking process in Theorem 4. Then, we compute the posterior distribution moments in terms of this formula in Proposition 7, Corollary 8. Consistency of the posterior distribution is stated in Proposition 9, noting the full support property of the process in Proposition 10. In Proposition 11, we discuss asymptotics of the process with respect a 'strength' parameter.

A main part of this work is also to consider the use and behavior of the Markovian stick-breaking process as a prior for inference of histograms. In this context, the generator G can be thought as a priori belief of weights or affinities in a 'network' of categories. For instance, in categorical data, one may believe that affinities between categories differ depending on the pair, and also that they may be directed in hierarchical situations. In using Dirichlet priors, there is an implicit assumption that the network connecting categories is complete and the affinity between two categories cannot depend on both categories. However, in using a Markovian stick-breaking prior, one can build into the prior a belief about the weight structure on the network by specification of the generator G. In simple experiments, we show interesting behaviors of the posterior distribution from these Markovian stick-breaking priors, in comparison to Dirichlet priors.

The structure of the paper is to define carefully the Markovian stick-breaking process in Section 2. Then, in Section 3, we state results on their moments, posterior distribution, and consistency. In Section 4, we discuss the use of these processes as priors, present simple numerical experiments, and provide some context with previous literature.

2. Definition of the Markovian stick-breaking process

We take as convention empty sums are 0, empty products of scalars are 1, empty products of matrices are the identity, and that a product of matrices is computed as

$$\prod_{j=1}^{n} M_j = M_n \cdot M_{n-1} \cdot \quad \cdots \quad \cdot M_1.$$

For a set $A \subseteq \mathfrak{X}$, we define $D(A)$ as the diagonal square matrix over $\mathfrak{X}$ with entries $D_{xx}(A) = \delta_x(A)$. For $x \in \mathfrak{X}$, let $e_x \in \mathbb{R}^{\mathfrak{X}}$ be the column vector with a 1 in the xth entry and 0's in other entries. Let also $\vec{1}$ be the column vector of all 1's.

We now give a precise definition of the well-known GEM 'Griffiths-Engel-McCloskey' residual allocation sequence, which apportions a unit resource into infinitely many parts.

Definition 1 (GEM). Let $\mathbf{X} = (X_j)_{j=1}^{\infty}$ be an i.i.d. collection of Beta$(1, \theta)$ variables for some positive constant θ. Define $\mathbf{P} = (P_j)_{j=1}^{\infty}$ by $P_1 = X_1$ and

$P_j = X_j\big(1 - \sum_{i=1}^{j-1} P_i\big)$ for $j \geq 2$, which leads to the formula

$$P_j = X_j \prod_{i=1}^{j-1}(1 - X_i).$$

We say $\mathbf{P}$ has GEM(θ) distribution.

To define the 'Markovian stick-breaking' process on the discrete space $\mathfrak{X}$, we now state carefully the definition of a Generator kernel or matrix.

DEFINITION 2 (Generator). We call a real-valued matrix $G = (G_{xy})_{x,y\in\mathfrak{X}}$, a generator matrix over $\mathfrak{X}$, if

(1) For each pair $x, y \in \mathfrak{X}$ with $x \neq y$, then $G_{x,y} \geq 0$.
(2) For each $x \in \mathfrak{X}$, $G_{xx} = -\sum_{y\in\mathfrak{X}-\{x\}} G_{xy}$.
(3) $\theta^G := \sup_{x\in\mathfrak{X}} |G_{xx}| < \infty$

If μ is a stochastic vector over $\mathfrak{X}$ and $\mu^T G = \vec{0}$, we call μ a stationary distribution of G.

Note that if $\theta \geq \theta^G$ for a generator matrix G, then

$$Q = I + G/\theta,$$

where I is the identity kernel, is a stochastic matrix over $\mathfrak{X}$. All such Q's share stationary distributions and communication classes. As such, we refer to the stationary distributions and irreducibility properties of G and Q's interchangeably.

We now define the 'Markovian stick-breaking' measure (MSB) as follows.

DEFINITION 3 (MSB(G)). Let G be an irreducible positive recurrent generator matrix over a discrete space $\mathfrak{X}$, with stationary distribution μ. Let $\theta \geq \theta^G$, and stochastic matrix $Q = I + G/\theta$. Let $\mathbf{P} \sim$ GEM(θ), and let $\mathbf{T} = (T_j)_{j=1}^{\infty}$ be a stationary, homogeneous Markov chain in $\mathfrak{X}$ with transition kernel Q and independent of $\mathbf{P}$. Define the random measure ν over $\mathfrak{X}$ by

$$\nu = \sum_{j=1}^{\infty} P_j \delta_{T_j} \ .$$

We say ν has Markovian stick-breaking distribution with generator G, and the pair (ν, T_1) has MSB(G) distribution.

We note, in this definition, the distribution of ν does not depend on the choice of $\theta \geq \theta^G$, say as its moments by Corollary 5 below depend only on G; see also [**11**] for more discussion. We remark also, when Q is a 'constant' stochastic matrix with common rows μ, then $\{T_j\}_{j\geq 1}$ is an i.i.d. sequence with common distribution μ and so the Markovian stick-breaking measure ν reduces to the Dirichlet distribution with parameters (θ, μ); see [**11**] for further remarks.

3. Results on moments, posterior distribution, and consistency

We now compute in the next formulas certain moments of the Markovian stick-breaking measure with respect to generator G, which identify the distribution of ν.

THEOREM 4. *Let G be an irreducible positive recurrent generator matrix on $\mathfrak{X}$, and let $(\nu, T_1) \sim MSB(G)$. Let also $(A_j)_{j=1}^n$ be a collection of disjoint subsets of $\mathfrak{X}$, $x \in \mathfrak{X}$, and $\vec{k} \in \{0,1,2,\ldots\}^n$. Then,*

$$\mathcal{E}\left[\prod_{j=1}^n \nu(A_j)^{k_j} \middle| T_1 = x\right] = \left(\#\mathbb{S}(\vec{k})\right)^{-1} \sum_{\sigma \in \mathbb{S}(\vec{k})} e_x^T \left[\prod_{j=1}^k \left((I - G/j)^{-1} D(A_{\sigma_j})\right)\right] \vec{1}$$

where $k = \sum_{j=1}^n k_j$, $\mathbb{S}(\vec{k})$ is the collection of distinct permutations of k-lists of k_1 many 1's, k_2 many 2's, and so on to k_n many n's, and $\#\mathbb{S}(\vec{k})$ is the cardinality of this set.

The proof of Theorem 4 is given in the Section 5.

COROLLARY 5. *In the context of the previous theorem, suppose $A_j = \{x_j\}$. Then,*

$$\mathcal{E}\left[\prod_{j=1}^n \nu(x_j)^{k_j}\right] = \left(\#\mathbb{S}(\vec{k})\right)^{-1} \sum_{\sigma \in \mathbb{S}(\vec{k})} \mu_{x_{\sigma_k}} \prod_{j=1}^{k-1} (I - G/j)^{-1}_{x_{\sigma_{j+1}}, x_{\sigma_j}}$$

We remark that Corollary 5 is an improvement of a corresponding formula in [**12**] found, by different means, when $\mathfrak{X}$ is finite and G has no nonzero entries.

These formulas will be of help to identify the posterior distribution, if the Markovian stick-breaking measure is used as a prior. In the case of the Dirichlet process, the posterior distribution is again in the class of Dirichlet processes: Namely given ν, let $Y_1, \ldots, Y_n$ be i.i.d. random variables with distribution ν. Then, the distribution of ν given $Y^n = \{Y_j\}_{j=1}^n$ is a Dirichlet process with parameters $(\theta, \mu + \sum_{j=1}^n \delta_{Y_j})$. However, when ν is a general Markovian stick-breaking measure, such a neat correspondence is not clear. But, later in Proposition 7, we write the posterior moments in terms of 'size-biased' moments with respect to the prior.

We now give a representation of a sequence $Y_1, \ldots, Y_n$, conditional on a sample ν from the Markovian stick-breaking process, which is i.i.d. with common distribution ν. This representation is standard with respect to the Dirichlet process and relatives such as species sampling processes (cf. Ch. 14 [**17**]).

PROPOSITION 6. *Consider the Markovian stick-breaking process ν built from $\mathbf{P}$ and $\mathbf{T}$. For $n \geq 1$, let $(J_i)_{i=1}^n$ be a collection of positive integer valued random variables such that $\mathcal{P}(J_i = j_i : 1 \leq i \leq n | \mathbf{P}, \mathbf{T}) = \prod_{i=1}^n P_{j_i}$. Define the sequence $Y^n = (Y_i)_{i=1}^n$ where $Y_i = T_{J_i}$ for $1 \leq i \leq n$. Then, $Y^n \big| \nu, T_1$ is a collection of i.i.d. variables taking values in $\mathfrak{X}$ with common distribution ν.*

PROOF. Compute, noting $\nu(x) = \sum_{j \geq 1} P_j \mathbb{1}(T_j = x)$, that

$$\begin{aligned}\mathcal{P}(Y^n = y^n | \mathbf{P}, \mathbf{T}) &= \sum_{j_1=1}^\infty \cdots \sum_{j_n=1}^\infty \mathcal{P}(J_i = j_i, T_{j_i} = y_j : 1 \leq i \leq n | \mathbf{P}, \mathbf{T}) \\ &= \sum_{j_1=1}^\infty \cdots \sum_{j_n=1}^\infty \prod_{i=1}^n P_{j_i} \mathbb{1}(T_{j_i} = y_i) = \prod_{i=1}^n \sum_{j=1}^\infty P_j \mathbb{1}(T_j = y_i) = \prod_{i=1}^n \nu(y_i)\end{aligned}$$

Since (ν, T_1) is a function of $\mathbf{P}$ and $\mathbf{T}$ and $\mathcal{P}(Y^n = y^n | \nu, T_1) = \mathcal{E}[\mathcal{P}(Y^n = y^n | \mathbf{P}, \mathbf{T}) | \nu, T_1]$, the result follows. □

The following identifies the posterior distribution in terms of its moments, given as certain 'size-biased' expressions with respect to the prior.

PROPOSITION 7. *Let ν be a random probability measure taking values in the simplex $\Delta_{\mathfrak{X}} = \{\mathbf{p} : \sum_{x\in\mathfrak{X}} p_x = 1, 0 \le p_x \le 1\}$, and let T be an $\mathfrak{X}$-valued random variable. For $n \ge 1$, let $Y^n = (Y_1, \ldots, Y_n)$ be a sequence of random variables such that $Y^n\big|\nu, T$ are i.i.d. with common distribution ν. Let also $\vec{k} = \vec{k}^n$ denote the frequencies of y^n, that is, for each $x \in \mathfrak{X}$, $k_x = \#\{j : 1 \le j \le n,\ y_j = x\}$. In addition, let $\vec{l} \in \{0, 1, 2, \ldots\}^{\mathfrak{X}}$, where $l_x = 0$ except for finitely many $x \in \mathfrak{X}$. Then, for events $A \in \sigma(T)$, we have*

$$\mathcal{E}\left[\prod_{x\in\mathfrak{X}} \nu(x)^{l_x}\Big|Y^n = y^n, A\right] = \frac{\mathcal{E}\Big[\prod_{x\in\mathfrak{X}} \nu(x)^{k_x+l_x}\Big|A\Big]}{\mathcal{E}\Big[\prod_{x\in\mathfrak{X}} \nu(x)^{k_x}\Big|A\Big]}$$

PROOF. Define m additional random variables $Y_{n+1}, Y_{n+2}, \ldots, Y_{n+m}$, by augmenting the probability space if necessary, such that together $Y_1, Y_2, \ldots, Y_n, Y_{n+1}, \ldots, Y_{n+m}|\nu, T$ are i.i.d. with common distribution ν. In particular, for $y^{n+m} \in \mathfrak{X}^{n+m}$, we have $\mathcal{P}\big(Y^{n+m} = y^{n+m}\big|\nu, T\big) = \prod_{j=1}^{n+m} \nu(y_j)$.

Recall now $\vec{k}$ the frequencies of y^n. Let $(y_{n+1}, y_{n+2}, \ldots, y_{n+m}) \in \mathfrak{X}^m$ be any sequence with frequencies $\vec{l}$. Then, we compute

$$\begin{aligned}
\mathcal{E}\left[\prod_{x\in\mathfrak{X}} \nu(x)^{l_x}\Big|Y^n = y^n, A\right] &= \mathcal{E}\left[\frac{\mathcal{P}\big(Y^{n+m} = y^{n+m}\big|\nu, T\big)}{\mathcal{P}\big(Y^n = y^n\big|\nu, T\big)}\Big|Y^n = y^n, A\right] \\
&= \mathcal{E}\Big[\mathcal{P}\big(Y^{n+m} = y^{n+m}\big|Y^n = y^n, \nu, A\big)\Big|Y^n = y^n, A\Big] \\
&= \mathcal{P}\big(Y^{n+m} = y^{n+m}\big|Y^n = y^n, A\big) \\
&= \frac{\mathcal{P}\big(Y^{n+m} = y^{n+m}\big|A\big)}{\mathcal{P}\big(Y^n = y^n\big|A\big)} = \frac{\mathcal{E}\Big[\prod_{x\in\mathfrak{X}} \nu(x)^{k_x+l_x}\Big|A\Big]}{\mathcal{E}\Big[\prod_{x\in\mathfrak{X}} \nu(x)^{k_x}\Big|A\Big]},
\end{aligned}$$

and the result follows. □

Returning to the Markovian stick-breaking process ν, given the 'data' $Y_1, \ldots, Y_n$ conditional on ν and T_1, we may evaluate the posterior moments of the Markovian stick-breaking measure as a case of Proposition 7.

COROLLARY 8. *Let ν be a Markovian stick-breaking process, and $Y_1, \ldots, Y_n|\nu, T_1$ be i.i.d. with common distribution ν (say, as in Proposition 6). Let also $\vec{k} = (k_x)_{x\in\mathfrak{X}}$ be such that $k_x = \#\{j : 1 \le j \le n \text{ and } y_j = x\}$ for each $x \in \mathfrak{X}$. Let in addition $\vec{l} \in \{0, 1, 2, \ldots\}^{\mathfrak{X}}$ be a vector with only finitely many non-zero entries, and $m = \sum_{x\in\mathfrak{X}} l_x$. Then, for each $x \in \mathfrak{X}$, we have*

$$\begin{aligned}
&\mathcal{E}\left[\prod_{w\in\mathfrak{X}} \nu(w)^{l_w}\Big|Y^n = y^n, T_1 = x\right] \\
&\quad = \frac{\Big(\#\mathbb{S}(\vec{k})\Big) \sum_{\sigma\in\mathbb{S}(\vec{k}+\vec{l})} (I - G/(n+m))^{-1}_{x,\sigma_{n+m}} \prod_{j=1}^{n+m-1} (I - G/j)^{-1}_{\sigma_{j+1},\sigma_j}}{\Big(\#\mathbb{S}(\vec{k}+\vec{l})\Big) \sum_{\sigma\in\mathbb{S}(\vec{k})} (I - G/n)^{-1}_{x,\sigma_n} \prod_{j=1}^{n-1} (I - G/j)^{-1}_{\sigma_{j+1},\sigma_j}}
\end{aligned}$$

and

$$\mathcal{E}\left[\prod_{w\in\mathfrak{X}} \nu(w)^{l_w}\Big|Y^n = y^n\right]$$
$$= \frac{\left(\#\mathbb{S}(\vec{k})\right)\sum_{\sigma\in\mathbb{S}(\vec{k}+\vec{l})}\mu_{\sigma_{n+m}}\prod_{j=1}^{n+m-1}(I-G/j)^{-1}_{\sigma_{j+1},\sigma_j}}{\left(\#\mathbb{S}(\vec{k}+\vec{l})\right)\sum_{\sigma\in\mathbb{S}(\vec{k})}\mu_{\sigma_n}\prod_{j=1}^{n-1}(I-G/j)^{-1}_{\sigma_{j+1},\sigma_j}}.$$

We now give a statement of 'consistency' with respect to the posterior distribution, in line with limits of 'Bayes estimators' in [**16**], by considering the moment expression in Proposition 7, when $\mathfrak{X}$ is finite. Consistency, in the case $\mathfrak{X}$ is countably infinite, may be pathological according to [**16**], and so we limit out discussion accordingly.

PROPOSITION 9. *Let ν be a Markovian stick-breaking process on a finite state space $\mathfrak{X}$. Let also $Y_1,\ldots,Y_n|\nu,T_1$ be i.i.d. with common distribution ν.*

Suppose, for each $x\in\mathfrak{X}$ as $n\uparrow\infty$, that $\frac{1}{n}\sum_{j=1}^n \mathbb{1}(Y_j=x)\to\eta_x$ a.s., where $\eta=\{\eta_x\}_{x\in\mathfrak{X}}\in\Delta_{\mathfrak{X}}$. Then, as $n\uparrow\infty$, the posterior distribution $\mu_n=\mathbf{P}(\nu\in\cdot|Y^n)$ converges a.s. to δ_η.

PROOF. Write,

$$\mu_n(B)=\mathbf{P}(\nu\in B|Y^n=y^n)=\frac{\mathbf{P}(Y^n=y^n,\nu\in B)}{\mathbf{P}(Y^n=y^n)}=\frac{\mathcal{E}\big[\prod_{j=1}^n\nu(y_j),\nu\in B\big]}{\mathcal{E}\big[\prod_{j=1}^n\nu(y_j)\big]}.$$

By Theorem 1 in [**16**], if η belongs to the support of ν, the desired convergence of μ_n to δ_η follows.

Hence, to finish, we note by Proposition 10 below that ν has full support on the simplex $\Delta_{\mathfrak{X}}$. □

The following is an improvement of a corresponding result in [**12**] when G has no zero entries, by directly considering the stick-breaking form of ν. See also [**5**] in this context which discusses support properties for a class of species sampling priors.

PROPOSITION 10. *For finite $\mathfrak{X}$, the Markovian stick-breaking measure ν with respect to irreducible G has full support on the simplex $\Delta_{\mathfrak{X}}$.*

PROOF. Let $r=|\mathfrak{X}|$. Since ν has the form $\sum_{j\geq 1}P_j\delta_{T_j}$, the idea is to consider a path of the Markov chain $\mathbf{T}$ with prescribed visits to states 1,2,…,r, and realizations of the GEM(θ) sequence $\mathbf{P}$ with values such that ν belongs to a small ϵ-ball around η.

Since Q is irreducible, there exists an integer $n\geq r$ and a path $(t_1,\ldots,t_n)\in\mathfrak{X}^n$ such that the chain $\mathbf{T}$ has positive probability of starting on the path $P(T_i=t_i: 1\leq i\leq n)>0$ and such that the path hits every state $x\in\mathfrak{X}$. For each state $x\in\mathfrak{X}$, define $i_x=\min\{i:t_i=x\}\in\{1,\ldots,n\}$ to be the first time the path hits state x.

Since $\mathbf{P}$ is distributed as a residual allocation model constructed from iid proportions each having full support on the unit interval, $\mathbf{P}$ has full support on Δ_∞. Thus, for each $\delta>0$, we have with positive probability that simultaneously $P_{i_x}>\eta_x-\delta$ for all $x\in\mathfrak{X}$. Noting the following containment of events

$$\begin{aligned}\big\{1\leq i\leq n: T_i=t_i;\ \forall x: P_{i_x}>\eta_x-\delta\big\}&\subseteq\big\{\forall x: -\delta<\nu_x-\eta_x<(r-1)\delta\big\}\\&\subseteq\big\{\forall x: |\nu_x-\eta_x|<(r-1)\delta\big\}\end{aligned}$$

and taking $\delta = \epsilon/(r-1)$, we then have

$\mathcal{P}(\forall x \in \mathfrak{X} : |\nu_x - \eta_x| < \epsilon) \geq \mathcal{P}(1 \leq i \leq n : T_i = t_i;\ \forall x \in \mathfrak{X} : P_{i_x} > \eta_x - \epsilon/(r-1)) > 0.$

Hence, ν is within ϵ of η with positive probability. □

When the stochastic matrix Q is fixed, the parameter θ in the representation of $G = \theta(Q - I)$ can be viewed as a type of 'strength' of the Markovian stick-breaking ν, as more discussed in the next section.

PROPOSITION 11. *Let Q be irreducible positive recurrent stochastic and define $G^\theta = \theta(Q - I)$. Then, the Markovian stick-breaking measure $\nu = \nu^{(\theta)}$ parametrized by G^θ converges in probability to the stationary vector μ of Q as $\theta \uparrow \infty$.*

PROOF. Suppose Q is aperiodic. Let $x \in \mathfrak{X}$. For all $\theta > 0$ we have $E[\nu(x)] = \mu_x$. As $\theta \uparrow \infty$, we have by Cor. 4.1 for each $n \in \mathbb{N}$ that

$$\begin{aligned}
&\lim_{\theta\to\infty} \mathcal{E}[\nu(x)^2] \\
&\quad = (I - G^\theta)^{-1}_{xx}\mu_x \\
&\quad = \lim_{\theta\to\infty} \frac{\mu_x}{\theta+1}\sum_{j=0}^{n-1}\left(\frac{\theta Q}{\theta+1}\right)^j_{xx} + \frac{\mu_x}{\theta+1}\left[\left(\frac{\theta Q}{\theta+1}\right)^n \sum_{j=0}^{\infty}\left(\frac{\theta Q}{\theta+1}\right)^j\right]_{xx} \\
&\quad = \lim_{\theta\to\infty} 0 + \mu_x\left[Q^n(I - G^\theta)^{-1}\right]_{xx} = \mu_x^2
\end{aligned}$$

since Q^n converges to a constant stochastic matrix with rows μ as $n \to \infty$ and $(I - G^\theta)^{-1}$ is stochastic. Therefore, $\nu(x)$ converges in probability to μ_x as $\theta \uparrow \infty$.

If Q is periodic, define aperiodic $Q' = 0.5(Q + I)$ and note $G^\theta = \theta(Q - I) = 2\theta(Q' - I)$. Since the proposition has been shown to apply to Q', the result holds also for Q. □

4. On use of the MSB(G) measure as a prior

We explore in this section the use of the Markovian stick breaking measure MSB(G) as a prior for multinomial probabilities. In a nutshell, with respect to such a prior, when G is in form $G = \theta(Q - I)$, the matrix Q specifies an affinity network which reflects prior beliefs of association among categories. Given observed data, the posterior mean histogram then computed will have the effect of 'smoothing' the empirical probability mass function (pmf) according to the affinity network, in that mass levels of related categories will tend be similar. The parameter θ as we will note will then represent a relative strength of this 'smoothing'. In particular, we consider, in simple examples, effects on the posterior mean histograms with respect to a few MSB(G) priors in relation to Dirichlet priors, which do not assert affinities among categories.

Of course, 'histogram smoothing' in the context of pmf estimation is an old subject with several Bayesian approaches. For instance, see Leonard [**23**], where multivariate logistic-normal priors are considered; Dickey and Jiang [**10**], where 'filtered' Dirichlet distributions are proposed; Wong [**28**], where generalized Dirichlet distributions are used; and more recently Demirhan and Demirhan [**9**]; see also the survey Agresti and Hitchcock [**2**], and books Agresti [**1**], Ghosal and Van der Vaart [**17**], and Congdon [**7, 8**] and references therein. We remark there is also a large body of work for 'histogram smoothing' with respect to Bayesian density estimation for continuous data, not unrelated to that for pmf inference. See, for instance, Petrone [**25**], Escobar and West [**13**], and Hellmayr and Gelfand [**19**], and references therein.

Similarly, categorical data may be viewed in terms of contingency tables with prior beliefs that certain factor outcomes are likely to co-occur or to occur separately, or that outcomes are likely to share a majority of factors. Again, there is considerable work on Bayesian inference in this vein. For instance, see Agresti and Hitchcock [**2**], and books Agresti [**1**], Ghosal and Van der Vaart [**17**], and Congdon [**7**, **8**] and references therein.

Histogram smoothing: Toy problem. We recall informally a basic 'toy problem', with respect to the inference of the distribution of say shoe sizes, to set-up the main ideas. Suppose a shoe seller is opening a new shop in town and wants to know the distribution of shoe sizes of the town population before stocking the shelves. Suppose that a person's shoe size is determined by their foot length, and that foot lengths are approximately Normal in distribution. Then, of course, we would expect that a histogram of shoe sizes would look approximately like a binned Normal histogram.

The shoe seller records the shoe sizes from a sample of individuals in town. In this multinomial data, categories are shoe sizes. We have some prior understanding of the context. Shoe sizes have a lower and upper bound, and presumably most people have shoe sizes relatively in the middle. Moreover, prior knowledge that shoe sizes arise from a continuous Normally distributed factor (foot length), would indicate that gaps in the shoe size sample histogram are likely not present in the true histogram.

One could use a Dirichlet prior, conveniently conjugate with multinomial data, though we will see shortly that such a prior cannot encompass all of the prior knowledge. Suppose there are d possible shoe sizes/categories, numbered $1, 2, \ldots, d$. We specify a Dirichlet prior with parameters $(\theta\mu)$ where $\mu \in \Delta_d$ is the best guess at the shoe size probability mass function, and concentration parameter $\theta > 0$ represents the level of confidence in the best guess μ. If there is no 'best guess', one could take $\mu = (1/d)(1, \ldots, 1)$, the uniform stochastic vector, and θ small.

Let $\vec{f} \in \{0, 1, 2, \ldots\}^d = (f_1, f_2, \ldots, f_d)$ be the count vector from the sample of size n collected, where f_i is the number of people in the sample with shoe size i. With this data in hand, one updates the prior belief by computing the posterior distribution. In the case, if the prior is Dirichlet$(\theta\mu)$, the posterior would be Dirichlet$(\theta\mu + \vec{f})$. Then, the posterior estimate of the population distribution of shoe sizes would be the posterior mean $(\theta\mu + \vec{f})/(\theta + n)$.

As an example, consider sample shoe size data collected from 15 Normal samples binned into $d = 16$ shoe sizes. Suppose we specify a so-called non-informative prior with $\mu = (1/d)\vec{1}$ and $\theta = 4$. In Figure 1, we see the prior estimate of the pmf in the left plot (i.e. μ) represented as a histogram. In the middle plot is the empirical pmf computed from the sample. The posterior mean histogram, a weighted average of the left and middle plots, is seen in the right plot.

The posterior mean histogram is 'smoother,' or less jagged, in that the two gaps in the data histogram have been partially filled. However, a Dirichlet prior does not allow too much control: There is no notion of association between categories built in to the prior. As such, one wouldn't be able to impose in some way that an

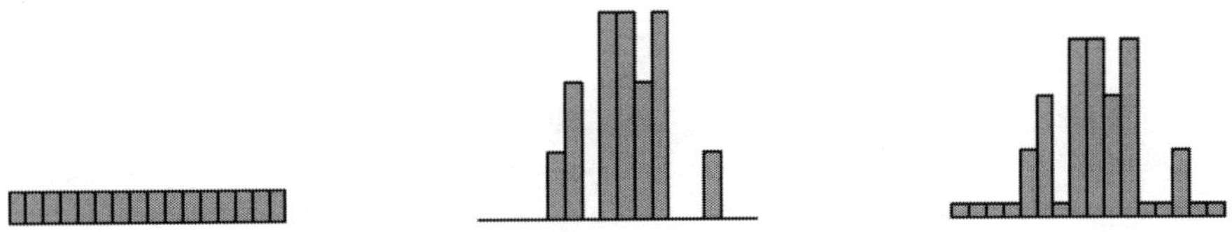

FIGURE 1. Left plot is the prior mean histogram for a Dirichlet prior with uniform mean; Middle plot is of data collected; Right plot is the posterior histogram.

empty bin between two 'tall' bars should be filled with a similarly 'tall' bar, or that an empty bin very far from any observed data should be left approximately empty.

In this context, we explore now use of a Markovian stick-breaking prior, which encodes associations between categories through specification of a network represented by the generator matrix G. In this general network, categories are nodes and edges, directed or undirected, specify affinity between categories. The adjacency matrix for this network is then formed into the generator matrix G by modifying diagonal entries appropriately to create generator matrix structure. Recall that the matrix G, in the form $G = \theta(Q - I)$, specifies the transition matrix Q for the Markovian sequence $\mathbf{T}$ as well as the parameter θ for the GEM sequence $\mathbf{P}$. Accordingly, counts in the different categories are associated not only with respect to the GEM $\mathbf{P}$ but also with respect to the Markovian $\mathbf{T}$.

We recall, in the Dirichlet$(\theta\mu)$ context, where $\mathbf{T}$ is an i.i.d. sequence with common distribution μ and $\mathbf{P}$ is GEM(θ), that the parameter θ is viewed as a 'strength', and can represent in a sense the number of data points equivalent to the prior 'belief'. The corresponding posterior mean mass function is the weighted average $(\theta\mu+\vec{f})/(\theta+n)$ where $(1/n)\vec{f}$ is the empirical data probability mass function. When $\theta \uparrow \infty$, the limit is the prior belief mean μ.

It is similar in the Markovian stick-breaking setting: If say the transition matrix Q representing the network is specified in advance, the parameter θ is also a sort of relative strength in that, as $\theta \uparrow \infty$, ν converges in probability to μ (Proposition 11).

Types of generators and associations. We now consider several ways, among others, in which a network or graph might be specified and an associated generator matrix G constructed.

In the context of this paper, graphs are connected, weighted, directed or undirected, and without self-loops. Weights should be nonnegative and the sum of weights of edges connected to (undirected) or coming into (directed) any one edge should have finite upper bound.

In general, once a graph has been specified, a generator matrix G is obtained from the adjacency matrix A of the graph by modifying the diagonal entries of A to give it a generator matrix structure. Note then that connectedness of the graph would ensure irreducibility of G. In the case of infinitely many categories, we would further demand that a graph result in a positive recurrent generator G.

Dirichlet graphs. The Dirichlet prior is a special case of the Markovian stick-breaking prior. For the purpose of comparison, we begin by specifying the graph or network associated with a Dirichlet$(\vec{\alpha})$ prior on d categories. The corresponding

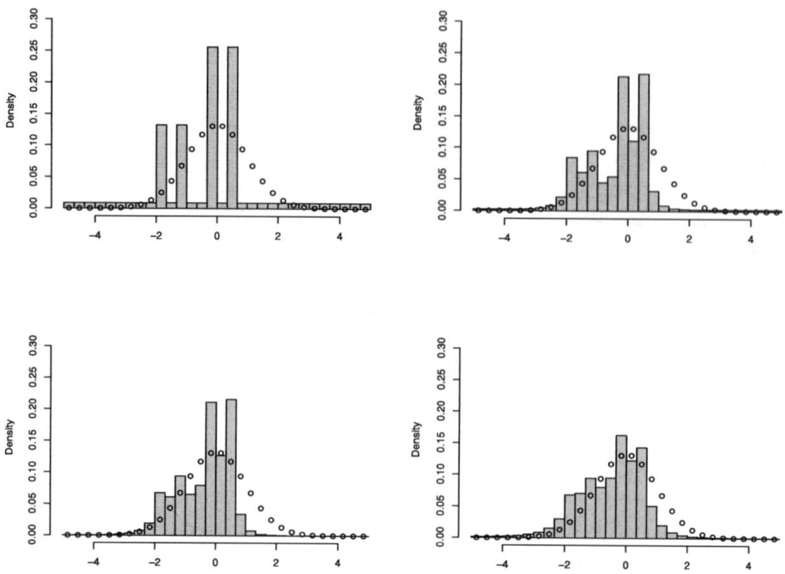

FIGURE 2. Binned Normal(0, 1) population pmf given by dotted curve; 6 samples across 30 bins in range $[-5, 5]$ (not pictured) with 1 point in bins 10, 12 and 2 points in bins 15, 17. Posterior mean mass functions from MSB(G) priors: Top left: G_1=Dirichlet$(w, \ldots, w)$, $w = 2/29$; Bottom left: G_2 =Tridiagonal with $w = 3$; Bottom right: G_3 =Tridiagonal with $w = 8$; Top right $G_4 = (G_1 + 2.5G_2)/3.5$.

graph on d nodes has a directed edge from node i to node j of weight α_j for each ordered pair of distinct nodes (i, j). Thus, for every node j, all incoming edges have weight α_j independent of the originating node, disallowing for special associations between pairs of nodes. The adjacency matrix A for this graph is constant with $A_{ij} = \alpha_j$, and the associated generator matrix G has the same off diagonal entries and diagonal entry $G_{jj} = \alpha_j - \sum_{i=1}^{d} \alpha_i$.

Geometric graphs. When categorical data arise from binning continuous data, categories come with a geometric arrangement. For ease, suppose the continuous data is real-valued data, and so categories (intervals in which the continuous data occur) come linearly ordered. This geometric arrangement can be reflected in a graph with categories represented by nodes and an undirected edge of weight w placed between each pair of adjacent categories, forming a line segment. The adjacency matrix A for such a graph has w in the first upper diagonal and first lower diagonal entries and zeros elsewhere. The associated generator matrix G has the same off diagonal entries as A and the necessary diagonal entries for generator structure. We will refer to this type of generator G as 'tridiagonal' with weight w. We mention that the prior MSB(G) mean, in this case, would be uniform. Moreover, the weight w represents a relative strength, and can be related to θ when G is put in form $G = \theta(Q - I)$. By increasing w, the 'smoothing' effect, relative to the geometry, on the posterior mean estimate of the pmf will strengthen.

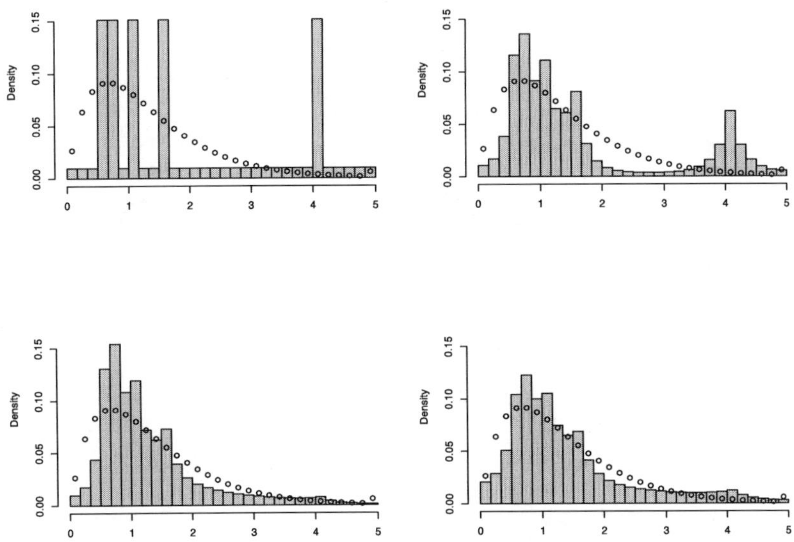

FIGURE 3. Binned Gamma(2, 1.5) population pmf given by dotted curve; 5 samples across 30 bins in range [0,8] (not pictured) with 1 point in bins $1, 2, 3, 7, 16$. Posterior mean mass functions from MSB(G) priors: Top left: G_1=Dirichlet$(w, \ldots, w)$, $w = 2/29$; Bottom left: G_2 =Tridiagonal with $w = 8$; Bottom right: G_3 =Tridiagonal with $w = 16$; Top right $G_4 = (G_1 + 2.5G_2)/3.5$.

There are of course other relevant settings. For instance, suppose the continuous data were angle data taking values on the circle and having full support. In such a case, the corresponding graph would be a cycle graph on d categories, and the corresponding 'wrapped' generator matrix would be obtained by modifying the tridiagonal generator with weight w to have entries $G_{d,1} = G_{1,d} = w$ and $G_{11} = G_{dd} = -2w$.

More complicated geometries can also be envisioned, for instance when the categories of interest are regions in a mesh of a many-dimensional setting.

Contingency tables. The Markovian stick-breaking prior might also be used for multi-factor categorical data, where a single data point is of the form $x = (s_1, s_2, \ldots, s_k) \in \mathfrak{X} = S_1 \times S_2 \times \cdots \times S_k$, representing k categorical factors observed, where S_i is the set of possible outcomes of factor i. As an example, one might simultaneously observe eye and hair color of individuals. Then $k = 2$ and $S = \{\text{eye colors}\} \times \{\text{hair colors}\}$, and a single observation might be (brown eyes, black hair). In certain contexts, such as genetics, we might have prior reason to believe that similar outcomes (differing by only a few factors) are similarly likely to occur in the population. Thus, a prior distribution on $\Delta_{\mathfrak{X}}$ should put more weight on distributions where similar outcomes have similar probabilities of occurring.

In specifying such an MSB prior on $\Delta_{\mathfrak{X}}$, we might translate the notion of similar outcomes into a network. For example: For two outcomes $x = (s_1, s_2, \ldots, s_k)$ and $y = (t_1, t_2, \ldots, t_k)$, place an undirected edge of weight w between them only if the outcomes are identical for all but one factor $s_j \neq t_j$. Such a prior associates any two

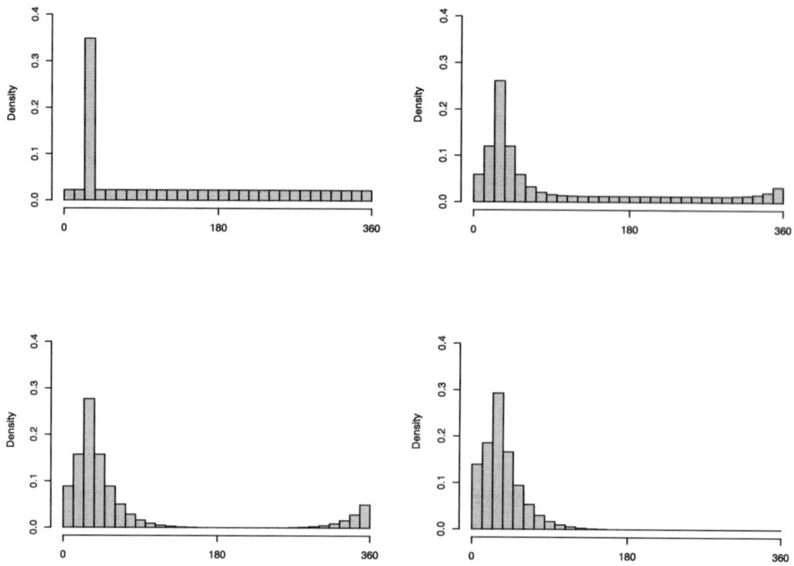

FIGURE 4. Wrapped vs unwrapped; 1 sample, across 30 bins in degree range [0, 360] (not pictured), in bin 3. Posterior mean mass functions from MSG(G) priors: Top left: G_1=Dirichlet$(w,\dots,w)$, $w = 2/29$; Bottom left: G_2 =Wrapped tridiagonal with $w = 3$; Bottom right: G_3 =(unwrapped) Tridiagonal with $w = 3$; Top right $G_4 = (G_1 + 2.5G_2)/3.5$.

outcomes differing only by a single factor. Interestingly, as the associated generator matrix is by construction lumpable according to each factor, this joint MSB prior on $\Delta_{\mathfrak{X}}$ has marginal Dirichlet$(w, w, \dots, w)$ prior on Δ_{S_i} for each factor. Similarly, we might specify a joint prior on $\Delta_{\mathfrak{X}}$ with pre-specified MSB$(G^{(i)})$ marginals on each Δ_{S_i} which encodes closeness of similar outcomes in $\mathfrak{X}$ by defining a joint generator matrix $G_{x,y} = G^{(j)}_{s_j,t_j}$ for $x = (s_1, s_2, \dots, s_k)$ and $y = (t_1, t_2, \dots, t_k)$ identical for all but one factor $s_j \neq t_j$, and $G_{x,y} = 0$ otherwise for $x \neq y$.

Directed vs undirected graphs. Since a graph with undirected edges corresponds to a symmetric adjacency matrix, the associated Markovian stick-breaking prior will correspond to a symmetric G with a uniform stationary vector. Necessarily then, an MSB, with non-uniform prior mean vector, corresponds to a directed graph. Note that some directed graphs also produce a uniform mean stationary vector, such as a directed cycle graph with equal weights.

One might envision using directed graphs in settings where there is a 'hierarchy', such as in employee data in different levels of management, for instance.

Simple numerical experiments. In Theorem 4, we have computed the posterior mean estimate of the probability mass function given that the prior is a Markovian stick-breaking measure with generator G and the empirical counts $\vec{k}$ of observed data. Specifically, let $\mathfrak{X}$ denote the set of categories and let $(\nu, T_1) \sim$ MSB(G), where ν is the prior. For a data vector $k = (k_w)_{w\in\mathfrak{X}}$ of non-negative integers and a category $x \in \mathfrak{X}$, define $v(k) = \big(v_x(k)\big)_{x\in\mathfrak{X}}$ by

$$v_x(k) = \mathcal{E}\left[\prod_{w\in\mathfrak{X}} \nu(w)^{k_w}\middle| T_1 = x\right] = \binom{n}{k}^{-1} \sum_{\sigma\in\mathbb{S}(k)} \left(I - \frac{G}{n}\right)^{-1}_{x,\sigma_n} \prod_{j=1}^{n-1} \left(I - \frac{G}{j}\right)^{-1}_{\sigma_{j+1},\sigma_j}$$

where $n = \sum_{w\in\mathfrak{X}} k_w$ and $\mathbb{S}(k)$ denotes the set of distinct permutations of a list containing precisely k_w many w's for each $w \in \mathfrak{X}$. Then, the posterior probability mass function, specified in terms of the posterior means, given the observed multinomial counts k, when evaluated at $x \in \mathfrak{X}$, is given by

$$p(x|k) = \frac{\mu^T v(k + e_x)}{\mu^T v(k)} = \frac{\sum_{w\in\mathfrak{X}} \mu_w v_w(k + e_x)}{\sum_{w\in\mathfrak{X}} \mu_w v_w(k)}$$

where μ is the stationary vector of G.

We consider now simple computational experiments to see how different generators G, with respect to Markovian stick-breaking priors, affect the posterior mean probability mass function, computed exactly from the above formulas with a small number of samples, in two types of data, one with Normal and the other with Gamma samples. We will consider G's, which are Dirichlet, tri-diagonal, and averages between these types, to see the effects.

In Figure 2, with respect to a generated Normal$(0, 1)$ sample histogram of 6 samples, across 30 bins from -5 to 5, with 1 point in bins 10 and 12 and 2 points in bins 15 and 17, posterior mean mass functions are plotted with respect to four Markovian stick-breaking priors. In the top left plot, the generator G_1 corresponds to a Dirichlet$(w, \ldots, w)$ where $w = 2/29$. In the bottom left and bottom right, the generators G_2 and G_3 are a tridiagonal matrices with $w = 3$ and $w = 8$ entries in the two off-diagonals respectively. In the top right, the generator G_4 is the average $G_4 = (G_1 + 2.5G_2)/(3.5)$.

Similarly, in Figure 3, with respect to a generated Gamma$(2, 1.5)$ sample histogram of 5 samples, again across 30 bins from 0 to 8, with 1 point in bins $1, 2, 3, 7$ and 16, posterior mean mass functions are plotted with respect to similar priors in the same locations as in Figure 2.

In Figure 4, the intent is to see the posterior mean mass function effects, with respect to one data point in bin 3, across 30 bins indexed by angles (degrees) of a circle, when the priors correspond to generators which are wrapped tri-diagonal G_2 with $w = 3$ in the bottom left, $G_1 =$Dirichlet$(w, \ldots, w)$ with $w = 2/29$ in the top left, their average $G_4 = (G_1 + 2.5G_2)/(3.5)$ in the top right, and an unwrapped tri-diagonal generator G_3 with $w = 3$ in the bottom right.

Discussion. Briefly, we were interested to see what effects might arise from using Markovian stick-breaking priors in probability mass function inference. We observe in Figures 2 and 3 that the posterior mean mass functions, computed from Markovian stick-breaking priors with tridiagonal G's in the bottom left and right, show clear effects due to the network affinities encoded in the generators in comparison to the posterior mean mass function with respect to the Dirichlet prior in the top left. The posterior mean mass function with respect to the prior built with the averaged G_4 generator incorporates a similarity structure with some positive weight between all categories, but with emphasis on neighbor categories. In Figure 4, one definitely sees the effect of wrapping in the bottom left, and also the averaging effect where all bins receive non-negligible mass in the top right.

It would seem that similarities between categories encoded in the generator G do affect the posterior distribution when the prior is a Markovian stick-breaking process with generator G. In terms of future work, there are of course several natural directions to pursue, among them to clarify more the scope and performance of these Markovian stick-breaking priors in various categorical network settings.

5. Proof of Theorem 4

We begin by enumerating some facts.

FACT 1. Let $\mathbf{P} \sim \mathrm{GEM}(\theta)$. Then $P_1 \sim \mathrm{Beta}(1,\theta)$ and

$$\mathcal{E}\big[(1-P_1)^j P_1^{k-j}\big] = \frac{\Gamma(1+\theta)}{\Gamma(1)\Gamma(\theta)}\frac{\Gamma(1+k-j)\Gamma(\theta+j)}{\Gamma(1+\theta+k)} = \frac{\theta\Gamma(k-j+1)\Gamma(\theta+j)}{\Gamma(\theta+k+1)}$$

$$\mathcal{E}\big[(1-P_1)^k\big] = \frac{\theta\Gamma(1)\Gamma(\theta+k)}{\Gamma(\theta+k+1)} = \frac{\theta}{\theta+k} \tag{1}$$

FACT 2. Let $G = \theta(Q-I)$ be a generator matrix and Q stochastic and $\theta > 0$. When $k > 0$:

$$\left(I - \frac{\theta Q}{\theta+k}\right)^{-1} = \frac{\theta+k}{k}(I-G/k)^{-1} \text{ and } Q(I-G/k)^{-1} = \frac{\theta+k}{\theta}\left((I-G/k)^{-1} - \frac{kI}{\theta+k}\right) \tag{2}$$

FACT 3. Consider the space $\{0,1,2,\ldots\}^n$ of non-negative integer n-vectors. For two vectors $\vec{k}, \vec{l} \in \{0,1,2,\ldots\}^n$, we say $\vec{l} < \vec{k}$ if for each $1 \le j \le n$, we have $l_j \le k_j$, and for some $1 \le j \le n$, in fact $l_j < k_j$. Note that this gives a strict partial ordering to all non-negative n-vectors; that the zero vector is strictly less than every other vector; and that each $\vec{k}$ is strictly greater than only finitely many n-vectors. Thus, for each n, the space is well-founded and an induction may be considered with respect to this partial ordering starting from 0.

FACT 4. For an n-vector $\vec{k}$ of non-negative integer entries, with $k = \sum_{j=1}^n k_j > 0$,

$$\#\mathbb{S}(\vec{k}) = \binom{k}{k_1, k_2, \ldots, k_n} = \frac{\Gamma(k+1)}{\prod_{j=1}^n \Gamma(k_i+1)} \tag{3}$$

The following proposition will help an induction in the proof of Theorem 4.

PROPOSITION 12. *Let G be an irreducible, positive recurrent generator matrix on $\mathfrak{X}$ and let $(\nu, T_1) \sim MSB(G)$. Then, for each $k \in \{0,1,2,\ldots\}$, $A \subseteq \mathfrak{X}$, and $x \in \mathfrak{X}$, we have*

$$\mathcal{E}\Big[\nu(A)^k \Big| T_1 = x\Big] = e_x^T \prod_{j=1}^k \Big((I-G/j)^{-1}D(A)\Big)\vec{1} \tag{4}$$

PROOF. Since (4) is a statement regarding the distribution of (ν, T_1), we can choose a particular instance of (ν, T_1) constructed from an independent pair $\mathbf{X} = (X_j)_{j=1}^\infty$ and $\mathbf{T} = (T_j)_{j=1}^\infty$ of, respectively, an i.i.d. sequence of $\mathrm{Beta}(1,\theta)$ variables and a stationary, homogeneous Markov chain with transition kernel Q, where $G = \theta(Q-I)$.

As usual, let $\mathbf{P}$ be defined with respect to $\mathbf{X}$ by $P_j = X_j \prod_{i=1}^{j-1}(1-X_i)$. For ease of notation, define the vector $v(k,A) = \big(v_x(k,A)\big)_{x\in\mathfrak{X}}$ by $v_x(k,A) = \mathcal{E}\big[\nu(A)^k \big| T_1 = x\big]$. We begin by finding a recursive (in k) formula for $v(k,A)$.

To this end, we define $\nu^* = \sum_{j=2}^{\infty}\Big[X_j \prod_{i=2}^{j-1}(1-X_i)\Big]\delta_{T_j}$ and note that ν^* is independent of $X_1 = P_1$ since $\mathbf{X}$ is i.i.d. and independent of $\mathbf{T}$. Furthermore, $\nu = P_1\delta_{T_1} - (1-P_1)\nu^*$. Write

$$
\begin{aligned}
e_x^T v(k,A) \;=\; v_x(k,A) \;&=\; \mathcal{E}\Big[\left(P_1\delta_x(A) + (1-P_1)\nu^*(A)\right)^k \left|T_1 = x\right.\Big] \\
&= \sum_{y\in\mathfrak{X}} \mathcal{P}(T_2 = y|T_1 = x)\mathcal{E}\Big[\left(P_1\delta_x(A) + (1-P_1)\nu^*(A)\right)^k \left|T_1 = x, T_2 = y\right.\Big] \\
&= \sum_{y\in\mathfrak{X}} Q_{xy}\bigg\{\mathcal{E}\Big[\left((1-P_1)\nu^*(A)\right)^k \big|T_1 = x, T_2 = y\Big] \\
&\qquad + \delta_x(A)\sum_{j=0}^{k-1}\binom{k}{j}\mathcal{E}\Big[P_1^{k-j}(1-P_1)^j\left(\nu^*(A)\right)^j\big|T_1 = x, T_2 = y\Big]\bigg\}
\end{aligned}
\tag{5}
$$

Clearly, as $\mathbf{X}$ and $\mathbf{T}$ are independent and $\mathbf{X}$ is i.i.d., $P_1 = X_1$ is independent of T_1, T_2, and ν^*. By the Markov property and since ν^* is not a function of T_1, we have $\nu^*|(T_2 = y)$ is independent of T_1. Furthermore, since $(X_j)_{j\geq 1} \overset{d}{=} (X_j)_{j\geq 2}$ as an i.i.d. sequence and $(T_j)_{j\geq 1} \overset{d}{=} (T_j)_{j\geq 2}$ as a stationary Markov chain, we have $(\nu, T_1) \overset{d}{=} (\nu^*, T_2)$, implying $\nu^*|(T_1 = x, T_2 = y) \overset{d}{=} \nu|(T_1 = y)$.

Thus, defining $\nu(0,A) = \vec{1}$, equation (5) becomes

$$
\begin{aligned}
&= \sum_{y\in\mathfrak{X}} Q_{xy}\Bigg[\mathcal{E}\left[(1-P_1)^k\right]\mathcal{E}\left[\nu(A)^k|T_1 = y\right] \\
&\qquad\qquad + \delta_x(A)\sum_{j=0}^{k-1}\binom{k}{j}\mathcal{E}[P_1^{k-j}(1-P_1)^j]\,\mathcal{E}\left[\nu(A)^j|T_1 = y\right]\Bigg] \\
&= \sum_{y\in\mathfrak{X}} Q_{xy}\left[\mathcal{E}\left[(1-P_1)^k\right]v_y(k,A) + \delta_x(A)\sum_{j=0}^{k-1}\binom{k}{j}\mathcal{E}\left[P_1^{k-j}(1-P_1)^j\right]v_y(j,A)\right] \\
&= e_x^T\left[\mathcal{E}\left[(1-P_1)^k\right]Qv(k,A) + D(A)Q\sum_{j=0}^{k-1}\binom{k}{j}\mathcal{E}\left[P_1^{k-j}(1-P_1)^j\right]v(j,A)\right],
\end{aligned}
$$

which, noting (1), equals

$$
= e_x^T\left[\frac{\theta}{\theta+k}Qv(k,A) + D(A)Q\sum_{j=0}^{k-1}\binom{k}{j}\frac{\theta\Gamma(k-j+1)\Gamma(\theta+j)}{\Gamma(\theta+k+1)}v(j,A)\right].
$$

Since the statement holds for every x, it follows that

$$v(k,A) = \frac{\theta}{\theta+k}Qv(k,A) + D(A)Q\sum_{j=0}^{k-1}\binom{k}{j}\frac{\theta\Gamma(k-j+1)\Gamma(\theta+j)}{\Gamma(\theta+k+1)}v(j,A)$$
$$= \frac{\theta}{\theta+k}Qv(k,A) + \frac{\theta\Gamma(k+1)}{\Gamma(\theta+k+1)}D(A)Q\sum_{j=0}^{k-1}\frac{\Gamma(\theta+j)}{\Gamma(j+1)}v(j,A)$$

and $\left(I-\frac{\theta Q}{\theta+k}\right)v(k,A) = \frac{\theta\Gamma(k+1)}{\Gamma(\theta+k+1)}D(A)Q\sum_{j=0}^{k-1}\frac{\Gamma(\theta+j)}{\Gamma(j+1)}v(j,A)$. Then,

$$v(k,A) = \frac{\theta\Gamma(k+1)}{\Gamma(\theta+k+1)}\left(I-\frac{\theta Q}{\theta+k}\right)^{-1}D(A)Q\sum_{j=0}^{k-1}\frac{\Gamma(\theta+j)}{\Gamma(j+1)}v(j,A)$$
$$= \frac{\theta\Gamma(k)}{\Gamma(\theta+k)}\left(I-\frac{G}{k}\right)^{-1}D(A)Q\sum_{j=0}^{k-1}\frac{\Gamma(\theta+j)}{\Gamma(j+1)}v(j,A), \tag{6}$$

where the last line follows from (2).

We now solve the recursion for $v(k,A)$ inductively. We have already specified $v(0,A)=\vec{1}$. By (6), we have

$$v(1,A) = \frac{\theta\Gamma(1)}{\Gamma(\theta+1)}\left(I-G/1\right)^{-1}D(A)Q\frac{\Gamma(\theta)}{\Gamma(1)}v(0,A)$$
$$= \left(I-G/1\right)^{-1}D(A)Q\vec{1} = \left(I-G/1\right)^{-1}D(A)\vec{1}$$

If, for $1\le j\le k-1$, $v(j,A) = \prod_{i=1}^{j}\left((I-G/i)^{-1}D(A)\right)\vec{1}$, then it follows from (6) that $v(k,A) = \frac{\theta\Gamma(k)}{\Gamma(\theta+k)}\left(I-G/k\right)^{-1}u_k$ where

$$u_k = D(A)Q\sum_{j=0}^{k-1}\frac{\Gamma(\theta+j)}{\Gamma(j+1)}\prod_{i=1}^{j}\left((I-G/i)^{-1}D(A)\right)\vec{1}.$$

We now claim that $u_k = w_k$ where

$$w_k = \frac{\Gamma(\theta+k)}{\theta\Gamma(k)}D(A)\prod_{i=1}^{k-1}\left((I-G/i)^{-1}D(A)\right)\vec{1}.$$

Indeed, if $u_k = w_k$, we would conclude that

$$v(k,A) = \frac{\theta\Gamma(k)}{\Gamma(\theta+k)}\left(I-\frac{G}{k}\right)^{-1}w_k = \prod_{i=1}^{k}\left((I-G/i)^{-1}D(A)\right)\vec{1},$$

finishing the proof of Proposition 12.

To verify the claim, observe that $u_1 = \Gamma(\theta)D(A)Q\vec{1} = \Gamma(\theta)D(A)\vec{1} = \big(\Gamma(\theta+1)/\theta\big)D(A)\vec{1} = w_1$. Suppose that $u_j = w_j$ for $j\le k$. Then,

$$u_{k+1}-w_{k+1} = u_k + \frac{\theta}{k}D(A)Q(I-G/k)^{-1}w_k - \frac{\theta+k}{k}D(A)(I-G/k)^{-1}w_k$$
$$= u_k - D(A)w_k = u_k - w_k = 0,$$

as $(D(A))^2 = D(A)$, finishing the proof. □

PROOF OF THEOREM 4. The theorem holds trivially for $\vec{k} = \vec{0}$. If $\vec{k} \neq \vec{0}$, without loss of generalization, we assume $\vec{k}$ has strictly positive entries. Otherwise, it may be represented as a vector $\vec{k}^\sharp$ of smaller length by omitting the zero entries, with $\vec{A}^\sharp$ the corresponding shortened vector of sets and $n^\sharp$ the new vector length.

As in the proof of the Proposition 12, we may choose a particular instance of (ν, T_1) constructed from an independent pair $\mathbf{X} = (X_j)_{j=1}^\infty$ and $\mathbf{T} = (T_j)_{j=1}^\infty$ of, respectively, an i.i.d. sequence of Beta$(1,\theta)$ variables and a stationary, homogeneous Markov chain with transition kernel Q, where $G = \theta(Q - I)$. Let $\mathbf{P}$ be defined with respect to $\mathbf{X}$ by $P_j = X_j \prod_{i=1}^{j-1}(1 - X_i)$.

Define now the vector $v(\vec{k}, \vec{A}) = \left(v_x(\vec{k}, \vec{A})\right)_{x\in\mathfrak{X}}$ by $v_x(\vec{k}, \vec{A}) = \mathcal{E}\left[\prod_{j=1}^n \nu(A_j)^{k_j} \middle| T_1 = x\right]$. We begin by finding a recursive (in $\vec{k}$ and n) formula for $v(\vec{k}, \vec{A})$, and then we solve the recursion using Lemma 13, stated at the end of the section.

To this end, recall the definition $\nu^* = \sum_{j=2}^\infty \left[X_j \prod_{i=2}^{j-1}(1-X_i)\right]\delta_{T_j}$. We compute

$$\begin{aligned} e_x^T v(\vec{k}, \vec{A}) &= \mathcal{E}\left[\prod_{j=1}^n (P_1\delta_{T_1}(A_j) + (1 - P_1)\nu^*(A_j))^{k_j} \middle| T_1 = x\right] \\ &= \sum_{y\in\mathfrak{X}} \mathcal{P}(T_2 = y | T_1 = x)\mathcal{E}\left[\prod_{j=1}^n (P_1\delta_x(A_j) + (1 - P_1)\nu^*(A_j))^{k_j} \middle| T_1 = x, T_2 = y\right], \end{aligned}$$

which equals, as the collection $\vec{A}$ consists of disjoint set so that $\delta_x(A_i)\delta_x(A_j) = 0$ for $i \neq j$,

$$\begin{aligned} (7)\quad \sum_{y\in\mathfrak{X}} Q_{x,y} \Bigg\{ &\mathcal{E}\left[\prod_{j=1}^n ((1 - P_1)\nu^*(A_j))^{k_j} \middle| T_1 = x, T_2 = y\right] \\ &+ \sum_{i=1}^n \delta_x(A_i) \sum_{l=0}^{k_i-1} \binom{k_i}{l} \mathcal{E}\Bigg[(1 - P_1)^l P_1^{k_i-l} \nu^*(A_i)^l \\ &\times \prod_{1\le j\le n;\ j\neq i} ((1 - P_1)\nu^*(A_j))^{k_j} \Bigg| T_1 = x, T_2 = y\Bigg]\Bigg\}. \end{aligned}$$

Recall the relations among P_1, ν, ν^*, T_1 and T_2 stated below (5). Then, by Fact 1, we have that (7) equals

$$\begin{aligned} &\sum_{y\in\mathfrak{X}} Q_{x,y} \Bigg[\mathcal{E}\left[(1 - P_1)^k\right] \mathcal{E}\left[\prod_{j=1}^n \nu(A_j)^{k_j} \middle| T_1 = y\right] \\ &\quad + \sum_{i=1}^n \delta_x(A_i) \sum_{l=0}^{k_i-1} \binom{k_i}{l} \mathcal{E}\left[(1 - P_1)^{k-k_i+l} P_1^{k_i-l}\right] \mathcal{E}\left[\nu(A_i)^l \prod_{1\le j\le n;\ j\neq i} (\nu(A_j))^{k_j} \middle| T_1 = y\right]\Bigg] \\ &= \sum_{y\in\mathfrak{X}} Q_{x,y} \Bigg[\frac{\theta}{\theta + k} v_y(\vec{k}, \vec{A}) + \\ &\quad + \sum_{i=1}^n \delta_x(A_i) \sum_{l=0}^{k_i-1} \binom{k_i}{l} \frac{\theta\Gamma(k_i - l + 1)\Gamma(\theta + k - k_i + l)}{\Gamma(\theta + k + 1)} v_y(\vec{k} + (l - k_i)e_i, \vec{A})\Bigg] \\ &= e_x^T \Bigg[\frac{\theta}{\theta + k} Q v(\vec{k}, \vec{A}) \\ &\quad + \sum_{i=1}^n \frac{\theta\Gamma(k_i + 1)}{\Gamma(\theta + k + 1)} D(A_i) Q \sum_{l=0}^{k_i-1} \frac{\Gamma(\theta + k - k_i + l)}{\Gamma(l + 1)} v(\vec{k} + (l - k_i)e_i, \vec{A})\Bigg] \end{aligned}$$

Since the above computation holds for every x, it may be written as a vector equation:

(8)

$$\begin{aligned} v(\vec{k},\vec{A}) &= \frac{\theta}{\theta+k}Qv(\vec{k},\vec{A}) + \sum_{i=1}^{n}\frac{\theta\Gamma(k_i+1)}{\Gamma(\theta+k+1)} \\ &\qquad \times D(A_i)Q\sum_{l=0}^{k_i-1}\frac{\Gamma(\theta+k-k_i+l)}{\Gamma(l+1)}v(\vec{k}+(l-k_j)e_j,\vec{A}) \\ &= \left(I-\frac{\theta Q}{\theta+k}\right)^{-1}\sum_{i=1}^{n}\frac{\theta\Gamma(k_i+1)}{\Gamma(\theta+k+1)} \\ &\qquad \times D(A_i)Q\sum_{l=0}^{k_i-1}\frac{\Gamma(\theta+k-k_i+l)}{\Gamma(l+1)}v(\vec{k}+(l-k_j)e_j,\vec{A}) \\ &= (I-G/k)^{-1}\sum_{i=1}^{n}\frac{\theta\Gamma(k_i+1)}{k\Gamma(\theta+k)} \\ &\qquad \times D(A_i)Q\sum_{l=0}^{k_i-1}\frac{\Gamma(\theta+k-k_i+l)}{\Gamma(l+1)}v(\vec{k}+(l-k_i)e_i,\vec{A}). \end{aligned}$$

The recursive formula (8) for $v(\vec{k},\vec{A})$ is in terms of the values of $v(\vec{l},\vec{A})$ only for $\vec{l}<\vec{k}$. If $\vec{k}$ has r zero entries and at least one positive entry, recall $\vec{k}^\sharp$, the reduction of $\vec{k}$ to a strictly positive $(n-r)$-vector by removal of zero entries, with $\vec{A}^\sharp$ corresponding. Then $v(\vec{k},\vec{A})=v(\vec{k}^\sharp,\vec{A}^\sharp)$. Thus, we consider simultaneously an induction on the value of n and, given n, an induction on the n-vector $\vec{k}$ according to the strict partial ordering from Fact 3.

When $n=1$, the theorem holds by Proposition 12. This is the base case for induction on n. Suppose by way of induction on n that, for each $1\le m<n$ and, given m, each non-negative integer m-vector $\vec{l}$ and m-vector $\vec{B}$ of disjoint subsets of $\mathfrak{X}$, we have that the theorem holds for $v(\vec{l},\vec{B})$.

Consider a non-negative integer n-vector $\vec{k}$ with at least one positive entry and an n-vector $\vec{A}$ of disjoint subsets of $\mathfrak{X}$. If $\vec{k}$ has any zero-entries, by the induction assumption on n,

$$v(\vec{k},\vec{A}) = v(\vec{k}^\sharp,\vec{A}^\sharp) = \left(\#\mathbb{S}(\vec{k})\right)^{-1}\sum_{\sigma\in\mathbb{S}(\vec{k})}\left[\prod_{r=1}^{k}\left((I-G/r)^{-1}D(A_{\sigma(j)})\right)\right]\vec{1}.$$

This is the base case for an induction on $\vec{k}\in\{0,1,2,\ldots\}^n$.

Suppose instead that $\vec{k}$ consists of positive integers. Given n, suppose by way of induction on $\vec{k}\in\{0,1,2,\ldots\}^n$ that for every $\vec{l}\in\{0,1,2,\ldots\}^n$ with $\vec{l}<\vec{k}$, the theorem holds for $v(\vec{l},\vec{A})$. Then, we have

$$\begin{aligned} v(\vec{k},\vec{A}) &= \frac{\theta}{k\Gamma(\theta+k)}(I-G/k)^{-1}\sum_{i=1}^{n}\Gamma(k_i+1) \\ &\qquad \times D(A_i)Q\sum_{l=0}^{k_i-1}\frac{\Gamma(\theta+k-k_i+l)}{\Gamma(l+1)}v(\vec{k}+(l-k_i)e_i,\vec{A}) \end{aligned}$$

equals, using (8),

$$\frac{\theta}{k\Gamma(\theta+k)}(I-G/k)^{-1}\sum_{i=1}^{n}\Gamma(k_i+1)D(A_i)Q\sum_{l=0}^{k_i-1}\frac{\Gamma(\theta+k-k_i+l)}{\Gamma(l+1)} \tag{9}$$
$$\times\left(\#\mathbb{S}(\vec{k}+(l-k_i)e_i)\right)^{-1}\sum_{\sigma\in\mathbb{S}(\vec{k}+(l-k_i)e_i)}\left[\prod_{r=1}^{k-k_i+l}((I-G/r)^{-1}D(A_{\sigma(j)}))\right]\vec{1}$$

Recalling (3), it then follows that (9) equals

$$\frac{\theta}{k\Gamma(\theta+k)}(I-G/k)^{-1}\sum_{i=1}^{n}\Gamma(k_i+1)D(A_i)Q\sum_{l=0}^{k_i-1}\frac{\Gamma(\theta+k-k_i+l)}{\Gamma(l+1)}$$
$$\times\frac{\Gamma(l+1)\prod_{j\neq i}\Gamma(k_j+1)}{\Gamma(k-k_i+l+1)}\sum_{\sigma\in\mathbb{S}(\vec{k}+(l-k_i)e_i)}\left[\prod_{r=1}^{k-k_i+l}((I-G/r)^{-1}D(A_{\sigma(j)}))\right]\vec{1}$$
$$=\frac{\prod_{j=1}^{n}\Gamma(k_j+1)}{k\Gamma(\theta+k)}(I-G/k)^{-1}\sum_{i=1}^{n}\theta D(A_i)Q\sum_{l=0}^{k_i-1}\frac{\Gamma(\theta+k-k_i+l)}{\Gamma(k-k_i+l+1)}$$
$$\times\sum_{\sigma\in\mathbb{S}(\vec{k}+(l-k_i)e_i)}\left[\prod_{r=1}^{k-k_i+l}((I-G/r)^{-1}D(A_{\sigma(j)}))\right]\vec{1}. \tag{10}$$

By Lemma 13, at the end of the section, (10) equals

$$\frac{\prod_{j=1}^{n}\Gamma(k_j+1)}{k\Gamma(\theta+k)}(I-G/k)^{-1}\sum_{i=1}^{n}\frac{\Gamma(\theta+k)}{\Gamma(k)}D(A_i)$$
$$\times\sum_{\sigma\in\mathbb{S}(\vec{k}-e_i)}\left[\prod_{j=1}^{k-1}((I-G/r)^{-1}D(A_{\sigma(j)}))\right]\vec{1}$$
$$=\frac{\prod_{j=1}^{n}\Gamma(k_j+1)}{\Gamma(k+1)}\sum_{\sigma\in\mathbb{S}(\vec{k})}\left[\prod_{j=1}^{k}((I-G/r)^{-1}D(A_{\sigma(j)}))\right]\vec{1}$$
$$=\left(\#\mathbb{S}(\vec{k})\right)^{-1}\sum_{\sigma\in\mathbb{S}(\vec{k})}\left[\prod_{j=1}^{k}((I-G/r)^{-1}D(A_{\sigma(j)}))\right]\vec{1}.$$

By induction on $\vec{k}$, the statement of the theorem holds for all $\vec{k}\in\{0,1,2,\ldots\}^n$. By induction on n, the theorem holds for all n as well. This completes the proof. □

We now state and prove the lemma referred to in the argument for Theorem 4.

LEMMA 13. *Let $m \geq 1$, $(A_j)_1^n$ be disjoint sets, $\vec{k} \in \{0,1,2,\ldots\}^{n-1}$, and $\tilde{k} = \sum_{j=1}^{n-1} k_j$. Then,*

$$\theta D(A_n) Q \sum_{l=0}^{m-1} \frac{\Gamma(\theta+\tilde{k}+l)}{\Gamma(\tilde{k}+l+1)} \sum_{\sigma\in\mathbb{S}(\vec{k},l)} \prod_{j=1}^{\tilde{k}+l} \left[(I-G/j)^{-1} D(A_{\sigma(j)})\right] \vec{1}$$
$$= \frac{\Gamma(\theta+\tilde{k}+m)}{\Gamma(\tilde{k}+m)} D(A_n) \sum_{\sigma\in\mathbb{S}(\vec{k},m-1)} \prod_{j=1}^{\tilde{k}+m-1} \left[(I-G/j)^{-1} D(A_{\sigma(j)})\right] \vec{1} \tag{11}$$

where $\mathbb{S}(\vec{k},l) = \mathbb{S}((k_1,\ldots,k_{n-1},l))$.

PROOF. We prove the lemma by induction. Define the left-hand side of (11) as u_m. Then, by (2) and that $D(A_i)D(A_j) = D(A_i)\delta_i(j)$, we have

$$\begin{aligned} u_1 &= \theta D(A_n) Q \frac{\Gamma(\theta+\tilde{k})}{\Gamma(\tilde{k}+1)} \sum_{\sigma\in\mathbb{S}(\vec{k},0)} \prod_{j=1}^{\tilde{k}+0} \left[(I-G/j)^{-1} D(A_{\sigma(j)})\right] \vec{1} \\ &= \frac{\theta+k}{\theta}\theta \frac{\Gamma(\theta+\tilde{k})}{\Gamma(\tilde{k}+1)} D(A_n) \sum_{\sigma\in\mathbb{S}(\vec{k},0)} \prod_{j=1}^{\tilde{k}+0} \left[(I-G/j)^{-1} D(A_{\sigma(j)})\right] \vec{1} \\ &= \frac{\Gamma(\theta+\tilde{k}+1)}{\Gamma(\tilde{k}+1)} D(A_n) \sum_{\sigma\in\mathbb{S}(\vec{k},0)} \prod_{j=1}^{\tilde{k}+0} \left[(I-G/j)^{-1} D(A_{\sigma(j)})\right] \vec{1}. \end{aligned}$$

By (2) and $D(A_i)D(A_j) = D(A_i)\delta_i(j)$ again,

$$\begin{aligned} u_{m+1} &= u_m + \theta D(A_n) Q \frac{\Gamma(\theta+\tilde{k}+m)}{\Gamma(\tilde{k}+m+1)} \sum_{\sigma\in\mathbb{S}(\vec{k},m)} \prod_{j=1}^{\tilde{k}+m} \left[(I-G/j)^{-1} D(A_{\sigma(j)})\right] \vec{1} \\ &= u_m + \frac{\theta+\tilde{k}+m}{\theta}\theta \frac{\Gamma(\theta+\tilde{k}+m)}{\Gamma(\tilde{k}+m+1)} D(A_n) \\ &\quad \times \left[\sum_{\sigma\in\mathbb{S}(\vec{k},m)} \prod_{j=1}^{\tilde{k}+m} \left[(I-G/j)^{-1} D(A_{\sigma(j)})\right] - \frac{\tilde{k}+m}{\theta+\tilde{k}+m} \right. \\ &\quad \left. \times \sum_{\sigma\in\mathbb{S}(\vec{k},m-1)} \prod_{j=1}^{\tilde{k}+m-1} \left[(I-G/j)^{-1} D(A_{\sigma(j)})\right] \right] \vec{1}, \end{aligned}$$

which further equals

$$\begin{aligned} &\frac{\Gamma(\theta+\tilde{k}+m+1)}{\Gamma(\tilde{k}+m+1)} D(A_n) \sum_{\sigma\in\mathbb{S}(\vec{k},m)} \prod_{j=1}^{\tilde{k}+m} \left[(I-G/j)^{-1} D(A_{\sigma(j)})\right] \vec{1} \\ &+ u_m - \frac{\Gamma(\theta+\tilde{k}+m)}{\Gamma(\tilde{k}+m)} D(A_n) \sum_{\sigma\in\mathbb{S}(\vec{k},m-1)} \prod_{j=1}^{\tilde{k}+m-1} \left[(I-G/j)^{-1} D(A_{\sigma(j)})\right] \vec{1}. \end{aligned}$$

We conclude the result via induction. □

Acknowledgments

We thank J. Sethuraman for reading and comments on a draft of this manuscript.

References

[1] Alan Agresti, *Analysis of ordinal categorical data*, 2nd ed., Wiley Series in Probability and Statistics, John Wiley & Sons, Inc., Hoboken, NJ, 2010, DOI 10.1002/9780470594001. MR2742515

[2] Alan Agresti and David B. Hitchcock, *Bayesian inference for categorical data analysis*, Stat. Methods Appl. **14** (2005), no. 3, 297–330, DOI 10.1007/s10260-005-0121-y. MR2211337

[3] J. Aitchison, *A general class of distributions on the simplex*, J. Roy. Statist. Soc. Ser. B **47** (1985), no. 1, 136–146. MR805071

[4] J. Aitchison and S. M. Shen, *Logistic-normal distributions: some properties and uses*, Biometrika **67** (1980), no. 2, 261–272, DOI 10.2307/2335470. MR581723

[5] Pier Giovanni Bissiri and Andrea Ongaro, *On the topological support of species sampling priors*, Electron. J. Stat. **8** (2014), no. 1, 861–882, DOI 10.1214/14-EJS912. MR3229100

[6] David Blackwell and James B. MacQueen, *Ferguson distributions via Pólya urn schemes*, Ann. Statist. **1** (1973), 353–355. MR362614

[7] Peter Congdon, *Bayesian models for categorical data*, Wiley Series in Probability and Statistics, John Wiley & Sons, Ltd., Chichester, 2005, DOI 10.1002/0470092394. MR2191351

[8] Peter Congdon, *Bayesian statistical modelling*, 2nd ed., Wiley Series in Probability and Statistics, John Wiley & Sons, Ltd., Chichester, 2006, DOI 10.1002/9780470035948. MR2281386

[9] Haydar Demirhan and Kamil Demirhan, *A Bayesian approach for the estimation of probability distributions under finite sample space*, Statist. Papers **57** (2016), no. 3, 589–603, DOI 10.1007/s00362-015-0669-z. MR3557362

[10] James M. Dickey and Thomas J. Jiang, *Filtered-variate prior distributions for histogram smoothing*, J. Amer. Statist. Assoc. **93** (1998), no. 442, 651–662, DOI 10.2307/2670116. MR1631349

[11] Dietz, Z., Lippitt, W., and Sethuraman, S. Stick-Breaking processes, Clumping, and Markov Chain Occupation Laws. *Sankhya A* (2021). `https://doi.org/10.1007/s13171-020-00236-x`

[12] Zach Dietz and Sunder Sethuraman, *Occupation laws for some time-nonhomogeneous Markov chains*, Electron. J. Probab. **12** (2007), no. 23, 661–683, DOI 10.1214/EJP.v12-413. MR2318406

[13] Michael D. Escobar and Mike West, *Bayesian density estimation and inference using mixtures*, J. Amer. Statist. Assoc. **90** (1995), no. 430, 577–588. MR1340510

[14] Thomas S. Ferguson, *A Bayesian analysis of some nonparametric problems*, Ann. Statist. **1** (1973), 209–230. MR350949

[15] J. Forster and A. Skene, *Calculation of marginal densities for parameters of multinomial distributions Stat. Comput.* bf 4(4) 279–286 (1994)

[16] David A. Freedman, *On the asymptotic behavior of Bayes' estimates in the discrete case*, Ann. Math. Statist. **34** (1963), 1386–1403, DOI 10.1214/aoms/1177703871. MR158483

[17] Subhashis Ghosal and Aad van der Vaart, *Fundamentals of nonparametric Bayesian inference*, Cambridge Series in Statistical and Probabilistic Mathematics, vol. 44, Cambridge University Press, Cambridge, 2017, DOI 10.1017/9781139029834. MR3587782

[18] C. Goutis, *Bayesian estimation methods for contingency tables, J. Ital. Statist. Soc.* **2**(1) 35–54 (1993)

[19] C. Hellmayr and A.E. Gelfand, *A partition Dirichlet process model for functional data analysis, Sankhya Ser. B* `https://doi.org/10.1007/s13571-019-00221-x`.

[20] R. King and S. P. Brooks, *Prior induction in log-linear models for general contingency table analysis*, Ann. Statist. **29** (2001), no. 3, 715–747, DOI 10.1214/aos/1009210687. MR1865338

[21] Michael Lavine, *Some aspects of Pólya tree distributions for statistical modelling*, Ann. Statist. **20** (1992), no. 3, 1222–1235, DOI 10.1214/aos/1176348767. MR1186248

[22] Thomas Leonard and John S. J. Hsu, *Bayesian methods*, Cambridge Series in Statistical and Probabilistic Mathematics, vol. 5, Cambridge University Press, Cambridge, 2001. An analysis for statisticians and interdisciplinary researchers; Reprint of the 1999 original. MR1847906

[23] T. Leonard, *A Bayesian method for histograms*, Biometrika **60** (1973), 297–308, DOI 10.1093/biomet/60.2.297. MR326902

[24] Peter Müller, Fernando Andrés Quintana, Alejandro Jara, and Tim Hanson, *Bayesian nonparametric data analysis*, Springer Series in Statistics, Springer, Cham, 2015, DOI 10.1007/978-3-319-18968-0. MR3309338

[25] Sonia Petrone, *Bayesian density estimation using Bernstein polynomials* (English, with English and French summaries), Canad. J. Statist. **27** (1999), no. 1, 105–126, DOI 10.2307/3315494. MR1703623

[26] Jim Pitman, *Some developments of the Blackwell-MacQueen urn scheme*, Statistics, probability and game theory, IMS Lecture Notes Monogr. Ser., vol. 30, Inst. Math. Statist., Hayward, CA, 1996, pp. 245–267, DOI 10.1214/lnms/1215453576. MR1481784

[27] Jayaram Sethuraman, *A constructive definition of Dirichlet priors*, Statist. Sinica **4** (1994), no. 2, 639–650. MR1309433

[28] Tzu-Tsung Wong, *Generalized Dirichlet distribution in Bayesian analysis*, Appl. Math. Comput. **97** (1998), no. 2-3, 165–181, DOI 10.1016/S0096-3003(97)10140-0. MR1643091

Biostatistics, University of Colorado Anschutz Medical Campus, Aurora, Colorado 80045
Email address: `william.lippitt@cuanschutz.edu`

Mathematics, University of Arizona, Tucson, Arizona 85721
Email address: `sethuram@math.arizona.edu`

Contemporary Mathematics
Volume **774**, 2021
https://doi.org/10.1090/conm/774/15572

Eulerian polynomials and Quasi-Birth-Death processes with time-varying-periodic rates

Barbara Margolius

ABSTRACT. A Quasi-Birth-Death (QBD) process is a stochastic process with a two dimensional state space, a level and a phase. An ergodic QBD with time-varying periodic transition rates will tend to an asymptotic periodic solution as time tends to infinity . Such QBDs are also asymptotically geometric. That is, as the level tends to infinity, the probability of the system being in state (k, j) at time t within the period tends to an expression of the form $f_j(t)\alpha^{-k}\Pi_j(k)$ where α is the smallest root of the determinant of a generating function related to the generating function for the unbounded (in the level) process, $\Pi_j(k)$ is a polynomial in k, the level, that may depend on j, the phase of the process, and $f_j(t)$ is a periodic function of time within the period which may also depend on the phase. These solutions are analogous to steady state solutions for QBDs with constant transition rates. If the time within the period is considered to be part of the state of the process, then they *are* steady-state solutions. In this paper, we consider the example of a two-priority queueing process with finite buffer for class-2 customers. For this example, we provide explicit results up to an integral in terms of the idle probability of the queue. We also use this asymptotic approach to provide exact solutions (up to an integral equation involving the probability the system is in level zero) for some of the level probabilities.

1. Introduction

In this paper we have two goals: to study asymptotic behavior of the level probabilities of ergodic, level independent Quasi-Birth-Death (QBD) processes (see [**8**] and [**12**]) with time-varying periodic rates, and to look in detail at an example involving priority queues. A QBD is a stochastic process with a two-dimensional state space, $\{Y(t), J(t)\}$, where $Y(t) \in \{0, 1, 2, \dots\}$ represents the level of the process and $J(t) \in \{0, 1, \dots, N\}$ the phase. We also consider the related unbounded or random walk process, $\{X(t), J(t)\}$, where $X(t) \in \mathbb{Z}$ represents the level of the process which no longer has a boundary at level zero. Such processes are pervasive in applications. Examples include call centers, distribution of packets of information over a network, emergency service delivery, care in a hospital, airplane flight schedules and many others. The literature surveys cited in what follows provide additional examples.

2020 *Mathematics Subject Classification.* Primary 60K25.
Key words and phrases. Queueing, Time-varying, Periodic.

The study of queues with time-varying transition rates has a long history dating back at least to Kolmogorov's [**7**] "Sur le problème d'attente" (On the problem of waiting). In 2018, Ward Whitt [**16**] provided a bibliography of research on the performance analysis of queueing systems with time-varying arrival rates with sections on numerical algorithms for Markov models, deterministic fluid models, infinite server queues, arrival process models, many server heavy traffic limits, and single-server queues, all with time-varying arrivals. In 2016, Defraeye and van Nieuwenhuyse [**3**] provided an extensive survey on staffing and scheduling for non-stationary demand for service. They provide a classification system with four elements, classification by: system assumptions, performance evaluation methods and metrics, optimization approach and application area. Their interest is in staffing and scheduling approaches for non-stationary demand. Also in 2016, Schwarz, et al [**14**] provide a survey that places articles on time-dependent queueing systems into categories based on whether the approach involves numerical and analytical solutions, piece-wise constant approximations, or approaches based on modified system characteristics. Within this framework, Schwarz, et al have characterized the approach used in this paper and earlier related papers by the author as "semi-analytical, semi-numerical". We obtain explicit results up to an integral equation over a single period.

In discussing systems with time-varying parameters, similar phrasing is used in the literature to mean different things. Time-dependent is sometimes used to refer to the transient solution for a system and sometimes used to refer to systems with transition rates which vary as a function of time. Here, for the most part, we are not referring to the transient solution of the system. Rather we are seeking the asymptotic periodic solution of the system in the sense of Breuer [**2**]. We do obtain the transient solution as an interim step to obtaining the asymptotic (in the level) behavior of the asymptotic (in time) periodic solution of the system. Earlier work by the author related to both transient solutions of QBDs with time-varying rates and asymptotic periodic solutions can be found in [**9**] and [**10**]. The former article solves these systems in both the transient and asymptotic periodic cases and provides factorial moments. The latter article puts the solutions in the context of analogues of the matrices $\mathbf{R}$ and $\mathbf{G}$ of matrix analytic methods. A more recent article [**11**] explores the asymptotic behavior (in the level) of queue-length processes with time-varying periodic transition rates when the unbounded random walk process is scalar. The 2019 article also notes that the asymptotic (in the level) behavior of QBDs with time-varying periodic transition rates will be governed by zeros of the determinant of a generating function related to the unbounded process.

Our approach applies techniques of Flajolet and Sedgwick [**5**] and involves finding the roots of the determinant of a matrix related to the generating function for a two-dimensional random walk over a single time period. In [**11**] we treated the asymptotic behavior of scalar queueing processes with time-varying periodic transition rates and touched upon how those results could be extended to QBDs. Here after a brief review of the $M_t/M_t/1$ queue, we show in detail, how the results extend to QBDs. We illustrate the approach using the example of a single server priority queue with finite buffer for class-2 customers [**6**].

The paper is organized as follows: In section 2, we explain the general method for analyzing time-varying periodic ergodic QBDs. In section 3, we show how this

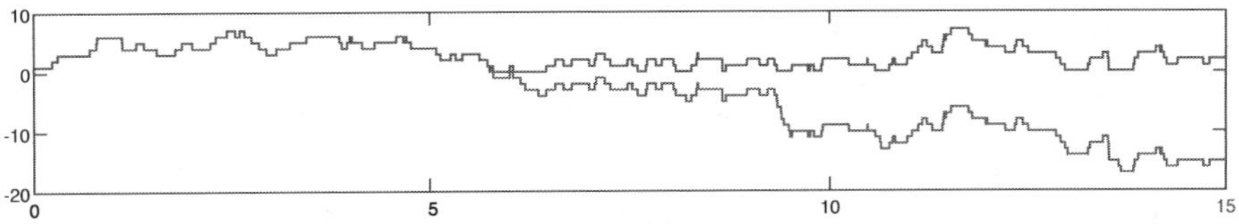

FIGURE 1. The graph shows the evolution of a random walk process in red, and the corresponding queue length process in blue (the darker line).

method applies in the trivial case of the single server queue with time-varying periodic transition rates. We then consider a two priority queue with finite buffer in section 4. Section 4 has subsections providing exact formulas for the level probabilities derived as coefficients on z^k from the generating function, providing a combinatorial argument for the formulas that we obtain in terms of generating functions for the Eulerian numbers, and asymptotic formulas for the level probabilities. The final section is a brief conclusion.

2. The approach

We define $\mathbf{q}_k(t)$ as a vector function with $N+1$ components which correspond to phases $j = 0, 1, \ldots, N$ in the quasi-birth-death process described above, k is the level of the process, and t is time. For an ergodic QBD with periodic transition rates, the level probabilities as the level number tends to infinity tend to periodic functions of time as the time t tends to infinity [**10**], that is $\lim_{n\to\infty} \mathbf{q}_k(t+n) = \mathbf{p}_k(t)$, $t \in [0, 1)$ for some periodic vector function $\mathbf{p}_k(t)$. Here and throughout the paper, the period is taken to be of length one, as we can always rescale time to make it so.

As the level k tends to infinity, $\mathbf{p}_k(t) \sim \sum_\ell \alpha_\ell^{-k} F_\ell(t)\Pi_\ell(n)$ where $F_\ell(t)$ is periodic and α_ℓ do not depend on time. In this expression, ℓ is the singularity number. $F_\ell(t)$ is an $N+1$ component periodic vector function with one component for each phase. $F_\ell(t)$ depends on singularity α_ℓ. α_ℓ and $\Pi_\ell(k)$ do not depend on time. The expressions $\Pi_\ell(k)$ are matrices of polynomials in k. This result is general for stable QBDs with time-varying periodic transition rates.

Random walks are closely connected to queueing processes. The graph shown in figure 1 shows a random walk path in blue (the darker line) and a queue length path in red. Initially when the random walk is positive the two paths are shared. Departures that take the random walk below zero are ignored in the queue length process. If we can understand random walks, then we can learn a great deal about the queueing processes that correspond to them.

Consider a random walk with two-dimensional state space $\{X(t), J(t)\}$ where $X(t) \in \mathbb{Z}$ gives the level of the process and $J(t) \in \{0, 1, \ldots, N\}$ gives the phase. We partition $\mathbf{p}(t)$ by levels into subvectors $\mathbf{p}_k(t)$, $k \in \mathbb{Z}$, where $\mathbf{p}_k(t)$ has $N+1$ components. The defining system satisfies the system of differential equations:

$$\frac{\partial}{\partial t}\mathbf{p}_k(t) = \mathbf{p}_{k-1}(t)\mathbf{A}_1(t) + \mathbf{p}_k(t)\mathbf{A}_0(t) + \mathbf{p}_{k+1}(t)\mathbf{A}_{-1}(t)$$

with the additional requirement that

$$\sum_{k=-\infty}^{\infty} \mathbf{p}_k(t)\mathbf{1} = 1$$

where $\mathbf{1}$ is an appropriately dimensioned vector of ones. $\mathbf{A}_i(t)$, $i = -1, 0, 1$ and $\mathbf{B}(t)$ are $(N+1) \times (N+1)$ matrix functions. The (i,j) component of the matrix gives the rate at which a transition occurs from phase i to phase j. The transition rates are periodic functions with period one.

Differential equations for the QBD process would include the random walk differential equations

$$\frac{\partial}{\partial t}\mathbf{p}_k(t) = \mathbf{p}_{k-1}(t)\mathbf{A}_1(t) + \mathbf{p}_k(t)\mathbf{A}_0(t) + \mathbf{p}_{k+1}(t)\mathbf{A}_{-1}(t) \tag{2.1}$$

for $k > 0$ and boundary condition

$$\frac{\partial}{\partial t}\mathbf{p}_0(t) = \mathbf{p}_0(t)\mathbf{B}(t) + \mathbf{p}_1(t)\mathbf{A}_{-1}(t). \tag{2.2}$$

We can solve the differential equations, (2.1) and (2.2) for the generating function of the system, $P(z,t) = \sum_{k=0}^{\infty} \mathbf{p}_k(t) z^k$, and get

$$P(z,t) = \int_s^t \mathbf{p}_0(u)\left(\mathbf{B}(u) - \mathbf{A}_0(u) - z^{-1}\mathbf{A}_{-1}(u)\right)\Phi(z,u,t)du + P(z,s)\Phi(z,s,t). \tag{2.3}$$

In writing $P(z,t)$ we are suppressing the dependence on the initial condition at time s from the notation. In the equation (2.3), $\Phi(z,s,t)$ is the generating function for the two-dimensional random walk. $\Phi(z,s,t)$ is a Laurent series in z. It satisfies the differential equations

$$\begin{aligned} \frac{\partial}{\partial t}\Phi(z,s,t) &= \Phi(z,s,t)\mathbf{A}(z,t), \\ \frac{\partial}{\partial s}\Phi(z,s,t) &= -\mathbf{A}(z,s)\Phi(z,s,t), \\ \Phi(z,t,t) &= \mathbf{I} \end{aligned} \tag{2.4}$$

where $\mathbf{A}(z,t) = z\mathbf{A}_1(t) + \mathbf{A}_0(t) + z^{-1}\mathbf{A}_{-1}(t)$. The coefficient on z^k is a matrix whose (i,j) component represents the probability of having a net change of k levels and transitioning from phase i at time s to phase j by time t.

When transition rates are periodic with period 1 and the system is ergodic, $P(z,t-1) = P(z,t)$ and so we may rewrite equation (2.3) as

$$P(z,t) = \int_{t-1}^t \mathbf{p}_0(u)\left(\mathbf{B}(u) - \mathbf{A}_0(u) - z^{-1}\mathbf{A}_{-1}(u)\right)\Phi(z,u,t)du \times \left(\mathbf{I} - \Phi(z,t-1,t)\right)^{-1}. \tag{2.5}$$

In what follows, equation (2.5) is our key equation.

For ergodic QBDs with time-varying periodic transition rates, we have the following theorem:

THEOREM 2.1. *The determinant of* $(\mathbf{I} - \Phi(z,t-1,t))$ *does not depend on* t, *that is,*

$$\det\left(\mathbf{I} - \Phi(z,0,1)\right) = \det\left(\mathbf{I} - \Phi(z,t-1,t)\right), \ \forall t.$$

PROOF. The random walk probability generating function satisfies the equation $\Phi(z,s,t) = \Phi(z,s,w)\Phi(z,w,t)$.

Also, by periodicity, we have that $\Phi(z,s,t) = \Phi(z,s-n,t-n)$, $n \in \mathbb{Z}$. In particular, $\Phi(z,t-1,s)\Phi(z,s,t) = \Phi(z,t-1,t)$ and $\Phi(z,s,t)\Phi(z,t-1,s) = \Phi(z,s-1,t-1)\Phi(z,t-1,s) = \Phi(z,s-1,s)$.

These facts together with the Sylvester Identity [**15**]:

$$\det(\mathbf{I} - AB) = \det(\mathbf{I} - BA)$$

prove that the determinant of $(\mathbf{I} - \Phi(z,t-1,t))$ does not depend on t. ■

Note that while it is true that

$$\det(\mathbf{I} - \Phi(z,s-1,s)) = \det(\mathbf{I} - \Phi(z,t-1,t))$$

it is not true in general that $(\mathbf{I} - \Phi(z,s-1,s))$ equals $(\mathbf{I} - \Phi(z,t-1,t))$ though the result does hold in the scalar case and in the priority queue example we present in this paper.

For general $\mathbf{A}_j(t)$, $j = -1,0,1$ it is not straightforward to write an explicit formula for the generating function. In section 4, we explore an example for which an explicit formula is available. In the next section, we will look at a simple scalar example.

3. Single-server queue

We begin with the single-server queue. For this QBD, the matrices $\mathbf{A}_{-1}(t) = \mu(t)$, $\mathbf{A}_0(t) = -\lambda(t) - \mu(t)$, $\mathbf{A}_1(t) = \lambda(t)$, and $\mathbf{B}(t) = -\lambda(t)$ are scalars. The level probabilities are also scalars. For this simple process, our key equation becomes

$$P(z,t) = \int_{t-1}^{t} p_0(u)\mu(u)(1-z^{-1})\mathrm{e}^{\int_u^t \lambda(\xi)(z-1)+\mu(\xi)(z^{-1}-1)d\xi}du \\ \times (1-\mathrm{e}^{\bar{\lambda}(z-1)+\bar{\mu}(z^{-1}-1)})^{-1}$$

where $\bar{\lambda} = \int_0^1 \lambda(u)du$ and $\bar{\mu} = \int_0^1 \mu(u)du$ are the average values of the transition rates during the time period.

The relevant zeros of the denominator are given by

$$\chi_\ell = \frac{1}{2\bar{\lambda}}\left(\bar{\lambda} + \bar{\mu} + 2\pi i\ell + \sqrt{(\bar{\lambda}+\bar{\mu}+2\pi i\ell)^2 - 4\bar{\lambda}\bar{\mu}}\right).$$

While the numbers

$$\frac{1}{2\bar{\lambda}}\left(\bar{\lambda} + \bar{\mu} + 2\pi i\ell - \sqrt{(\bar{\lambda}+\bar{\mu}+2\pi i\ell)^2 - 4\bar{\lambda}\bar{\mu}}\right)$$

are also zeros of the denominator, they are inside or on the unit circle and so, for an ergodic process, will also be zeros of the numerator. They are removable singularities of the generating function.

An exact formula for the level probabilities is then given by

$$(3.1)\quad p_k(t) = \\ \int_{t-1}^{t} p_0(u)\mu(u)\sum_{\ell=-\infty}^{\infty}\frac{(1-\chi_\ell^{-1})\chi_\ell^{-k}}{\sqrt{(\bar{\lambda}+\bar{\mu}+2\pi i\ell)^2 - 4\bar{\lambda}\bar{\mu}}}\mathrm{e}^{\int_u^t(\lambda(\xi)(\chi_\ell-1)+\mu(\xi)(\chi_\ell^{-1}-1))d\xi}du.$$

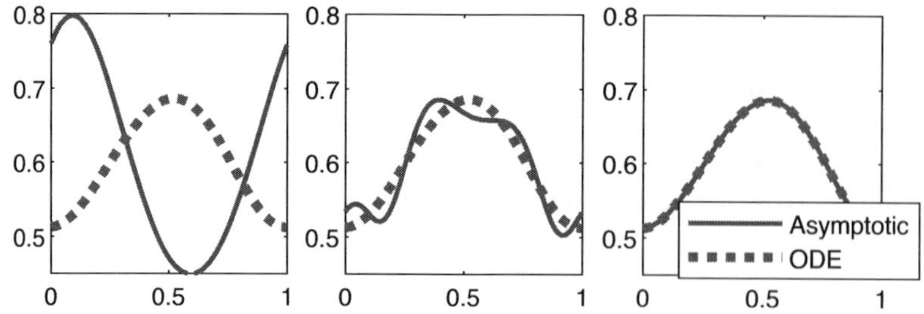

FIGURE 2.
$p_0(t)$, $M=0$ $p_0(t)$, $M=2$ $p_0(t)$, $M=7$
Comparison of ODE Solution and Asymptotic Periodic Solution for $p_0(t)$ for different values of M.

To prove this result, follow the approach in [**5**], pp. 258–259 and 268–269 and note that the coefficient integral from Cauchy's integral formula becomes the sum of all of the residues.

We truncate the infinite sum to estimate the probabilities, choosing a suitable value of M with

$$p_k(t) \approx \int_{t-1}^{t} p_0(u)\mu(u) \sum_{\ell=-M}^{M} \frac{(1-\chi_\ell^{-1})\chi_\ell^{-k}}{\sqrt{(\bar\lambda+\bar\mu+2\pi i\ell)^2-4\bar\lambda\bar\mu}} e^{\int_u^t (\lambda(\xi)(\chi_\ell-1)+\mu(\xi)(\chi_\ell^{-1}-1))d\xi} du. \tag{3.2}$$

EXAMPLE 3.1. Suppose that we have $\lambda(t) = 2 + \frac{2}{3}\cos(2\pi t)$ and $\mu(t) = 5 + \frac{5}{2}\sin(2\pi t)$. For this example, $\bar\lambda = 2$ and $\bar\mu = 5$, and so the

$$\chi_k = \frac{1}{4}\left(7 + 2\pi ik + \sqrt{9 + 28\pi ik - 4\pi^2k^2}\right).$$

Figures 2 and 3 show the probabilities computed using equation (3.2) truncated using various values of M. Note that convergence is rapid even for small numbers in the queue. For $p_{10}(t)$ we used only a single term ($M = 0$). For the approximation for $p_0(t)$, fifteen terms achieved a high degree of accuracy. The error in estimation of $p_k(t)$ can be bounded by the tail of the Riemann zeta function times a constant.

Other scalar examples can be found in [**11**]. The exact solution for a single-server queue using asymptotic analysis is new here and could easily be extended to the other examples in that paper.

4. Single-server priority queue with finite Buffer

We consider a single server pre-emptive priority queue with two classes of customer, with class-2 customers having finite waiting room. For this example,

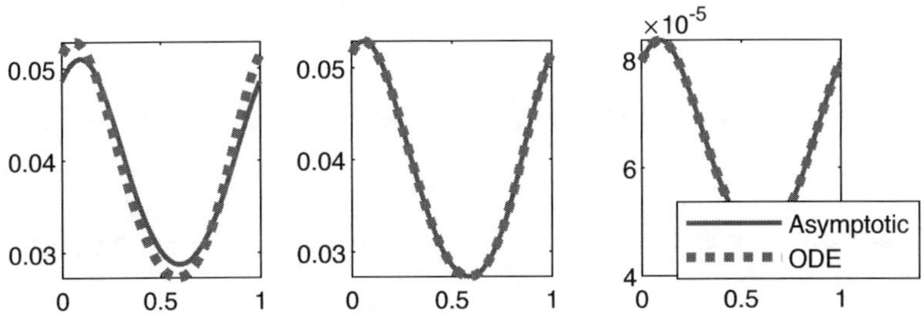

FIGURE 3.
$p_3(t)$, $M = 0$ $p_3(t)$, $M = 2$ $p_{10}(t)$, $M = 0$
Comparison of ODE Solution and Asymptotic Periodic Solution for $p_3(t)$ and $p_{10}(t)$ for different values of M.

$$\mathbf{B}(t) = \begin{bmatrix} -\lambda(t) & \lambda_2(t) & \cdot & \cdot & \cdot \\ \mu_2(t) & -\lambda(t)-\mu_2(t) & \lambda_2(t) & \cdot & \cdot \\ \cdot & \ddots & \ddots & \ddots & \cdot \\ \cdot & \ddots & \mu_2(t) & -\lambda(t)-\mu_2(t) & \lambda_2(t) \\ \cdot & \cdot & \ddots & \mu_2(t) & -\lambda_1(t)-\mu_2(t) \end{bmatrix},$$

$$\mathbf{A}_0(t) = \begin{bmatrix} -\lambda(t)-\mu_1(t) & \lambda_2(t) & \cdot & \cdot \\ \cdot & -\lambda(t)-\mu_1(t) & \lambda_2(t) & \cdot \\ \cdot & \ddots & \ddots & \cdot \\ \cdot & \ddots & -\lambda(t)-\mu_1(t) & \lambda_2(t) \\ \cdot & \ddots & \ddots & -\lambda_1(t)-\mu_1(t) \end{bmatrix},$$

$$\mathbf{A}_1(t) = \lambda_1(t)\mathbf{I}$$

and

$$\mathbf{A}_{-1}(t) = \mu_1(t)\mathbf{I}$$

are $(N+1) \times (N+1)$ sub-matrices where N is the size of the buffer for class-2 customers.

Let

$$a_j(s,t) = \frac{\left(\int_s^t \lambda_2(u)du\right)^j}{j!} \mathrm{e}^{-\int_s^t \lambda_2(u)du}$$

and

$$a_{>j}(s,t) = 1 - \sum_{k=0}^{j} \frac{\left(\int_s^t \lambda_2(u)du\right)^k}{k!} \mathrm{e}^{-\int_s^t \lambda_2(u)du}$$

represent the probability of j arrivals of class-2 customers occurring at time-varying rate $\lambda_2(t)$ during the time interval $[s,t)$.

Let $f = f(z,s,t) = \int_s^t \left(\lambda_1(u)z - (\lambda_1(u)+\mu_1(u)) + \mu_1(u)z^{-1}\right) du$ and observe that

$$\mathrm{e}^{f(z,s,t)} = \sum_{k=-\infty}^{\infty} P\{Y(s,t)=k\}z^k \tag{4.1}$$

where $Y(s,t)$ represents the net number of steps taken by a random walk governed by the time-varying transition rates $\lambda_1(t)$ and $\mu_1(t)$ during the time interval $[s,t)$. An explicit formula for $P\{Y(s,t)=k\}$ is

$$P\{Y(s,t)=k\} = \left(\frac{\int_s^t \lambda_1(u)du}{\int_s^t \mu_1(u)du}\right)^{k/2} \mathrm{e}^{-\int_s^t(\lambda_1(u)+\mu_1(u))du} I_k\left(2\sqrt{\int_s^t \lambda_1(u)du \int_s^t \mu_1(u)du}\right) \tag{4.2}$$

where $I_k(x)$ is the kth modified Bessel function [**4**].

We suppress the time interval in the notation, and write simply a_j for the arrival probabilities for class-2 customers and e^f for the random walk generating function for class-1 customers, then $\Phi(z,s,t)$ is given by

$$\Phi(z,s,t) = \mathrm{e}^f \begin{bmatrix} a_0 & a_1 & a_2 & \cdots & \cdots & a_{N-1} & a_{>N-1} \\ 0 & a_0 & a_1 & \ddots & \cdots & a_{N-2} & a_{>N-2} \\ 0 & 0 & \ddots & \ddots & \ddots & \vdots & \vdots \\ 0 & 0 & 0 & a_0 & \cdots & a_{N-i} & a_{>N-i} \\ 0 & 0 & 0 & 0 & \ddots & \ddots & \vdots \\ 0 & 0 & 0 & 0 & 0 & a_0 & a_{>0} \\ 0 & 0 & 0 & 0 & 0 & 0 & 1 \end{bmatrix}. \tag{4.3}$$

To prove this, differentiate the function with respect to t and note that it solves the differential equations given in equation (2.4).

Now consider $(\mathbf{I} - \Phi(z,t-1,t))^{-1}$ and recall that this may be written as the geometric series

$$(\mathbf{I} - \Phi(z,t-1,t))^{-1} = \sum_{n=0}^{\infty} \Phi^n(z,t-1,t) = \sum_{n=0}^{\infty} \Phi(z,t-1,t-1+n) \tag{4.4}$$

with $\Phi^0(z,0,1) = \Phi(z,t,t) = \mathbf{I}$. Since we are assuming that the transition rates are periodic with period 1, then for any of the transition rates, the integral of the rate over a single period is equal to its average value and we write, for example, $\bar{\lambda}_2 = \int_{t-1}^t \lambda_2(u)du$.

We compute $(\mathbf{I}-\Phi)^{-1}$ using the infinite sum. Clearly, the matrix will be upper triangular and other than the last column, also Toeplitz. We sum a component in the upper triangle

$$[(\mathbf{I} - \Phi(z,t-1,t))^{-1}]_{j,j+m} = \sum_{k=0}^{\infty} [\Phi(z,t-1,t+k)]_{j,j+m}$$

for $m \geq 0$. For $j+m < N$, $[\Phi(z,t-1,t)]_{j,j+m}$ is given by

$$[\Phi(z,t-1,t)]_{j,j+m} = \frac{\bar{\lambda}_2^m}{m!}\mathrm{e}^{-\bar{\lambda}_2}\mathrm{e}^{\bar{\lambda}_1(z-1)+\bar{\mu}_1(z^{-1}-1)},$$

and $[\Phi^n(z,t-1,t)]_{j,j+m} = [\Phi(z,t-1,t+n-1)]_{j,j+m}$ is given by

$$[\Phi(z,t-1,t+n-1)]_{j,j+m} = \frac{\bar{\lambda}_2^m n^m}{m!} \mathrm{e}^{-\bar{\lambda}_2 n} \mathrm{e}^{(\bar{\lambda}_1(z-1)+\bar{\mu}_1(z^{-1}-1))n}. \tag{4.5}$$

From equations (4.4) and (4.5), we have

$$[(\mathbf{I}-\Phi(z,t-1,t))^{-1}]_{j,j+m} = \frac{\bar{\lambda}_2^m}{m!} \sum_{n=1}^{\infty} n^m \mathrm{e}^{-\bar{\lambda}_2 n} \mathrm{e}^{(\bar{\lambda}_1(z-1)+\bar{\mu}_1(z^{-1}-1))n}.$$

We can write the final sum in closed form using the Carlitz identity [**13**]:

$$\frac{S_m(t)}{(1-t)^{m+1}} = \sum_{k=0}^{\infty} (k+1)^m t^k$$

where $S_m(t)$ is the mth Eulerian polynomial, that is the generating function for the Eulerian numbers. The triangular array of these numbers is given in the *Online Encyclopedia of Integer Sequences* as sequence **A008292**, [**1**]. Numerous additional references are available from that source.

We define $\phi(z,u,t) = \mathrm{e}^{\int_u^t (\lambda_1(\nu)(z-1)+\mu_1(\nu)(z^{-1}-1)-\lambda_2(\nu))d\nu}$, and, in the case when the integral is over a single period, $\phi(z) = \mathrm{e}^{\bar{\lambda}_1(z-1)+\bar{\mu}_1(z^{-1}-1)-\bar{\lambda}_2}$ so

$$[(\mathbf{I}-\Phi(z,t-1,t))^{-1}]_{j,j+m} = \frac{\bar{\lambda}_2^m}{m!} \frac{\phi(z) S_m(\phi(z))}{(1-\phi(z))^{m+1}}$$

for $m > 0$, and

$$[(\mathbf{I}-\Phi(z,t-1,t))^{-1}]_{j,j} = \frac{1}{1-\phi(z)}.$$

The matrix $(\mathbf{I}-\Phi)^{-1}$ is the upper triangular matrix given below. The entries in the right-most column are given by

$$\left[(\mathbf{I}-\Phi)^{-1}\right]_{i,N} = \frac{1}{1-\mathrm{e}^f} - \sum_{j=i}^{N-1} \left[(\mathbf{I}-\Phi)^{-1}\right]_{i,j},$$

so

$$(\mathbf{I}-\Phi)^{-1} =$$

$$\begin{bmatrix}
\frac{1}{1-\phi(z)} & \frac{\bar{\lambda}_2\phi(z)}{1!(1-\phi(z))^2} & \frac{\bar{\lambda}_2^2\phi(z)(1+\phi(z))}{2!(1-\phi(z))^3} & \cdots & \frac{\bar{\lambda}_2^{N-1}\phi(z)S_{N-1}(\phi(z))}{(N-1)!(1-\phi(z))^N} & \frac{1}{1-\mathrm{e}^f} - \sum_{j=i}^{N-1}\left[(\mathbf{I}-\Phi)^{-1}\right]_{0,j} \\
0 & \ddots & \ddots & \ddots & \vdots & \vdots \\
\vdots & \ddots & \frac{1}{1-\phi(z)} & \cdots & \frac{\bar{\lambda}_2^{N-i}\phi(z)S_{N-i}(\phi(z))}{(N-i)!(1-\phi(z))^{N-i+1}} & \frac{1}{1-\mathrm{e}^f} - \sum_{j=i}^{N-1}\left[(\mathbf{I}-\Phi)^{-1}\right]_{i-1,j} \\
\vdots & \cdots & \ddots & \ddots & \ddots & \vdots \\
\vdots & \cdots & \cdots & \ddots & \frac{1}{1-\phi(z)} & \frac{1}{1-\mathrm{e}^f} - \frac{1}{1-\phi(z)} \\
0 & \cdots & \cdots & \cdots & 0 & \frac{1}{1-\mathrm{e}^f}
\end{bmatrix}$$

4.1. Explicit formula for level probabilities up to an integral equation. It is possible to write an expression for

$$[z^k]P(z,t) = \mathbf{p}_k(t) =$$

$$[z^k] \int_{t-1}^{t} \mathbf{p}_0(u) \left(\mathbf{B}(u) - \mathbf{A}_0(u) - z^{-1}\mathbf{A}_{-1}(u)\right) \Phi(z,u,t) du \left(\mathbf{I}-\Phi(z,t-1,t)\right)^{-1}.$$

For this process, $\mathbf{B}(u) - \mathbf{A}_0(u) - z^{-1}\mathbf{A}_{-1}(u)$ simplify quite a bit and we have

$$\mathbf{B}(u) - \mathbf{A}_0(u) - z^{-1}\mathbf{A}_{-1}(u) = (\mu_1(u)(1-z^{-1}) - \mu_2(u))\mathbf{I} + \begin{bmatrix} \mu_2(u) & 0 & \cdots & 0 \\ \mu_2(t) & 0 & \cdots & 0 \\ \vdots & \ddots & \vdots & \vdots \\ 0 & 0 & \mu_2(t) & 0 \end{bmatrix}.$$

Define

$$\mathbf{v}(u) = \begin{bmatrix}(p_{00}(u) + p_{01}(u))\mu_2(u) & p_{02}(u)\mu_2(u) & \cdots & p_{0N}(u)\mu_2(u) & 0\end{bmatrix},$$

then

$$\begin{aligned}\mathbf{p}_k(t) = \int_{t-1}^{t} & \mathbf{p}_0(u)(\mu_1(u) - \mu_2(u)) \sum_{\ell=-\infty}^{\infty} \left([z^\ell]\Phi(z,u,t)\right)\left([z^{k-\ell}]\left(\mathbf{I} - \Phi\right)^{-1}\right) \\ & - \mathbf{p}_0(u)\mu_1(u) \sum_{\ell=-\infty}^{\infty} \left([z^\ell]\Phi(z,u,t)\right)\left([z^{k-\ell+1}]\left(\mathbf{I} - \Phi\right)^{-1}\right) \\ & + \mathbf{v}(u) \sum_{\ell=-\infty}^{\infty} \left([z^\ell]\Phi(z,u,t)\right)\left([z^{k-\ell}]\left(\mathbf{I} - \Phi\right)^{-1}\right).\end{aligned}$$

The matrices $[z^k]\Phi(z,u,t)$ and $[z^k](\mathbf{I} - \Phi(z,t-1,t))^{-1}$ are upper triangular. For $j+m \neq N$, $m = 0, \ldots, N-1$

$$\left[[z^k]\Phi(z,u,t)\right]_{j,j+m} = \frac{\left(\int_u^t \lambda_2(\nu)d\nu\right)^m}{m!} \mathrm{e}^{-\int_u^t \lambda_2(\nu)d\nu} P\{Y(u,t) = k\}$$

and

$$\begin{aligned}[z^k][(\mathbf{I} - \Phi(z,t-1,t))^{-1}]_{j,j+m} &= [z^k]\frac{\bar\lambda_2^m}{m!}\sum_{n=1}^{\infty} n^m \mathrm{e}^{(-\bar\lambda_2 + \bar\lambda_1(z-1) + \bar\mu_1(z^{-1}-1))n} \\ &= \frac{\bar\lambda_2^m}{m!}\sum_{n=1}^{\infty} n^m \mathrm{e}^{-\bar\lambda_2 n} \sum_{\ell=0}^{\infty} \frac{\bar\lambda_1^{k+\ell}\bar\mu_1^\ell n^{2\ell+k}}{(k+\ell)!\ell!} \mathrm{e}^{-(\bar\lambda_1+\bar\mu_1)n} \\ &= \frac{\bar\lambda_2^m}{m!}\sum_{\ell=0}^{\infty} \frac{\bar\lambda_1^{k+\ell}\bar\mu_1^\ell}{(k+\ell)!\ell!} \sum_{n=1}^{\infty} n^{2\ell+k+m} \mathrm{e}^{-(\bar\lambda_1+\bar\mu_1+\bar\lambda_2)n} \\ &= \mathrm{e}^{-(\bar\lambda_1+\bar\mu_1+\bar\lambda_2)} \frac{\bar\lambda_2^m}{m!}\sum_{\ell=0}^{\infty} \frac{\bar\lambda_1^{k+\ell}\bar\mu_1^\ell}{(k+\ell)!\ell!} \frac{S_{2\ell+k+m}\left(\mathrm{e}^{-(\bar\lambda_1+\bar\mu_1+\bar\lambda_2)}\right)}{(1-\mathrm{e}^{-(\bar\lambda_1+\bar\mu_1+\bar\lambda_2)})^{2\ell+k+m+1}} \\ &= \frac{\mathrm{e}^{-(\bar\lambda_1+\bar\mu_1+\bar\lambda_2)}\bar\lambda_1^k}{(1-\mathrm{e}^{-(\bar\lambda_1+\bar\mu_1+\bar\lambda_2)})^{m+k+1}} \frac{\bar\lambda_2^m}{m!} \\ &\quad \times \sum_{\ell=0}^{\infty} \binom{2\ell+k}{\ell} \frac{\bar\lambda_1^\ell\bar\mu_1^\ell}{(\bar\lambda_1+\bar\mu_1)^{2\ell}} \frac{S_{2\ell+k+m}\left(\mathrm{e}^{-(\bar\lambda_1+\bar\mu_1+\bar\lambda_2)}\right)(\bar\lambda_1+\bar\mu_1)^{2\ell}}{(2\ell+k)!(1-\mathrm{e}^{-(\bar\lambda_1+\bar\mu_1+\bar\lambda_2)})^{2\ell}}.\end{aligned}$$

This form of the formula for the level probabilities is not felicitous for computation. Other inconvenient forms are also possible. For example, we could express $[z^k][(\mathbf{I} - \Phi(z,t-1,t))^{-1}]_{j,j+m}$ as a hadamard product. Instead, we employ an asymptotic approach.

In subsection 4.2, we develop a combinatorial argument for why the Eulerian numbers appear in the formula for the level probabilities. This subsection may be skipped without loss of continuity.

4.2. A combinatorial argument for the appearance of Eulerian numbers in the formula for the level probabilities. The Eulerian polynomial $S_m(t)$ is a generating function for the number of descents in permutations of the integers $\{1, \ldots, m\}$. The coefficient on t^k gives the number of permutations of $\{1, \ldots, m\}$ with k descents. The Eulerian polynomial $S_4(t) = 1 + 11t + 11t^2 + t^3$ because there is one permutation with no descents (the sequence 1234), there are 11 each with one or two descents, and there is one permutation with 3 descents, (the sequence 4321), for the integers $\{1, 2, 3, 4\}$.

For example, the permutation 5367124 has two descents. These occur from 5 to 3 and from 7 to 1.

Note that $\sum_{n=0}^{\infty}(n+1)^m t^n$ is the generating function for the number of ways to place m distinct balls into $n+1$ boxes. If we place the boxes in a fixed order, there is a natural way to associate a permutation of the integers $\{1, \ldots, m\}$ with the placement of balls in boxes. If there is more than one ball in a box, we list the balls in order by their labels. We represent the partition of the balls into boxes with vertical bars, so for our permutation 5367124, we must have at least one bar at each descent, and we may have more than one bar there, there may be any number of bars placed between other numbers in the permutation. One possible placement of four bars would be 5||367|124|. This represents a placement of seven balls in five boxes. The first box has ball 5, the second box (represented by adjacent bars) is empty; the third box has balls 3, 6, and 7; the fourth box has balls 1, 2, and 4 and the fifth box is empty.

The generating function for the number of bars that may be placed in a gap with a descent (since at least one bar must be placed there) is $t+t^2+t^3+\cdots = \frac{t}{1-t}$. The generating function for a gap with no descent is $1+t+t^2+t^3+\cdots = \frac{1}{1-t}$ since any number of bars may be placed there including no bar at all. The generating function for a permutation of length m with n descents is $\frac{t^n}{(1-t)^{m+1}}$, the product of these generating functions. That is, the product of $m+1$ generating functions, one for each gap. n of the generating functions will have a t in the numerator and the rest will have a 1 in the numerator. All will have $1-t$ in the denominator. In general, for permutation w, we will have the generating function $\frac{t^{\text{des}w}}{(1-t)^{m+1}}$ where des$w$ is the number of descents in permutation w. To get the generating function over all permutations of length m, we sum

$$\sum_{w \in S_m} \frac{t^{\text{des}w}}{(1-t)^{m+1}} = \frac{S_m(t)}{(1-t)^{m+1}},$$

but this is the same as the generating function for the number of ways to place m balls into $n+1$ boxes, so

$$\frac{S_m(t)}{(1-t)^{m+1}} = \sum_{n=0}^{\infty}(n+1)^m t^n.$$

This proof of the Carlitz identity is given in Petersen as exercise 1.14, pp. 17–18 and 366–368 [**13**].

The components of the matrix $(\mathbf{I}-\Phi)^{-1}$, are given by

$$[(\mathbf{I}-\Phi(z,t-1,t))^{-1}]_{j,j+m} = \frac{\bar{\lambda}_2^m}{m!}\frac{\phi(z)S_m(\phi(z))}{(1-\phi(z))^{m+1}}$$

because of the combinatorial interpretation of the generating function represented there. Let us think of each period as a day. In this interpretation, the boxes correspond to days, and the balls to arrivals of class-2 customers. The labeling of the balls is the time stamp for the time within the day when they arrived. The function $\phi^n(z)$ is the generating function for the net change in the number of class-1 arrivals over the course of n days. The m class-2 arrivals are distributed over the n days. The order of arrival of the class-2 customers is not important, so we divide by $m!$. As in the balls and boxes case, our generating function can be constructed as the product of generating functions. Where there is a descent, there is at least one arrival of a class-2 customer, so those generating functions are expressed as $\frac{\bar{\lambda}_2\phi(z)}{1-\phi(z)}$. Where there is no descent, there may be any number of arrivals of class-2 customers including zero, and so those generating functions are expressed as $\frac{\phi(z)}{1-\phi(z)}$. The product of these for a given permutation w is $\frac{\bar{\lambda}_2^m\phi^{\operatorname{desw}}(z)}{1-\phi(z)}$. Summing over all permutations yields the generating function $\frac{\bar{\lambda}_2^m}{m!}\frac{\phi(z)S_m(\phi(z))}{(1-\phi(z))^{m+1}}$.

4.3. Asymptotic results. The asymptotic behavior of the generating function is governed by its singularities.

THEOREM 4.1. *[Flajolet and Sedgewick, Theorem IV. 10, p.258* **[5]***] Let $f(z)$ be a function meromorphic at all points of the closed disc $|z| \le R$, with poles at points $\alpha_1, \alpha_2, \ldots, \alpha_m$. Assume that $f(z)$ is analytic at all points of $|z| = R$ and at $z=0$. Then there exist m polynomials $\{\Pi_\ell(x)\}_{\ell=1}^m$ such that:*

$$f_k \equiv [z^k]f(z) = \sum_{\ell=1}^m \Pi_\ell(k)\alpha_\ell^{-k} + O(R^{-k}).$$

Furthermore the degree of Π_ℓ is equal to the order of the pole of f at α_ℓ minus one.

The probability generating function for a time-varying, periodic, level independent QBD is a vector of meromorphic functions. The mth component of $P(z,t)$ is the generating function for being in level k and phase m, so $[P(z,t)]_m = \sum_{k=0}^\infty p_{k,m}(t)z^k$. At $z=0$, we have $[P(0,t)]_m = p_{0,m}(t)$, $m=0,\ldots,N$. The singularities in the probability generating function occur where the determinant of the matrix $\mathbf{I}-\Phi(z,t-1,t)$ is zero.

In our application of the preceding theorem, we will use the notation $\Pi_\ell(k)$ to represent matrices of polynomials in k, one such matrix for each pole of $P(z,t)$, that is for each root of the determinant of $\mathbf{I}-\Phi(z,t-1,t)$.

In our example, since $\mathbf{I}-\Phi(z,t-1,t)$ is triangular, the determinant is the product of the diagonal elements and is given by $(1-\phi(z))^N(1-e^f)$. Roots occur whenever $\phi(z)=1$ or $\mathrm{e}^f=1$. $\phi(z)=1$ whenever $\bar{\lambda}_1(z-1)+\bar{\mu}_1(z^{-1}-1)-\bar{\lambda}_2 = 2\pi i\ell$ for $\ell\in\mathbb{Z}$. $\mathrm{e}^f=1$ whenever $f=2\pi i\ell$.

Given ℓ, there are two roots to $(1-\phi(z))$ and two more for $(1-e^f)$. For fixed ℓ, the roots for $(1-\phi(z))$ are

$$\begin{aligned}
\alpha_{\ell+} &= \frac{1}{2\bar{\lambda}_1}\left(\bar{\lambda}_1+\bar{\lambda}_2+\bar{\mu}_1+2\pi i\ell+\sqrt{\left(\bar{\lambda}_1+\bar{\lambda}_2+\bar{\mu}_1-2\pi i\ell\right)^2-4\bar{\lambda}_1\bar{\mu}_1}\right),\\
\alpha_{\ell-} &= \frac{1}{2\bar{\lambda}_1}\left(\bar{\lambda}_1+\bar{\lambda}_2+\bar{\mu}_1+2\pi i\ell-\sqrt{\left(\bar{\lambda}_1+\bar{\lambda}_2+\bar{\mu}_1-2\pi i\ell\right)^2-4\bar{\lambda}_1\bar{\mu}_1}\right).
\end{aligned}$$

For $(1-e^f)$, the roots occur at

$$\begin{aligned}
\beta_{\ell+} &= \frac{\bar{\lambda}_1+\bar{\mu}_1-2\pi i\ell+\sqrt{(\bar{\lambda}_1+\bar{\mu}_1-2\pi i\ell)^2-4\bar{\lambda}_1\bar{\mu}_1}}{2\bar{\lambda}_1},\\
\beta_{\ell-} &= \frac{\bar{\lambda}_1+\bar{\mu}_1-2\pi i\ell-\sqrt{(\bar{\lambda}_1+\bar{\mu}_1-2\pi i\ell)^2-4\bar{\lambda}_1\bar{\mu}_1}}{2\bar{\lambda}_1}.
\end{aligned}$$

For $\ell=0$, these roots are $\frac{\bar{\mu}_1}{\bar{\lambda}_1}$ and 1. The roots $\alpha_{\ell-}$ and $\beta_{\ell-}$ occur on or inside the unit circle, so they must also be zeros of the numerator. Otherwise, the probability generating function would not converge and the QBD would not be ergodic. We are interested in the singularities that occur at the roots $\alpha_{\ell+}$ and $\beta_{\ell+}$.

Near the root at $z=\alpha_{\ell+}$, we have $1-\phi(z)$ has a zero when $\{\bar{\lambda}_1(z-1)+\bar{\mu}_1(z^{-1}-1)-\bar{\lambda}_2\}=2\pi i\ell$, for $\ell\in\mathbb{Z}$. This expression has the two roots given above. Observe that

$$\begin{aligned}
&-\left(\bar{\lambda}_1(z-1)+\bar{\mu}_1(z^{-1}-1)-\bar{\lambda}_2-2\pi i\ell\right)\\
&\qquad\qquad\qquad = \frac{1}{z}\bar{\lambda}_1(\alpha_{\ell+}-z)(z-\alpha_{\ell-})\\
&\qquad\qquad\qquad\qquad = \frac{\alpha_{\ell+}}{z}\bar{\lambda}_1(1-z/\alpha_{\ell+})(z-\alpha_{\ell-}),
\end{aligned}$$

so

$$\frac{1}{1-\phi(z)}\approx\frac{1}{c_\ell(1-z/\alpha_{\ell+})}$$

for z near $\alpha_{\ell+}$, where

$$c_\ell=\frac{\alpha_{\ell+}}{z}\bar{\lambda}_1(z-\alpha_{\ell-})\bigg|_{z=\alpha_{\ell+}}=\bar{\lambda}_1(\alpha_{\ell+}-\alpha_{\ell-})=\sqrt{(\bar{\lambda}_1+\bar{\mu}_1+\bar{\lambda}_2+2\pi i\ell)^2-4\bar{\lambda}_1\bar{\mu}_1}.$$

The exponential $\phi(z)$ is equal to one for $z=\alpha_{\ell+}$, so $S_m(\phi(\alpha_{\ell+}))=m!$. Hence for $j+m<N$, we may approximate

$$[(\mathbf{I}-\Phi(z,t-1,t))^{-1}]_{j,j+m}\approx\sum_{\ell=-\infty}^{\infty}\frac{\bar{\lambda}_2^m}{c_\ell^{m+1}}\sum_{k=0}^{\infty}\binom{k+m}{k}\alpha_{\ell+}^{-k}z^k$$

where $\alpha_{\ell+}$ is the larger root of $\bar{\lambda}_1(z-1)+\bar{\mu}_1(z^{-1}-1)-\bar{\lambda}_2-2\pi i\ell$ for z near $\alpha_{\ell+}$.

An asymptotic formula for the coefficient on z^k in $[(\mathbf{I}-\Phi(z,t-1,t))^{-1}]_{j,j+m}$ is

$$[z^k][(\mathbf{I}-\Phi(z,t-1,t))^{-1}]_{j,j+m} = [z^k]\frac{\bar{\lambda}_2^m}{m!}\frac{\phi(z)S_m(\phi(z))}{(1-\phi(z))^{m+1}} \approx$$
$$[z^k]\sum_{\ell=-\infty}^{\infty}\frac{\bar{\lambda}_2^m}{c_\ell^{m+1}}\sum_{k=0}^{\infty}\binom{k+m}{k}\alpha_{\ell+}^{-k}z^k$$
$$= \sum_{\ell=-\infty}^{\infty}\frac{\bar{\lambda}_2^m}{c_\ell^{m+1}}\binom{k+m}{k}\alpha_{\ell+}^{-k}$$

The level probabilities when there are no class-2 customers are

$$p_{k,0}(t) = \int_{t-1}^{t} p_{00}(u)\mu_1(u)\sum_{\ell=-\infty}^{\infty}\frac{(1-\alpha_{\ell+}^{-1})}{c_\ell}\phi(\alpha_{\ell+},u,t)du\alpha_{\ell+}^{-k}.$$

These probabilities are exact. The proof is essentially the same as for equation (3.1). This formula is just the sum of the residues of the Cauchy integral formula for the coefficients.

The level probabilities when there is one class-2 customer are approximately

$$(4.6)\quad p_{k,1}(t) \approx$$
$$(k+1)\sum_{\ell=-\infty}^{\infty}\frac{\alpha_{\ell+}^{-k}\bar{\lambda}_2}{c_\ell^2}\int_{t-1}^{t}(p_{00}(u)\mu_1(u)(1-\alpha_{\ell+}^{-1})+p_{01}(u)\mu_2(u))\phi(\alpha_{\ell+},u,t)du.$$

More generally,

$$(4.7)\quad p_{k,j}(t) \approx$$
$$\binom{k+j}{k}\sum_{\ell=-\infty}^{\infty}\frac{\alpha_{\ell+}^{-k}\bar{\lambda}_2^j}{c_\ell^{j+1}}\int_{t-1}^{t}(p_{00}(u)\mu_1(u)(1-\alpha_{\ell+}^{-1})+p_{01}(u)\mu_2(u))\phi(\alpha_{\ell+},u,t)du.$$

This result is asymptotic in the level and for small k may yield a poor estimate. The estimates improve as the level increases. This is the dominant term in the asymptotic expansion because the singularity is of greatest multiplicity for this term. Figures 4 and 5 illustrate this behavior for phase 1 (one class-2 customer, shown in figure 4) and phase 2 (two class-2 customers shown in figure 5) and three different levels corresponding to three, thirteen and twenty-three class-1 customers.

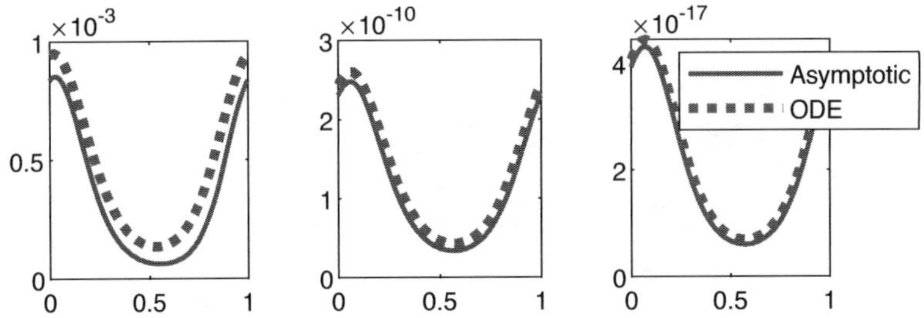

FIGURE 4.

$p_{3,1}(t)$ $p_{13,1}(t)$ $p_{23,1}(t)$

Comparison of ODE solution and asymptotic periodic solution for three different numbers of class-1 customers and one class-2 customer. The asymptotic estimates are computed using equation (4.6).

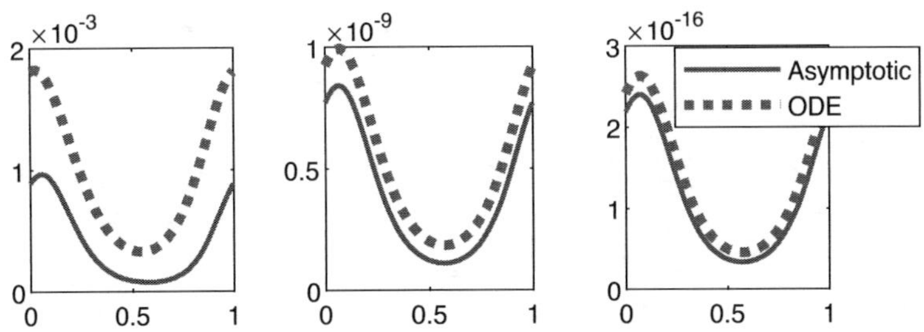

FIGURE 5.

$p_{3,2}(t)$ $p_{13,2}(t)$ $p_{23,2}(t)$

Comparison of ODE solution and asymptotic periodic solution for three different numbers of class-1 customers and two class-2 customers. The asymptotic estimates are computed using equation (4.7).

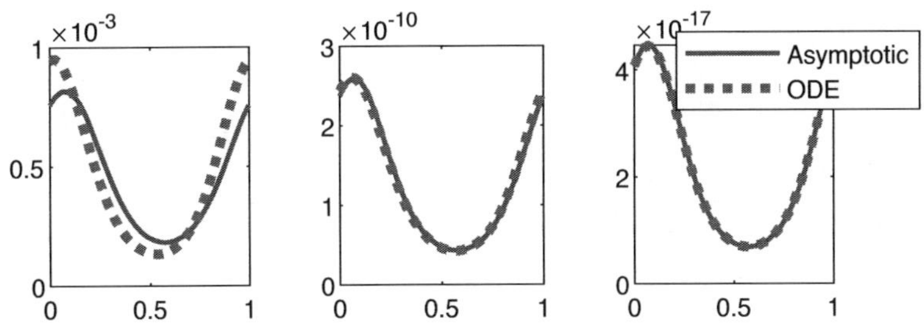

FIGURE 6.

$p_{3,1}(t)$ $p_{13,1}(t)$ $p_{23,1}(t)$

Comparison of ODE Solution and asymptotic periodic solution for three different numbers of class-1 customers and one class-2 customer using the formula given in equation (4.8), but including only the term $\ell = 0$ from the infinite sum.

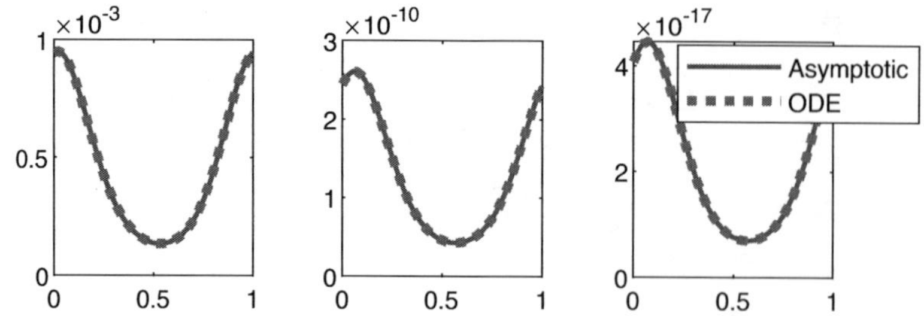

FIGURE 7.

$p_{3,1}(t)$ $\qquad$ $p_{13,1}(t)$ $\qquad$ $p_{23,1}(t)$

Comparison of ODE Solution and asymptotic periodic solution for three different numbers of class-1 customers and one class-2 customer using the formula given in equation (4.8), but including only terms $\ell = -2, -1, 0, 1, 2$ from the infinite sum.

It is possible to use the asymptotic approach to obtain exact formulas for any of the phases, simply by summing the residues. For phase 1, the exact formula is

$$
\begin{aligned}
(4.8)\quad & p_{k,1}(t) = \\
& (k+1)\sum_{\ell=-\infty}^{\infty} \frac{\alpha_{\ell+}^{-k}}{c_\ell^2} \int_{t-1}^{t} (p_{00}(u)\mu_1(u)(1-\alpha_{\ell+}^{-1}) + p_{01}u)\mu_2(u))\bar{\lambda}_2\phi(\alpha_{\ell+},u,t)du \\
& + \sum_{\ell=-\infty}^{\infty} \frac{\alpha_{\ell+}^{-k}}{c_\ell} \int_{t-1}^{t} \left(p_{00}(u)\mu_1(u)(1-\alpha_{\ell+}^{-1}) + p_{01}(u)\mu_2(u)\right)\phi(\alpha_{\ell+},u,t)du \\
& + \sum_{\ell=-\infty}^{\infty} \frac{\alpha_{\ell+}^{-k}}{c_\ell} \int_{t-1}^{t} \left(p_{00}(u)\mu_1(u)(1-\alpha_{\ell+}^{-1}) + p_{01}(u)\mu_2(u)\right) \int_u^t \lambda_2(\nu)d\nu\phi(\alpha_{\ell+},u,t)du \\
& + \sum_{\ell=-\infty}^{\infty} \frac{\alpha_{\ell+}^{-k-1}}{c_\ell^3}\bar{\lambda}_2 \int_{t-1}^{t} p_{00}(u)\mu_1(u)(-2\pi i\ell - \bar{\lambda}_1 - \bar{\lambda}_2 + \bar{\mu}_1)\phi(\alpha_{\ell+},u,t)du \\
& + \sum_{\ell=-\infty}^{\infty} \frac{\alpha_{\ell+}^{-k-1}}{c_\ell^3}\bar{\mu}_1\bar{\lambda}_2 2p_{01}(u)\mu_2(u)\phi(\alpha_{\ell+},u,t)du \\
& + \sum_{\ell=-\infty}^{\infty} \frac{\alpha_{\ell+}^{-k}}{c_\ell^2} \int_{t-1}^{t} \left[\left(\int_u^t (\mu_1(\nu)\alpha_{\ell+}^{-1} - \lambda_1(\nu)\alpha_{\ell+})d\nu\right)\bar{\lambda}_2(p_{00}(u)\mu_1(u)(1-\alpha_{\ell+}^{-1})\right. \\
& \qquad \left. + p_{01}(u)\mu_2(u))\phi(\alpha_{\ell+},u,t)\right]du.
\end{aligned}
$$

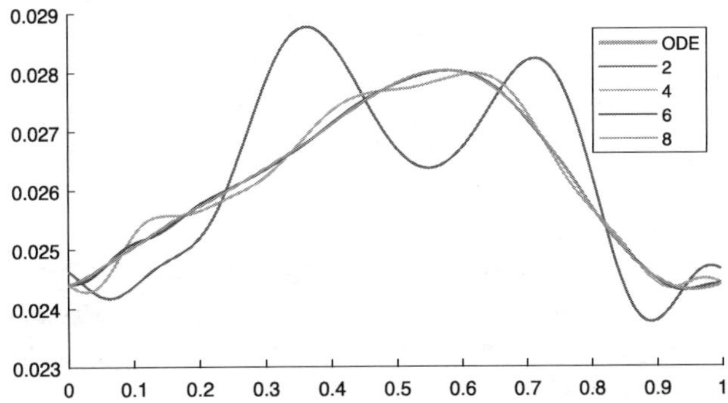

FIGURE 8. ODE solution versus asymptotic periodic solutions with varying values of M for $p_{0,1}(t)$ using equation (4.8).

Equation (4.8) is computed from the phase 1 component of the key equation (2.5), $[P(z,t)]_1$.

$$\lim_{z\to\alpha_{\ell+}} \left(1-\frac{z}{\alpha_{\ell+}}\right)^2 [P(z,t)]_1 = \frac{\alpha_{\ell+}^{-k}}{c_\ell^2}\int_{t-1}^{t} (p_{00}(u)\mu_1(u)(1-\alpha_{\ell+}^{-1}) + p_{01}u)\mu_2(u))\bar{\lambda}_2\phi(\alpha_{\ell+},u,t)du = g_1(\alpha_{\ell+},t)$$

and yields the first term of equation (4.8). We subtract this singularity and compute the following limit to get the remaining terms:

$$\lim_{z\to\alpha_{\ell+}} \left(1-\frac{z}{\alpha_{\ell+}}\right)([P(z,t)]_1 - \frac{g_1(\alpha_{\ell+},t)}{\left(1-\frac{z}{\alpha_{\ell+}}\right)^2}) = g_2(\alpha_{\ell+},t).$$

Then equation (4.8) can be written as

$$[P(z,t)]_1 = \sum_{\ell=-\infty}^{\infty}\left[\frac{g_1(\alpha_{\ell+},t)}{\left(1-\frac{z}{\alpha_{\ell+}}\right)^2} + \frac{g_2(\alpha_{\ell+},t)}{\left(1-\frac{z}{\alpha_{\ell+}}\right)}\right].$$

A similar approach can be used to get exact solutions for each phase. Note that the single term asymptotic estimate given in equation (4.7) is close for bigger values of k.

In general for time-varying, periodic QBDs, we have the following theorem:

THEOREM 4.2. *Let $P(z,t)$ be the probability generating function for a level independent ergodic QBD with time-varying periodic transition rates, meromorphic at all points of the closed disc $|z| < R$, with poles at points $\alpha_1, \alpha_2, \ldots, \alpha_m$.*

$$\mathbf{p}_k(t) \equiv [z^k]P(z,t) = \sum_{\ell=1}^{m} \alpha_\ell^{-k} F_\ell(t)\Pi_\ell(k) + O(R^{-k})\mathbf{e}$$

where

$$F_\ell(t) = \int_{t-1}^{t} \mathbf{p}_0(u)\left(\mathbf{B}(u) - \mathbf{A}_0(u) - \alpha_\ell^{-1}\mathbf{A}_{-1}(u)\right)\Phi(\alpha_\ell,u,t)du,$$

$\Pi_\ell(k)$ is a matrix of polynomials in k that depend on the pole, α_ℓ, and $\mathbf{e}$ *is a row vector of ones. Furthermore, the degree of the highest order polynomials in the matrix $\Pi_\ell(k)$ is equal to the order of the pole of $P(z,t)$ minus one.*

PROOF. This follows directly from Theorem 4.1. ∎

5. Conclusion

In this paper, we have shown how to extend the results of [**11**] to quasi-birth-death processes by providing a detailed example of a two-priority queue with finite buffer. The example has interesting combinatorial interpretations in terms of generating functions related to the Eulerian numbers.

In addition, we provide an exact formula for the asymptotic periodic level probabilities of the single-server queue in terms of an integral equation.

Acknowledgment

We thank Timothy Clos for helpful discussions on earlier versions of this paper.

References

[1] *The On-Line Encyclopedia of Integer Sequences, Sequence A008292*, 2020 (accessed July 30, 2020).

[2] Lothar Breuer, *The periodic BMAP/PH/c queue*, Queueing Syst. **38** (2001), no. 1, 67–76, DOI 10.1023/A:1010872128919. MR1839239

[3] M Defraeye and I van Nieuwenhuyse, *Staffing and scheduling under nonstationary demand for service: a literature review*, *Omega*, 58:4–25, 2016.

[4] William Feller, *An introduction to probability theory and its applications. Vol. I*, Third edition, John Wiley & Sons, Inc., New York-London-Sydney, 1968. MR0228020

[5] Philippe Flajolet and Robert Sedgewick, *Analytic combinatorics*, Cambridge University Press, Cambridge, 2009, DOI 10.1017/CBO9780511801655. MR2483235

[6] Winfried K. Grassmann and Steve Drekic, *Multiple eigenvalues in spectral analysis for solving QBD processes*, Methodol. Comput. Appl. Probab. **10** (2008), no. 1, 73–83, DOI 10.1007/s11009-007-9036-4. MR2394036

[7] A Kolmogorov, *Sur le problème d'attente*, *MatematicheskiiSbornik*, 38:101–106, 1931.

[8] G. Latouche and V. Ramaswami, *Introduction to matrix analytic methods in stochastic modeling*, ASA-SIAM Series on Statistics and Applied Probability, Society for Industrial and Applied Mathematics (SIAM), Philadelphia, PA; American Statistical Association, Alexandria, VA, 1999, DOI 10.1137/1.9780898719734. MR1674122

[9] B. H. Margolius, *Transient and periodic solution to the time-inhomogeneous quasi-birth process*, Queueing Syst. **56** (2007), no. 3-4, 183–194, DOI 10.1007/s11134-007-9027-8. MR2336105

[10] B. H. Margolius, *The matrices R and G of matrix analytic methods and the time-inhomogeneous periodic quasi-birth-and-death process*, Queueing Syst. **60** (2008), no. 1-2, 131–151, DOI 10.1007/s11134-008-9090-9. MR2452753

[11] Barbara Margolius, *Asymptotic estimates for queueing systems with time-varying periodic transition rates*, Lattice path combinatorics and applications, Dev. Math., vol. 58, Springer, Cham, 2019, pp. 307–326. MR3930461

[12] Marcel F. Neuts, *Matrix-geometric solutions in stochastic models*, Johns Hopkins Series in the Mathematical Sciences, vol. 2, Johns Hopkins University Press, Baltimore, Md., 1981. An algorithmic approach. MR618123

[13] T. Kyle Petersen, *Eulerian numbers*, Birkhäuser Advanced Texts: Basler Lehrbücher. [Birkhäuser Advanced Texts: Basel Textbooks], Birkhäuser/Springer, New York, 2015. With a foreword by Richard Stanley, DOI 10.1007/978-1-4939-3091-3. MR3408615

[14] Justus Arne Schwarz, Gregor Selinka, and Raik Stolletz, *Performance analysis of time-dependent queueing systems: Survey and classification*, *Omega*, 63:170–189, 2016.

[15] Jan Vrbik and Paul Vrbik, *Yet Another Proof of Sylvester's Determinant Identity*, *arXiv e-prints*, page `arXiv:1512.08747`, Dec 2015.

[16] W Whitt, *Time-varying queues*, *Queueing Models and Service Management*, 1(2):79–164, 2018.

Cleveland State University, Cleveland, Ohio 44115-2214, United States
Email address: b.margolius@csuohio.edu

Contemporary Mathematics
Volume **774**, 2021
https://doi.org/10.1090/conm/774/15573

Random measure algebras

Jason H. J. Park

ABSTRACT. In this article, we introduce algebras of random measures. Algebra is a vector space V over a field F with a multiplication satisfying the property: 1) distribution and 2) $c(x \cdot y) = (cx) \cdot y = x \cdot (cy)$ for every $c \in F$ and $x, y \in V$. The first operation is a trivial addition operation. For the second operation, we present three different methods 1) a convolution by covariance method, 2) O-dot product, 3) a convolution of bimeasures by Morse-Transue integral. With those operations, it is possible to build three different algebras of random measures.

1. Introduction

A random measure is a vector-valued set function of which codomain is a function space on probability measurable space. It is a direct analog of a stochastic process. Constructing algebras of random measures will enrich the study of stochastic processes since we can apply many algebra theorems to the stochastic analysis.

A main difficulty of constructing algebras of random measures is defining the second operation between random measures. To build an algebra, we need two operations. First one is the trivial addition. For the second operation, the usual multiplication cannot work since a multiplication of two random measures is not necessarily a random measure. Previously, J.E. Huneycutt [**H**] has shown products and convolutions of vector measures. Also, D. Dehay [**D**] has shown a Fubini theorem type of product between harmonizable time series. M.M. Rao [**R**] has suggested three possible operations, which are 1) a convolution by using covariance functions, 2) O-dot product, and 3) a convolution by strict Morse-Transue integral. J.H.J Park has constructed random measure algebras by using those definitions suggested by M.M. Rao. In this article, section 2 mainly consists of preliminary results and backgrounds of random measures. In section 3, 4 and 5, we show the definitions of second operations and the algebras of random measures by using those operations, respectively.

REMARK 1.1. J. Chang, H.S. Chung, and D. Skoug [**CCS**] introduced a convolution product on Wiener Space. Their convolution is the convolution of nonrandom functionals on a Wiener Space, which is a vector space of nonrandom integrable functionals relative to the Wiener measure which is translation invariant in an infinite dimensional space $\mathbb{R}^\infty$. However, there is no randomness involved in the

---

2020 *Mathematics Subject Classification.* Primary 60G57, 28B05; Secondary 16S99.
Key words and phrases. random, measure, probability, algebra.

functions. It is an interesting functional analysis problem, but our interest is in the convolution of stochastic processes and measures of various types, and is distinct from their works.

2. Preliminaries

DEFINITION 2.1. Let $(G, \mathcal{G})$ be a measurable space, where G is a locally compact abelian group and $\mathcal{G}$ is σ-algebra of G. Let Z be a vector-valued σ-additive set function such that $Z : \mathcal{G} \to L^2(\Omega, \Sigma, P)$, where (Ω, Σ, P) is a probability measure space. Then Z is termed a random measure.

In this article, we focus on second order random measures. An output of Z is a random variable with second moment. If the range of Z is L^p-space, then Z would be a pth order random measure. A random variable with a second moment has its importance because its variance exists. If we consider a stochastic process $\{X_t\}_{t \in I}$ with each X_t is a random variable with a second moment, then a covariance function exists for any two random variables X_t, X_s, where $t, s \in I$.

DEFINITION 2.2. (1) A mapping $\beta : \mathcal{G} \times \mathcal{G} \to \mathbb{C}$ is called a bimeasure if it is separately additive, that is, if $\beta(E, \cdot)$ and $\beta(\cdot, F)$ are (scalar-valued) additive measures for every $E, F \in \mathcal{G}$.

(2) A bimeasure $\beta : \mathcal{G} \times \mathcal{G} \to \mathbb{C}$ is said to be σ-additive if it is separately σ-additive.

(3) Let β be a bimeasure. For every pair $(E, F) \in \mathcal{G} \times \mathcal{G}$, we define the (Vitali) variation $|\beta|(E, F)$ of β on (E, F) by following equality:

$$|\beta|(E, F) = \sup \Sigma_{i \in I} \Sigma_{j \in J} |\beta(E_i, F_j)| \leq +\infty,$$

where the supremum is taken for all finite families $\{E_i\}_{i \in I}$ of disjoint sets from $\mathcal{G}$ with $\cup_{i \in I} E_i = E$ and all families $\{F_j\}_{j \in J}$ of disjoint sets from $\mathcal{G}$ with $\cup_{i \in I} F_i = F$.

(4) A bimeasure is called positive definite if

$$\sum_{i=1}^{n} \sum_{j=1}^{n} a_i \bar{a}_j \beta(E_i, E_j) \geq 0$$

for all $a_i, a_j \in \mathbb{C}, E_i, E_j \in \mathcal{G}$ and $1 \leq i, j \leq n$ where $n \in \mathbb{N}$.

A bimeasure is an 'analog' of a covariance function. The following lemma states that there is always a corresponding bimeasure β to Z and vice versa.

LEMMA 2.3. [**R**]

(1) *Let $Z : \mathcal{G} \to \mathbb{H}$ (a Hilbert Space) be a (bounded) vector measure and $\langle \cdot, \cdot \rangle_{\mathbb{H}}$ be the inner product of $\mathbb{H}$. The mapping $\beta : \mathcal{G} \times \mathcal{G} \to \mathbb{C}$ defined by*

$$\beta(E, F) = \langle Z(E), Z(F) \rangle_{\mathbb{H}}, \qquad E, F \in \mathcal{G}$$

is a (bounded) positive definite bimeasure and it is the bimeasure induced by Z.

(2) *Conversely, it follows from (basic) properties of the reproducing kernel Hilbert space that for a (bounded) positive definite bimeasure $\beta : \mathcal{G} \times \mathcal{G} \to \mathbb{C}$ there exist a Hilbert space $\mathbb{H}_\beta$ and a (bounded) vector measure $Z_\beta : \mathcal{G} \to \mathbb{H}_\beta$ such that*

$$\langle Z_\beta(E), Z_\beta(F) \rangle_{\mathbb{H}_\beta} = \beta(E, F) \qquad E, F \in \mathcal{G}, \qquad \overline{span(Z_\beta)} = \mathbb{H}_\beta,$$

where $span(Z_\beta)$ *denotes the linear span of* $\{Z(E)|E \in \mathcal{G}\}$.

EXAMPLE 2.4. Let $\{W_t\}_{t\in\mathbb{R}^+\cup\{0\}}$ be a Wiener Process with a positive diffusion coefficient σ, that is,

(1) Each increment $W(s+t) - W(s)$ is $N(0, \sigma^2 t)$
(2) For every pair of disjoint time intervals $(t_1, t_2], (t_3, t_4]$ with $0 \le t_1 < t_2 \le t_3 < t_4$, the increments $W(t_4) - W(t_3)$ and $W(t_2) - W(t_1)$ are independent random variables, and similarly for n disjoint time intervals, where n is an arbitrary positive integer.
(3) $W(0) = 0$ and $W(t)$ is continuous as a function of t.

Let $\mathcal{G}$ be a σ-algebra of Borel subsets of $\mathbb{R}^+\cup\{0\}$. If $(t, s) \in \mathcal{G}$, where $0 \le t < s \le \infty$, then we define a random measure $Z : \mathcal{G} \to L^2(\Omega, \Sigma, P)$ by $Z((t, s)) = W(s) - W(t)$. This is an example of a stochastic process that can be written in terms of a random measure.

DEFINITION 2.5. Given a centered $L_0^2(P)$-stochastic process $\{X(t) : t \in \mathbb{R}^+ \cup \{0\}\}$, its covariance function, or kernel is given by $C(t, s) = Cov(X(t), X(s))$.

LEMMA 2.6. [**P1**] *For a Wiener Process* $\{W(t) : t \in \mathbb{R}^+ \cup \{0\}\}$, *its covariance function is* $Cov(W(s), W(t)) = \sigma^2 \min\{s, t\}$ *for* $s, t \ge 0$.

3. A convolution by covariance method

The Morse-Transue integral by M. Morse and W. Transue [**MT**] is essential in our work. The definition of Morse-Transue integral is as follows:

DEFINITION 3.1. [**MT**] If $(G_i, \mathcal{G}_i), i = 1, 2$ are measurable spaces with $(G, \mathcal{G})$(i.e $G = G_1 \times G_2$ and $\mathcal{G} = \mathcal{G}_1 \otimes \mathcal{G}_2$) and $f_i : G_i \to \mathbb{C}$ (measurable relative to $\mathcal{G}_i, i = 1, 2$) are given, then the pair (f_1, f_2) is called *strictly* β-integrable where β is a bimeasure on $\mathcal{G}_1 \times \mathcal{G}_2$, provided the following two conditions hold:

(1) f_1 is $\beta(\cdot, B)$-integrable (L-S) for each $B \in \mathcal{G}_2$ and f_2 is $\beta(A, \cdot)$-integrable (L-S) for each $A \in \mathcal{G}_1$ such that $\tilde{\beta}(A, F) = \int_F f_2(w_2)\beta(A, dw_2)$ is σ-additive in $A \in \mathcal{G}_1$ for each $F \in \mathcal{G}_2$ and $\tilde{\beta}_2(E, B) = \int_E f_1(w_1)\beta(dw_1, B)$ is σ-additive in $B \in \mathcal{G}_2$ for each $E \in \mathcal{G}_1$; [L-S for Lebesgue Stieltjes]
(2) f_1 is $\tilde{\beta}_1(\cdot, F)$-integrable (L-S), f_2 is $\tilde{\beta}_2(E, \cdot)$-integrable (L-S) and

$$\int_E f_1(w_1)\tilde{\beta}_1(dw_1, F) = \int_F f_2(w_2)\tilde{\beta}_2(E, dw_2), E \in \mathcal{G}_1, F \in \mathcal{G}_2 \tag{3.1}$$

The common value in (3.1) is denoted $\int_E \int_F (f_1, f_2) d\beta$. It is called the *strict Morse-Transue integral*, if (3.1) holds each pair (E, F) as above.

A bimeasure induces a bilinear form. The definition of bilinear form is as follows:

DEFINITION 3.2. (1) Let $(G_i, \mathcal{G}_i), i = 1, 2$ be measurable spaces with G_i as a locally compact Hausdorff space and $\mathcal{G}_i$ as its Borel σ-algebra. Let $\beta : \mathcal{G}_1 \times \mathcal{G}_2 \to \mathbb{C}$ be a bimeasure, $C_c(G_i)$ be the space of scalar continuous functions with compact support. Then we define the corresponding bilinear form $B : C_c(G_1) \times C_c(G_2) \to \mathbb{C}$ as follows with Morse-Transue integral:

$$B(f_1, f_2) = \int_{G_2} \int_{G_1} f_1(w_1) f_2(w_2) \beta(dw_1, dw_2), f_i \in C_c(G_i)$$

Conversely, if a bilinear form B is given, then its corresponding bimeasure β is defined by $\beta(A, B) = B(\chi_A, \chi_B)$.

(2) A bilinear form B is bounded if there is a constant C (C depends on B) such that

$$|B(f_1, f_2)| \leq C||f_1||||f_2||,$$

for all $f_i \in C_c(G_i)$, where $||\cdot||$ is the uniform norm.

If a bilinear form B is correspondent to a bimeasure β, then we have following properties. B is bounded if and only if β is bounded, and B is positive definite if and only if β is positive definite.

LEMMA 3.3. [**R**] *Let $(G_i, \mathcal{G}_i), i = 1, 2$ be measurable spaces where each G_i is a locally compact Hausdorff space and each $\mathcal{G}_i$ denotes the corresponding σ-algebra of Borel sets. Let $\beta : \mathcal{G}_1 \times \mathcal{G}_2 \to \mathbb{C}$ be a bimeasure, $C_c(G_i)$ be the space of scalar continuous functions on G_i with a compact support, and $B : C_c(G_1) \times C_c(G_2) \to \mathbb{C}$ be the corresponding bilinear form of the bimeasure β. Then we have the following:*

(1) *β is bounded if and only if B is bounded.*
(2) *β is positive definite if and only if B is positive definite.*

Now, we introduce the definition of Fourier Transform of a bilinear form by C. Graham and B. Schreiber [**GS**].

DEFINITION 3.4. [**GS**] Let G_i be LCA groups with character (or dual) groups $\Gamma_i, i = 1, 2$. Set $G = G_1 \times G_2$ and $\Gamma = \Gamma_1 \times \Gamma_2$. For each character $(\lambda_1, \lambda_2) \in \Gamma_1 \times \Gamma_2$ (so $\lambda_i \in \Gamma_i, i = 1, 2$), consider

$$\hat{B}(\lambda_1, \lambda_2) = \langle \bar{\lambda_1} \otimes \bar{\lambda_2}, B \rangle = \int_{G_2} \int_{G_1} \bar{\lambda}_1 \bar{\lambda}_2 d\beta(\lambda_1, \lambda_2), \qquad \lambda_1 \in \Gamma_1, \lambda_2 \in \Gamma_2,$$

where β is the corresponding bimeasure of B and the integral is the MT-integral.

The space $S(\Gamma_1, \Gamma_2)$ consists of bounded, uniformly continuous functions on $\Gamma_1 \times \Gamma_2$, and it is an algebra.

DEFINITION 3.5. [**GS**] If Γ_1, Γ_2 are LCA groups and $V_1(\cdot), V_2(\cdot)$ are strongly continuous unitary representatives of Γ_1, Γ_2 on a Hilbert space $\mathbb{H}$, then

$$S(\Gamma_1, \Gamma_2) = \{\alpha : \Gamma_1 \times \Gamma_2 \to \mathbb{C} | \langle V_1(\lambda_1)\xi, V_2(\lambda_2)\eta \rangle_{\mathbb{H}} = \alpha(\lambda_1, \lambda_2)\},$$

for some $\xi, \eta \in \mathbb{H}$. (Here ξ, η are arbitrarily fixed.)

LEMMA 3.6. [**GS**] *The set $S(\Gamma_1, \Gamma_2)$ of Definition 3.5 is closed under pointwise products, sums and complex conjugation. Hence, it is an algebra.*

There is an one-to-one correspondence between $\hat{B}$ and $\alpha \in S(\Gamma_1, \Gamma_2)$.

THEOREM 3.7. [**GS**] *Let G_1, G_2 be locally compact abelian (LCA) groups with Γ_1, Γ_2 as their dual groups. Let $\tilde{B}(G_1, G_2)$ be the space of all bounded bilinear forms [obtained from bounded bimeasures through the MT-integration as before] and $S(\Gamma_1, \Gamma_2)$ be the corresponding function space of Definition 3.5. Then we obtain the following conclusions:*

(1) *For each $B \in \tilde{B}(G_1, G_2)$, its transform $\hat{B}$ exists, and satisfies $\hat{B} \in S(\Gamma_1, \Gamma_2)$, and for each $\alpha \in S(\Gamma_1, \Gamma_2)$ there is a unique $B \in \tilde{B}(G_1, G_2)$ satisfying $\alpha = \hat{B}$.*

(2) *For each* $\alpha \in S(\Gamma_1, \Gamma_2)$ *and* $B \in \tilde{B}(G_1, G_2)$ *of (1) so that* $\alpha = \hat{B}$, *we have* $||B|| \le ||\xi|| ||\eta||$, *where* (ξ, η) *defines* α *as in Definition 3.5.*

Because $S(\Gamma_1, \Gamma_2)$ is an algebra, there exists an element $\hat{B} = \hat{B}_1 \cdot \hat{B}_2$ in $S(\Gamma_1, \Gamma_2)$. And we can define an operation $*$ by $B_1 * B_2 = B$ such that $\hat{B}$ is correspondent to B.

DEFINITION 3.8. [**R**] Let $B_1, B_2 \in \tilde{B}(G_1, G_2)$ and $\hat{B}_1, \hat{B}_2 \in S(\Gamma_1, \Gamma_2)$ as in Theorem 3.7. We define the convolution $B_1 * B_2 \in \tilde{B}(G_1, G_2)$ by the equation $(B_1 * B_2)^\wedge = \hat{B}_1 \cdot \hat{B}_2$ which is unambiguously defined in the (complex-valued) space $S(\Gamma_1, \Gamma_2)$.

And $B_1 * B_2$ is also bounded by the following lemma.

LEMMA 3.9. [**R**] *If* B_1, B_2 *are bounded bilinear forms on* $C_0(G_1) \times C_0(G_2)$, *then their composition* $B_1 * B_2$, *of Definition* 3.8 *is continuous and satisfies*

$$||B_1 * B_2|| \le K_G^2 ||B_1|| ||B_2||, \tag{3.2}$$

where $K_G > 0$ *is the Grothendieck constant which is known to satisfy* $\frac{\pi}{2} < K_G \le \pi(2\log(1+\sqrt{2}))^{-1} = 1.782\cdots$.

We are now ready to define the convolution of random measures. Let $(G, \mathcal{G})$ be a measurable space, G be a locally compact abelian group (LCA), and $\mathcal{G}$ be a σ-algebra of Borel sets of G. Consider a second order random measure $Z : \mathcal{G} \to L_0^2(\Omega, \Sigma, P)$, where (Ω, Σ, P) is a probability measurable space. Let $\beta : \mathcal{G} \times \mathcal{G} \to \mathbb{C}$ be the corresponding bimeasure defined by

$$\beta(A, B) = E[Z(A)\overline{Z(B)}] = \int_\Omega Z(A)\overline{Z(B)} dP.$$

Let $B : C_0(G) \times C_0(G) \to \mathbb{C}$ be the bilinear form of β defined by

$$B(f, g) = \int_G \int_G f(g_1) g(g_2) d\beta(g_1, g_2)$$

A convolution of random measures is defined as follows:

THEOREM 3.10. [**R**] *Let* $(G, \mathcal{G})$ *be a measurable space, where* G *is a LCA group and* $\mathcal{G}$ *is the Borel* σ*-algebra of* G. *Suppose* $Z_i : \mathcal{G} \to L_0^2(P), i = 1, 2$ *be* σ*-additive random measures, and* B_1, B_2 *be the corresponding bounded bilinear forms on* $C_0(G)$ *so that their composition* $B_1 * B_2$ *is well-defined (whence* (3.2) *holds). Then there exists a unique random measure* Z *on* $\mathcal{G}$ *with values in* $L_0^2(P)$ *such that its covariance bimeasure determines a bounded bilinear form* B *which is precisely* $B_1 * B_2$ *and if* Z *is defined as* $Z = Z_1 * Z_2 : \mathcal{G} \to L_0^2(P)$ *where the probability space* (Ω, Σ, P) *can be taken rich enough to support all this structure, then* Z *has its covariance bimeasure form as* B.

Let $RM(G)$ be the set of second ordered random measures $Z : \mathcal{G} \to L_0^2(\Omega, \Sigma, P)$. Let $BM(G)$ be the set of bimeasures induced from random measures from $RM(G)$. Let $BL(G)$ be the set of bilinear forms induced from $BM(G)$. If a random measure Z has its corresponding bimeasure β, we will use the notation $Z \sim \beta$. And if β induces the bilinear form B, then we will write $\beta \sim B$ for convenience.

We have following elementary results for the basic structure of $RM(G)$, $BM(G)$, and $BL(G)$ from J.H.J Park [**P2**].

THEOREM 3.11. **[P2]**

(1) $BM(G, +)$ *is an abelian group.*
(2) $BM(G, +)$ *is a unitary* $\mathbb{C}$*-module.*
(3) $BL(G, +)$ *is an abelian group.*
(4) $BL(G, +)$ *is a unitary* $\mathbb{C}$*-module.*
(5) $RM(G, +)$ *is an abelian group.*
(6) $RM(G, +)$ *is a unitary* $\mathbb{C}$*-module.*

We have following results by J.H.J. Park **[P2]**.

DEFINITION 3.12. **[P2]** Suppose B_1, B_2 are bounded bilinear forms and β_1, β_2 are their bimeasures as defined in Theorem 3.2. Let $\beta_1 * \beta_2$ denote the corresponding bimeasure of $B_1 * B_2$ as defined in Definition 3.8, and $Z_1 * Z_2$ denote the induced random measure of the bimeasure $\beta_1 * \beta_2$, as given by Theorem 3.10.

LEMMA 3.13. **[P2]** *Given* $\alpha \in \mathbb{C}$ *and* $\beta_1, \beta_2 \in BM(G)$ *with corresponding bilinear forms* B_1, B_2. *We have:*

(1) $\beta_1 + \beta_2$ *induces the bilinear form* $B_1 + B_2$. *(i.e.* $\beta_1 + \beta_2 \sim B_1 + B_2$*)*
(2) $\alpha\beta$ *induces the bilinear form* αB. *(i.e.* $\alpha\beta \sim \alpha B$*)*

LEMMA 3.14. **[P2]** *Let* B_1, B_2 *be bounded bilinear forms and* $\hat{B}_1, \hat{B}_2$ *be their Fourier transforms. Then* $\widehat{B_1 + B_2} = \hat{B}_1 + \hat{B}_2$.

LEMMA 3.15. **[P2]** *Suppose* G *is an LCA group. Then the algebra* $S(\Gamma, \Gamma)$ *has an identity element* $Id(\cdot, \cdot) \in S(\Gamma, \Gamma)$ *(under pointwise multiplication), which is defined by* $Id(\lambda_1, \lambda_2) = 1$ *for all* $\lambda_1, \lambda_2 \in \hat{G}$.

LEMMA 3.16. **[P2]** *Suppose* B *is a bounded bilinear form and* $\hat{B}$ *is the corresponding Fourier transform. For each* $c \in \mathbb{C}$, *the bounded bilinear form* cB *has the corresponding Fourier transform* $c\hat{B}$. *(i.e If* $B \sim \hat{B}$, *then* $cB \sim c\hat{B}$.*)*

LEMMA 3.17. **[P2]** *Let* $\alpha \in \mathbb{C}$, $Z : G \to L^2(\Omega, \Sigma, P)$ *be a random measure, and* β *be the corresponding bimeasure of* Z *(i.e.* $\beta(A, B) = E[Z(A)\overline{Z(B)}]$. *Then* αZ *has the bimeasure* $|\alpha|^2\beta$. *(i.e. If* $Z \sim \beta$ *then* $\alpha Z \sim |\alpha|^2\beta$*)*

$BL(G)$ is an algebra and it is shown in **[GS]**.

THEOREM 3.18. **[P2]** *Let* $BM(G)$ *be a set of all bimeasures induced from a second ordered random measure over the locally compact abelian group* G, *with the operation* $(+, *)$. *Then*

(1) $BM(G)$ *is a commutative ring with identity.*
(2) $BM(G)$ *is a* $\mathbb{C}$*-algebra.*

THEOREM 3.19. **[P2]** $BL(G)$ *is a ring and an algebra over* $\mathbb{C}$.

THEOREM 3.20. **[P2]** *Let* $RM(G)$ *be a set of all second ordered random measures over* G, *with the operation* $(+, *)$. *Then*

(1) $RM(G)$ *is a commutative ring.*
(2) $RM(G)$ *is a ring with identity.*
(3) $RM(G)$ *is a normed-ring, whose norm* $|| \cdot ||$ *is the semi-variation.*
(4) $RM(G)$ *is a normed* $\mathbb{C}$*-Algebra with its norm semi-variation.*
(5) $RM(G)$ *is a linear space.*

4. O-dot product and convolution of bimeasures

In this section, we introduce the O-dot product and build an algebra by using O-dot product as the second operation. The following proposition leads to the definition of O-dot product.

PROPOSITION 4.1. **[R]** *Let $\beta_i : \mathcal{G}_i \times \mathcal{G}_i \to \mathbb{C}, i = 1, 2$ be a pair of positive definite kernels and $\beta = \beta_1 \cdot \beta_2 : (\mathcal{G}_1 \times \mathcal{G}_1) \times (\mathcal{G}_2 \times \mathcal{G}_2) \to \mathbb{C}$ as their pointwise product. Then β is positive definite. If we let $\mathbb{H}_\beta, \mathbb{H}_{\beta_1}, \mathbb{H}_{\beta_2}$ the corresponding reproducing kernel Hilbert (or Aronszajn) spaces, then $\mathbb{H}_\beta = \mathbb{H}_{\beta_1} \otimes \mathbb{H}_{\beta_2}$, so that $\mathbb{H}_\beta$ is a tensor product of $\mathbb{H}_{\beta_1}$ and $\mathbb{H}_{\beta_2}$.*

Proposition 4.1 illustrates that the product of two bimeasures is a pointwise multiplication of two bimeasures, which leads to the following definition.

DEFINITION 4.2. **[R]** Let $(G, \mathcal{G})$ be a measurable space and $Z_i : \mathcal{G} \to L_0^2(P)$ be a pair $(i = 1, 2)$ of random measures into $L_0^2(P)$ the Hilbert space of (equivalence classes of) centered (complex) random variables on a probability space (Ω, Σ, P) with covariance bimeasures $\beta_i : \mathcal{G} \times \mathcal{G} \to \mathbb{C}$ given by $\beta_i(A, B) = \langle Z_i(A), Z_i(B)\rangle$ using the inner product notation. Let $\beta = \beta_1 \cdot \beta_2 : (\mathcal{G} \times \mathcal{G}) \times (\mathcal{G} \times \mathcal{G}) \to \mathbb{C}$ be the product, pointwise as in Proposition 4.1.

The product $\beta = \beta_1 \cdot \beta_2$ in Definition 4.2 has the domain $(\mathcal{G} \times \mathcal{G}) \times (\mathcal{G} \times \mathcal{G})$, where β_1, β_2 has the domain $\mathcal{G} \times \mathcal{G}$. However, we can take a diagonal of $(\mathcal{G} \times \mathcal{G}) \times (\mathcal{G} \times \mathcal{G})$ which is isomorphic to $\mathcal{G} \times \mathcal{G}$. Let $\tilde{\beta} : (\mathcal{G} \times \mathcal{G}) \times (\mathcal{G} \times \mathcal{G}) \to \mathbb{C}$ be such that

$$\tilde{\beta} = \begin{cases} \beta & \text{on diagonal set of } (\mathcal{G} \times \mathcal{G}) \times (\mathcal{G} \times \mathcal{G}) \\ 0 & \text{otherwise.} \end{cases}$$

The diagonal set of $(\mathcal{G} \times \mathcal{G}) \times (\mathcal{G} \times \mathcal{G})$ is $\{(A \times B, A \times B) | A \times B \in \mathcal{G} \times \mathcal{G}\}$. Therefore, $\tilde{\beta}$ is defined on isomorphic copy of $\mathcal{G} \times \mathcal{G}$. We rewrite

$$\tilde{\beta}(A, B) = \beta(A \times B, A \times B).$$

The following lemma shows $\tilde{\beta}$ is positive definite and σ-additive.

LEMMA 4.3. **[P3]** *Suppose $\tilde{\beta}$ is defined as above.*

(1) *$\tilde{\beta}$ is positive definite*

(2) *$\tilde{\beta} : \mathcal{G} \times \mathcal{G} \to \mathbb{C}$ is separately σ-additive.*

We also define O-dot product of random measures Z_1 and Z_2. Random measures Z_1, Z_2 are corresponding random measures of β_1, β_2 by Lemma 2.3.

DEFINITION 4.4. **[P3]** Suppose the $\beta = \beta_1 \cdot \beta_2 : (\mathcal{G} \times \mathcal{G}) \times (\mathcal{G} \times \mathcal{G}) \to \mathbb{C}$ as in Definition 4.2. Let $\tilde{\beta} = \beta$ on the diagonal of $(\mathcal{G} \times \mathcal{G}) \times (\mathcal{G} \times \mathcal{G})$, and 0 otherwise. Define O-dot product $\odot$ of bimeasures by $\tilde{\beta} = \beta_1 \odot \beta_2$. Therefore, $\tilde{\beta} : \mathcal{G} \times \mathcal{G} \to \mathbb{C}$ is defined by $\tilde{\beta}(A, B) = \beta_1 \odot \beta_2(A, B) = \beta_1(A, B) \cdot \beta_2(A, B)$, where $A \times B \in \mathcal{G} \times \mathcal{G}$.

Moreover, there exist a reproducing kernel Hilbert space, $\mathbb{H}$ of $\tilde{\beta}$ and a random measure Z such that $\tilde{\beta}(A, B) = E[Z(A)\overline{Z(B)}]$. If Z_1, Z_2 and $\mathbb{H}_1, \mathbb{H}_2$ are the corresponding random measures and reproducing kernel Hilbert spaces for bimeasures β_1, β_2, then define the O-dot product of Z_1 and Z_2 as $Z = Z_1 \odot Z_2$, whose bimeasure is $\tilde{\beta}$.

REMARK 4.5. Note that there is a slight change from the definition of Rao's O-dot product **[R]**. In this article, we have restricted domain of the product bimeasure β so it can have the same domain of β_1, β_2.

THEOREM 4.6. [**P3**]

(1) *$BM(G, \odot)$ is a ring.*
(2) *$BM(G, \odot)$ is an algebra over $\mathbb{C}$.*

The multiplicative identity of $BM(G)$ is not trivial. One can think of a bimeasure $\delta(A, B) = 1$ for all $A, B \in \mathcal{G}$. However, this δ will not have the additive property of bimeasure.

THEOREM 4.7. [**P3**]

(1) *$RM(G, \odot)$ is a ring.*
(2) *$RM(G, \odot)$ is an algebra over $\mathbb{C}$.*

Let Z_W be a random measure that represents a Wiener process. The product $Z = Z_W \odot Z_W$ is a random measure with its bimeasure $\tilde{\beta}(A, B) = \sigma^4 \mu(A \cup B)^2$, where μ is the Lebesgue measure. The proof is illustrated in [**P3**].

LEMMA 4.8. [**P3**] *Suppose $Z_W : \mathcal{G} \to L^2(P)$ is a random measure that represents Wiener Process, where $\mathcal{G}$ is a σ-algebra of Borel subsets of $\mathbb{R}^+$. Suppose β_W is the corresponding bimeasure of Z_W (i.e. β_W is a scalar bimeasure induced from a Wiener process). If $Z = Z_W \odot Z_W : \mathcal{G} \to L^2(P)$, then Z has the bimeasure $\tilde{\beta} = \beta_W \odot \beta_W : \mathcal{G} \times \mathcal{G} \to \mathbb{C}$ such that $\tilde{\beta}(A, B) = \sigma^4 \mu(A \cap B)^2$, where σ is a positive diffusion coefficient of the Wiener Process and μ is a Lebesgue measure.*

$Z \odot Z$ has the covariance bimeasure $\tilde{\beta} = \sigma^4(\mu(A \cap B))^2$, where Z is a Wiener random measure. There exist a unique Gaussian Process corresponding to a given bimeasure. However, $Z \odot Z$ itself will not be a Wiener measure in the classical sense of Wiener's.

5. Convolution by strict Morse-Transue integral

The third and last convolution operation comes from the convolution of bimeasures by using Morse-Transue integral. The following proposition also can be considered as a definition.

PROPOSITION 5.1. [**R**] *Let $\mathcal{B}([0, 1])$ be a σ-algebra of Borel subsets of $[0, 1]$. If $Z_i : \mathcal{B}([0, 1]) \to L^2(P), i = 1, 2$ are random measures and β_i are their induced bimeasures respectively, then a convolution of bimeasures β_1 and β_2 is defined by*

$$(5.1) \qquad (\beta_1 * \beta_2)(A, B) = \int_0^1 \int_0^1 \beta_1(A - x, B - y) \beta_2(dx, dy), A, B \in \mathcal{B}([0, 1]),$$

and $(\beta_1 * \beta_2)(\cdot, \cdot)$ is a well-defined positive definite bimeasure on $\mathcal{B}([0, 1]) \times \mathcal{B}([0, 1])$. Also, it induces a random measure $Z : \mathcal{B}([0, 1]) \to L^2(P)$ with a finite Vitali variation.

The integration in above proposition is Morse-Transue integral [**MT**]. The convolution of bimeasures is well-defined, and it is commutative.

LEMMA 5.2. [**P3**] *The convolution products of positive definite bimeasures are commutative.*

With the newly defined convolution of bimeasures, we can define the convolution of random measures.

DEFINITION 5.3. [**R**] Let $Z_i : \mathcal{B}_0([0,1]) \to L^2(P), i = 1, 2$, are a pair of random measures. Define a convolution product of their induced bimeasures β_i by

$$(\beta_1 * \beta_2)(A, B) = \int_0^1 \int_0^1 \beta_1(A - x, B - y)\beta_2(dx, dy), A, B \in \mathcal{B}_0([0,1]).$$

$(\beta_1 * \beta_2)(\cdot, \cdot)$ is a well-defined positive definite bimeasure on $\mathcal{B}_0([0,1]) \times \mathcal{B}_0([0,1])$ and there is a random measure $Z : \mathcal{B}_0([0,1]) \to L^2(P)$ whose induced bimeasure is $(\beta_1 * \beta_2)(\cdot, \cdot)$. Define $Z = Z_1 * Z_2$. Then Z is well-defined with its codomain is $L_0^2(P)$.

We used the same notation $*$ for the convolution operation. However, the convolution $*$ in this section is different from $*$ in section 3. In this section, the convolution is defined by using Morse-Transue integral, whereas the convolution in section 3 is defined by the covariance method.

Let's denote $BM([0,1])$ as a set of positive definite bimeasures $\beta : \mathcal{B}_0([0,1]) \times \mathcal{B}_0([0,1]) \to \mathbb{C}$, and $RM([0,1])$ as a set of random measures $Z : \mathcal{B}_0([0,1]) \to L^2(\Omega, \Sigma, P)$. We investigate the algebraic structure of $BM([0,1])$ and $RM([0,1])$ with the convolution.

THEOREM 5.4. [**P3**]

(1) *The set of bimeasure $BM([0,1])$ with convolution is a ring with identity.*
(2) *$BM([0,1], *)$ is a $\mathbb{C}$-algebra.*

We move on to the structure of random measure algebra, which is our main interest.

THEOREM 5.5. [**P3**]

(1) *$RM([0,1], *)$ is a ring with identity.*
(2) *$RM([0,1], *)$ is a $\mathbb{C}$-algebra.*

Suppose Z is a random measure such that $Z : \mathcal{B}([0,1]) \to L_0^2(\Omega, \Sigma, P)$ such that $Z([t_i, t_{i+1}]) = W_{t_i} - W_{t_{i+1}}$, where $\{W_t\}_{t \in \mathbb{R}^+ \cup \{0\}}$ is a Wiener process. Let β be the bimeasure of Z, that is $\beta(A, B) = E[Z(A)\overline{Z(B)}]$. We explicitly compute $\beta * \beta(A, B)$ if $A = [t_1, t_2], B = [s_1, s_2]$ with $s_1 < t_1 < s_2 < t_2$.

THEOREM 5.6. [**P3**] *Let $\{W_t\}_{t \in [0,1]}$ be a Wiener process, and Z be the associated random measure, that is $Z : \mathcal{B}([0,1]) \to L_0^2(\Omega, \Sigma, P)$ such that $Z([t_i, t_{i+1}]) = W_{t_i} - W_{t_{i+1}}$. Let β be the bimeasure of Z, that is $\beta(A, B) = E[Z(A)\overline{Z(B)}]$ with $A = [t_1, t_2], B = [s_1, s_2]$ with $s_1 < t_1 < s_2 < t_2$. Then*

$$\begin{aligned}\beta * \beta(A, B) \quad &= -\tfrac{s_1^2 t_1^2}{4} + \tfrac{s_2 t_1^3}{4} - \tfrac{t_1^4}{12} + \tfrac{s_2}{3} + \tfrac{t_1 s_2}{2} + \tfrac{t_1^2 s_2}{2} + \tfrac{t_1^3 s_2}{6} - \tfrac{s_2^2}{2} - \tfrac{t_1 s_2^2}{2} - \tfrac{t_1^2 s_2^2}{4} \\ &+ \tfrac{t_1 s_2^3}{6} - \tfrac{s^4}{12} + \tfrac{t_2}{3} + \tfrac{s_1 t_2}{2} + \tfrac{s_1^2 t_2}{2} + \tfrac{s_1^3 t_2}{6} + \tfrac{s_2 t_2}{2} + \tfrac{s_2^2 t_2}{2} + \tfrac{s_2^3 t_2}{6} - \tfrac{t_2^2}{2} \\ &- \tfrac{s_1 t_2^2}{2} - \tfrac{s_1^2 t_2^2}{4} - \tfrac{s_2 t_2^2}{2} - \tfrac{s_2^2 t_2^2}{4} + \tfrac{t_2^3}{3} + \tfrac{s_1 t_2^3}{6} + \tfrac{s_2 t_2^3}{6} - \tfrac{t_2^4}{12}.\end{aligned}$$

There exists a unique (centered) Gaussian Process corresponding to a given covariance function, since a Gaussian process is determined by its mean and covariance functions. However, with such a complex representation, it is not trivial to express the exact representation of the Gaussian Process related to the covariance function above.

References

[CCS] Hyun Soo Chung, David Skoug, and Seung Jun Chang, *Relationships involving transforms and convolutions via the translation theorem*, Stoch. Anal. Appl. **32** (2014), no. 2, 348–363, DOI 10.1080/07362994.2013.877350. MR3177075

[D] Dominique Dehay, *On the product of two harmonizable time series*, Stochastic Process. Appl. **38** (1991), no. 2, 347–358, DOI 10.1016/0304-4149(91)90099-X. MR1119989

[GS] Colin C. Graham and Bertram M. Schreiber, *Bimeasure algebras on LCA groups*, Pacific J. Math. **115** (1984), no. 1, 91–127. MR762204

[H] James E. Huneycutt Jr., *Products and convolutions of vector valued set functions*, Studia Math. **41** (1972), 119–129, DOI 10.4064/sm-41-2-119-129. MR302855

[MT] Marston Morse and William Transue, *C-bimeasures Λ and their integral extensions*, Ann. of Math. (2) **64** (1956), 480–504, DOI 10.2307/1969597. MR86116

[P1] Jason Hong Jae Park, *Random Measure Algebras Under Convolution*, ProQuest LLC, Ann Arbor, MI, 2015. Thesis (Ph.D.)–University of California, Riverside. MR3427327

[P2] Jason Hong Jae Park, *A random measure algebra under convolution*, J. Stat. Theory Pract. **10** (2016), no. 4, 768–779, DOI 10.1080/15598608.2016.1224745. MR3558401

[P3] J.H.J. Park, *Random Measure Algebras Under O-dot Product and Morse-Transue Integral Convolution*, International J. of Stats. and Prob., **8(6)**, (2019), 73–81.

[R] M. M. Rao, *Random and vector measures*, Series on Multivariate Analysis, vol. 9, World Scientific Publishing Co. Pte. Ltd., Hackensack, NJ, 2012. MR2840012

Department of Mathematics, Univeristy of California, Riverside, Riverside, California, 92521

Current address: Department of Mathematical Sciences, Univeristy of Nevada, Las Vegas, Las Vegas, Nevada, 89154

Email address: jason.park@unlv.edu

Contemporary Mathematics
Volume **774**, 2021
https://doi.org/10.1090/conm/774/15574

From additive to second-order processes

M. M. Rao and R. J. Swift

ABSTRACT. The familiar Poisson process is a member of a class of stochastic processes known as additive processes. This broad class also contains the birth-death processes. Second-order processes are processes with two moments finite. The class of second-order processes includes the well-known weakly stationary as well as harmonizable processes.

A natural evolution of concepts linking the class of additive processes and the class of second-order processes will be detailed. The connection arises via stable processes and random measures

In the writing of the second edition of *Probability Theory with Applications* [**8**], a conversation occurred regarding the presentation of the material that would ultimately comprise Chapter 8 - A Glimpse of Stochastic Processes. We decided, for pedagogical reasons, on the order of topics that are presented in that chapter. However, there is a path of ideas we discussed that shows the connection of additive processes with second order processes that illustrates a natural connection to random measures. In this article, we follow that path and uncover some connections between familiar topics in the study of stochastic processes.

1. Counting processes

Let (Ω, Σ, P) be the underlying probability space and consider a nonnegative integer-valued process

$$\{N_t, t \geq 0\}$$

with independent increments. We can think of N_t as the total number of events that have occurred up to time t. Such a process is often termed a *counting process.*

If one lets $N_0 = 0$, and assumes that the probability of an event occurring during an interval of length t depends upon t, the familiar Poisson processes arises. In particular assume, as originally done by Poisson, that

$$P[N_{\Delta t} = 1] = \lambda \Delta t + o(\Delta t). \tag{1.1}$$

For a small value of Δt, equation (1.1) gives

$$P[N_{\Delta t} \geq 2] = o(\Delta t) \tag{1.2}$$

2020 *Mathematics Subject Classification.* Primary 60GXX.

Key words and phrases. Additive process, stable process, stationary process, harmonizable process.

and that events in nonoverlapping time intervals are independent. Letting

$$P_n(t) = P[N_t = n | N_0 = 0]$$

be the conditional probability of n events at time t given that there were none initially. It follows from the Kolmogorov forward equations that

$$P'_n(t) = -\lambda P_n(t) + \lambda P_{n-1}(t) \quad \text{for } n \geq 1. \tag{1.3}$$

The assumption that $N_0 = 0$ gives $P[N_0 = n] = 0$ for $n \geq 1$ so recursively solving (1.3) it follows that

$$P_n(t) = e^{-\lambda t}\frac{(\lambda t)^n}{n!} \qquad \text{for } n \geq 0, \tag{1.4}$$

which is the *Poisson process* with rate parameter $\lambda > 0$.

Alternately, a Poisson process N_t process can be obtained by letting X be an exponentially distributed random variable so that

$$P[X < x] = 1 - e^{-\lambda x}, x \geq 0, \lambda > 0.$$

If $X_1, \ldots, X_n$ are independent with the same distribution as X, let

$$S_n = \sum_{k=1}^{n} X_k,$$

be the partial sum and for $t \geq 0$, set

$$N_t = \sup\{n \geq 1 : S_n \leq t\}$$

so that N_t is the last time before the sequence $\{S_n, n \geq 1\}$ crosses the level $t \geq 0$, where as usual $\sup(\emptyset) = 0$. Then N_t is an integer valued random variable, and its distribution is easily obtained.

In fact, since S_n has a gamma distribution whose density is given by

$$f_{S_n}(x) = \frac{\lambda^n x^{n-1}}{\Gamma(n)} e^{-\lambda x}, \quad x \geq 0, n \geq 1, \lambda > 0,$$

we have for $n = 0, 1, 2, \ldots$ (set $S_0 = 0$), since

$$[N_t \geq n] = [S_n \leq t],$$

so that

$$\begin{aligned}
P[N_t = n] &= P[S_n \leq t, S_{n+1} > t] \\
&= \int\!\!\int_{[S_n \leq t, X_{n+1} + S_n > t]} f_{S_n}(x) f_{X_{n+1}}(y)\, dx\, dy \\
&\qquad (\text{ since } S_n, X_{n+1} \text{ are independent}) \\
&= \int_0^t f_{S_n}(x) dx \int_{[X_{n+1} > t - x]} f_{X_{n+1}}(y)\, dy \\
&= \int_0^t f_{S_n}(x)\, dx\, P[X_{n+1} > t - x] = e^{-\lambda t} \cdot \frac{(\lambda t)^n}{n!}. \qquad (1.5)
\end{aligned}$$

Thus $\{N_t, t \geq 0\}$ is a Poisson process. Moreover, it has the properties for $\omega \in \Omega$:

(a) $N_0(\omega) = 0$, $\lim_{t \to \infty} N_t(\omega) = \infty$,
(b) integer valued, nondecreasing, right continuous.

These properties characterize the Poisson process in the sense that such a process has independent stationary increments as well as the distribution given by equation (1.5).

THEOREM 1.1. *Let* $\{X_t, t \geq 0\}$ *be a nonnegative integer valued nondecreasing right continuous process with jumps of size 1 and support* $\mathbb{Z}^+ = \{0, 1, 2, \ldots\}$. *Then the following are equivalent conditions:*

1. the process is given by $X_t = \max\{n : S_n \leq t\}$, *where* $S_n = \sum_{k=1}^n X_k$, *with the* X_k *as i.i.d and exponentially distributed, i.e.,* $P[X > x] = e^{-\lambda x}, x \geq 0, \lambda > 0$,

2. the process has independent stationary increments, each of which is Poisson distributed, so that (1.5) *holds for* $0 < s < t$ *in the form*

$$P[X_t - X_s = n] = e^{-\lambda(t-s)} \frac{[\lambda(t-s)]^n}{n!}, \quad \lambda \geq 0, n = 0, 1, 2, \ldots,$$

3. the process has no fixed discontinuities, and satisfies the Poisson (conditional) postulates: for each $0 < t_1 < \ldots < t_k$; *and* $n_k \in \mathbb{Z}^+$ *one has for a* $\lambda \geq 0$ *as* $h \searrow 0$

(i) $P[X_{t_k+h} - X_{t_k} = 1 | X_{t_j} = n_j, j = 0, 1, \ldots k] = \lambda h + o(h)$

(ii) $P[X_{t_k+h} - X_{t_k} \geq 2 | X_{t_j} = n_j, j = 0, 1, \ldots k] = o(h)$.

1.1. Independent increment processes. The preceding generalizes by letting $\{X_t, t \in [0, 1]\}$ be a process with independent increments and consider the corresponding characteristic function (ch.f.) of the process.

That is, for $0 < s < t < 1$, let $\phi_{s,t}$ be the ch.f. of $X_t - X_s$. Then if

$$0 < t_1 < t_2 < t_3 < 1,$$

by the independence of $X_{t_3} - X_{t_2}$ and $X_{t_2} - X_{t_1}$, it follows that

$$\phi_{t_1,t_3}(u) = \phi_{t_1,t_2}(u)\phi_{t_2,t_3}(u), \quad u \in \mathbb{R}. \tag{1.6}$$

Now if the process is stochastically continuous; i.e., for each $\varepsilon > 0$,

$$\lim_{t \to s} P[X_t - X_s| \geq \varepsilon] = 0, \quad s \in (0, 1),$$

then

$$\lim_{t \to s} \phi_{s,t}(u) = 1$$

uniformly in u and s, t in compact intervals.

Hence if $0 \leq s < t_0 < t_1 < \cdots < t_n < t \leq 1$, with $t_k = s + (k(t-s)/n)$, we have

$$\phi_{s,t}(u) = \prod_{i=0}^{n-1} \phi_{t_i,t_{i+1}}(u), \quad u \in \mathbb{R}, \tag{1.7}$$

so that $\phi_{s,t}$ is infinitely divisible.

Using the Lévy-Khintchine representation (c.f. Rao & Swift [**8**]) with $s = 0 < t < 1, u \in \mathbb{R}$, gives

$$\phi_{0,t}(u) = \exp\left\{ i\gamma_t u + \int_{\mathbb{R}} \left(e^{iuv} - 1 - \frac{iuv}{1+v^2} \right) \frac{1+v^2}{v^2} \, dG_t(v) \right\} \tag{1.8}$$

for a unique pair $\{\gamma_t, G_t\}$.

Thus for a subinterval $[s,t] \subset [0,1]$, using (1.8), we obtain a pair $\gamma_{s,t}$ and $G_{s,t}$ in (1.8) for $\phi_{s,t}$. Using (1.6) applied to $0 < s < t < 1$, so that $\phi_{0,t} = \phi_{0,s} \cdot \phi_{s,t}$ we have

$$\text{Log}\phi_{s,t}(u) = \text{Log}\phi_{0,t}(u) - \text{Log}\phi_{0,s}(u). \tag{1.9}$$

Substituting (1.8) in (1.9), we obtain

$$\gamma_{s,t} = \gamma_t - \gamma_s \text{ and } G_{s,t} = G_t - G_s.$$

Thus
(1.10)

$$\phi_{s,t}(u) = \exp\left\{i(\gamma_t - \gamma_s)u + \int_{\mathbb{R}} \left(e^{iuv} - 1 - \frac{iuv}{1+v^2}\right) \frac{1+v^2}{v^2}\, d(G_t - G_s)(u)\right\},$$

A simple form occurs when G_t has no jump (so $\sigma^2 = 0$) at the origin, so

$$\varphi(t) = \exp\{i\gamma t + \int_0^\infty (e^{itx} - 1)\, dN(x)\}, \quad t \in \mathbb{R},$$

where $N(\{0\}) = 0, \gamma$ is a constant and $N(\cdot)$ is nondecreasing with

$$\int_{0+}^{2} u^2\, dN(u) < \infty.$$

Now if we rewrite the Poisson probabilities (1.4) as:

$$\pi_\lambda(\cdot) = e^{-\lambda} \sum_{n=0}^{\infty} \frac{\lambda^n}{n!} \delta_n(\cdot) \tag{1.11}$$

where $t = 1$ and $\delta_n(\cdot)$ is the Dirac point measure and $\pi_0 = \delta_0$, supp $(\pi_\lambda) = \{0, 1, 2, \ldots\} = \boldsymbol{Z}^+$. Then $\pi_\lambda(\cdot)$ is a measure on $\mathbb{P}(\boldsymbol{Z}^+)$, and if $\lambda_1, \lambda_2 \geq 0$ one has the convolution

$$(\pi_{\lambda_1} * \pi_{\lambda_2})(A) = \int_{\boldsymbol{Z}^+} \pi_{\lambda_1}(A - x)\pi_{\lambda_2}(dx)$$

and its ch.f.

$$\hat{\pi}_\lambda(t) = \int_{\boldsymbol{Z}^+} e^{itx}\pi_\lambda(dx) = e^{-\lambda} \sum_{n=0}^{\infty} e^{itn} \frac{\lambda^n}{n!} = e^{\lambda(e^{it}-1)},$$

which gives

$$\widehat{(\pi_{\lambda_1} * \pi_{\lambda_2})}(t) = \hat{\pi}_{\lambda_1}(t)\hat{\pi}_{\lambda_2}(t) = \hat{\pi}_{\lambda_1+\lambda_2}(t)$$

so that $\{\pi_\lambda, \lambda \geq 0\}$ is a semi-group of probability measures.

1.2. An extension. The previous work motivates the following extension. Let $(S, \mathbb{B}, \nu)$ be a finite measure space and $0 < c = \nu(S) < \infty$. Let $\tilde{\nu}(\cdot) = \frac{1}{c}\nu(\cdot)$, then $(S, \mathbb{B}, \tilde{\nu})$ is a probability space different from (Ω, Σ, P).

Let $X_j : S \to \mathbb{R}$ be independent identically distributed random variables relative to $\tilde{\nu}$. Then $\delta_{X_j} : \mathbb{R} \to \mathbb{R}^+$ is a random measure on $(\mathbb{R}, \mathbb{R})$, the Borelian line, in the sense that for each $s \in S$, $\delta_{X_j(s)}(\cdot)$ is the Dirac point measure.

If N is a Poisson random variable with intensity $c (= \nu(s))$ so that

$$P[N = n] = \frac{c^n}{n!} e^{-c}$$

then the measure π_c in (1.11) can be considered as a compound variable by:

$$\tilde{\pi}(B) = \sum_{j=1}^{N} \delta_{X_j}(B), \quad B \in \mathbb{B}, \tag{1.12}$$

where N is the Poisson random variable with $\nu(B)$ as intensity noted above.

Here N and X_j are independent. As a composition of N and X_j, all at most countable, $\tilde{\pi}(\cdot)$ is a random variable. In fact,

$$[\tilde{\pi}(B) = n] = \bigcup_{m \geq n} [\sum_{j=1}^{m} \delta_{X_j}(B) = n] \cap [N = n],$$

for each integer $n \geq 1$ so that $\tilde{\pi}(B)$ is measurable for Σ, and thus is a random element for all $B \in \mathbb{B}$.

THEOREM 1.2. *For each $B \in \mathbb{B}$, $\tilde{\pi}(B)$ is Poisson distributed with intensity $c \cdot \tilde{\nu}(B) = \nu(B)$, implying that $\tilde{\pi(\cdot)}$ is pointwise a.e. σ-additive.*

2. Random measures

Hereafter we write $\pi(\cdot)$ for $\tilde{\pi}(\cdot)$ to simplify notation. Now if we abstract the idea of a Poisson measure as given in the previous section, we have the following definition.

DEFINITION 2.1. Let $L^0(P)$ be the space of all real random variables on a probability space (Ω, Σ, P) and $(S, \mathbb{B})$ be a measurable space. A mapping $\mu : \mathbb{B} \to L^0(P)$ is called a **random measure**, if the following hold:

(i) $A_n \in \mathbb{B}, n = 1, 2, \ldots,$ disjoint, implies $\{\mu(A_n), n \geq 1\}$ is a mutually independent family of infinitely divisible random variables,

(ii) for A_n as above, $\mu(\bigcup_{n=1}^{\infty} A_n) = \sum_{n=1}^{\infty} \mu(A_n)$, the series converges in P-measure.

An important subclass of the infinitely divisible distribution functions (d.f.'s) is the *stable family*. In the present context, these are called *stable random measures*, and they include the Poisson case.

Recall that a stable random variable $X : \mathbb{R} \to L^0(P)$ has its characteristic function

$$\varphi(t) = E(e^{itX}) = \int_{\Omega} e^{itX} dP$$

to be given (by the Lévy formula) as:

$$\varphi(t) = \exp\{i\gamma t - c|t|^{\alpha}(1 - i\beta \text{sgnt} \cdot \varpi(t, \alpha))\}, \tag{2.1}$$

where

$$\gamma \in \mathbb{R}, |\beta| \leq 1, c \geq 0, 0 < \alpha \leq 2$$

and

$$\varpi(t, \alpha) = \begin{cases} \tan \frac{\pi\alpha}{2}, & \text{if } \alpha \neq 1 \\ -\frac{2}{\pi} \log |t|, & \text{if } \alpha = 1. \end{cases}$$

Here α is the characteristic exponent of φ (or X), and $\alpha > 2$ implies $c = 0$, to signify that X is a constant.

Once can show that the ch.f. φ of a stable random measure $\mu : \mathbb{B} \to L^0(P)$, (2.1), has the following form:

$$\begin{aligned}
\varphi_A(t) &= E(e^{it\mu(A)}) = \int_\Omega e^{it\mu(A)}\, dP, \\
&= \exp\{i\gamma(A)t - c(A)|t|^\alpha(1 - i\beta(A)\text{sgnt}\ \varpi(t,\alpha))\}, \\
&\qquad \text{for } 0 < \alpha \leq 2, \\
&= \exp\{-\psi(A,t)\},
\end{aligned} \tag{2.2}$$

for all $A \in \mathbb{B}, \nu(A) < \infty$ where $\nu : \mathbb{B} \to \overline{\mathbb{R}}^+$ is a σ-finite measure.

The function $\psi(\cdot,\cdot)$ is often called the *characteristic exponent* which is uniquely determined by the parameters (γ, c, α, and β) and conversely determines them to make (2.3) the Lévy formula.

The Poisson random measure $\pi : \mathbb{B} \times \Omega \to \mathbb{R}^+$ is a function of a set and a point, so that $\pi(A,\omega)(= \pi(A)(\omega))$ is a nonnegative number which is σ-additive in the first variable and a measurable (point) function in the second.

In the classical literature (Zygmund, [**11**]), the Poisson kernel is utilized to define a Poisson integral which is used to study the continuity, differentiation and related properties of functions representable as Poisson integrals. This leads us to a bit of harmonic analysis.

3. Harmonic analysis as a bridge

For a Lebesgue integrable $f : [-\pi, \pi] \to \mathbb{R}$, using the orthonormal system

$$\frac{1}{2}, \cos nx, \sin nx, n = 1, 2, \ldots,$$

consider the Fourier coefficients a_k, b_k given by

$$a_k = \frac{1}{\pi}\int_{-\pi}^{\pi} f(x)\cos kx\, dx, \qquad b_k = \frac{1}{\pi}\int_{-\pi}^{\pi} f(x)\sin kx\, dx$$

and for $0 \leq r < 1$, set

$$f_r(x) = \frac{1}{2}a_0 + \sum_{k=1}^{\infty}(a_k \cos kx + b_k \sin kx)r^k.$$

Then the Poisson kernel $P(\cdot,\cdot)$ is given by

$$P(r,x) = \frac{1}{2}\sum_{k=1}^{\infty} r^k \cos kx \left(= \frac{1}{2}\frac{1-r^2}{1-2r\cos x + r^2}\right) \geq 0,$$

with

$$\frac{1}{\pi}\int_{-\pi}^{\pi} P(r,x)\, dx = 1,$$

and $f_r(\cdot)$ representable as the convolution:

$$(Tf)(r,x) = f_r(x) = \frac{1}{\pi}\int_{-\pi}^{\pi} f(x)P(r,u-x)\, du, \quad 0 \leq r < 1. \tag{3.1}$$

Classically, this results asserts that $f_r(x) \to f(x)$, for all continuous periodic functions f, uniformly as $r \to 1$. Thus, T is a continuous linear mapping on $L^1(-\pi, \pi]$.

Replacing $P(r,x)\,dx$ by $\pi(\omega, ds)$ or more inclusively $\mu(ds)(\omega)$ of the above Definition, one could consider the corresponding analysis for random functions or processes (or sequences) that admit integral representation, modeling that of (3.1).

Here the Lebesgue interval $[-\pi,\pi]$ is replaced by $(S,\mathbb{B},\nu)$ and ω (in lieu of r) varies in (Ω,Σ,P). Such a general study has been undertaken by P. Lévy, [**4**],[**5**] when μ is a stable random measure. The resulting class of processes is now called *Lévy processes.*

From this, we can now define an integral of a scaler function relative to a stable random measure $\mu : \mathbb{B} \to L^B(P)$. In the simple case of a Poisson random measure, the intensity measure $\nu : \mathbb{B} \to \overline{\mathbb{R}}^+$ defines the triple $(S,\mathbb{B},\nu)$. In the general case of a stable random measure, we have $\gamma(\cdot), c(\cdot)$ and $\beta(\cdot)$ as set functions, with σ-additivity properties but are not related to ν of the triple.

A simplification here is to *assume* that $\gamma(\cdot)$ and $c(\cdot)$ are proportional to ν and β is a constant. Thus, let $\gamma(A) = a\nu(A), (a \in \mathbb{R})\; c(A) = c\nu(A), (c \geq 0)$, and $|\beta| \leq 1$ is a constant. The characteristic exponent $\psi(\cdot,\cdot)$ becomes for $a \in \mathbb{R}, 0 < \alpha \leq 2$, $A \in \mathbb{B}_0, t \in \mathbb{R}$,

$$\psi(A,t) = ia\nu(A)t - c\nu(A)|t|^\alpha\{1 - i\beta \operatorname{sgn} t \cdot \varpi(t,\alpha)\}. \tag{3.2}$$

It can be shown that $\exp\{-\psi(A,\cdot)\}$ is a characteristic function. Using this, one can establish the existence of an α-stable random measure into $L^0(P)$ on a probability space (Ω,Σ,P). This gives that the random measure $\mu : \mathbb{B}_0 \to L^0(P)$ is "controlled" by ν in the sense that $\mu(A) = 0$, a.e. $[P]$ holds whenever $\nu(A) = 0$, and μ is governed by the quadruple (a,c,β,ν).

4. Stable processes

Recall that a process $\{X_t, t \in I\}$ is *strictly stationary* if for each finite set of indices $t_1,\ldots,t_n \in I$ with $t_1 + s,\ldots,t_n + s \in I$ for any $s \in I$ (for any index set I with such an algebraic structure), all the distributions of $(X_{t_1},\ldots,X_{t_n})$ and $(X_{t_1+s},\ldots,X_{t_n+s})$ are identical.

Equivalently, their ch.f.'s satisfy

$$E(\exp[i\sum_{j=1}^n u_j X_{t_j}]) = E(\exp[i\sum_{j=1}^n u_j X_{t_j+s}]), \quad u_j \in \mathbb{R}. \tag{4.1}$$

We now consider this property for a class of α-stable processes. For simplicity we treat here only the *symmetric* α-stable class. Thus, a process $\{X_t, t \in I\}$ is termed α-stable if each finite linear combination $\sum_{j=1}^n a_j X_{t_j}$ is α-stable.

For each $n \geq 1$, the finite dimensional ch.f. of $X_{t_1},\ldots,X_{t_n}$ is representable as:

$$\varphi_{t_1,\ldots,t_n}(u_1,\ldots,u_n) = \exp\{-\int_{\mathbb{R}^n} \left|\sum_{j=1}^n u_j e^{i<t_j,\lambda>}\right|^\alpha dG_n(\lambda)\} \tag{4.2}$$

where the support of G_n is the unit sphere.

The G_n measure is defined on the space $(\mathbb{R}^n,\mathbb{B}_n)$ and as n varies, the system of measure spaces

$$\{(\mathbb{R}^n,\mathbb{B}_n,G_n), n \geq 1\}$$

changes. The consistency of the finite dimensional distributions of the process implies there is a unique measure G on the cylinder σ-algebra $\mathbb{B}$ of $\mathbb{R}^I$ whose projection, or n-dimensional marginal, satisfies $G_n = G \circ \pi_n^{-1}$ where $\pi_n : \mathbb{R}^I \to \mathbb{R}^n$

is the coordinate projection. If such a G exists, it is called the *spectral measure* of the α-stable process.

An α-stable symmetric process for which

$$\varphi_{t_1,\ldots,t_n}(u_1,\ldots,u_n) = \exp\{-\int_{\mathbb{R}^n}\left|\sum_{j=1}^{n} u_j e^{i<t_j,\lambda>}\right|^{\alpha} dG_n(\lambda),$$

holds with $G_n = G\circ\pi_n^{-1}$ is called a *strongly stationary α-stable* process. These processes are automatically strictly stationary.

Since the measure G is obtained through an application of the Kolmogorov-Bochner theorem, one may hope that all symmetric strictly stationary α-stable processes are also strongly stationary. However, it is shown by Marcus and Pisier [**6**] that the inclusion is proper unless $\alpha = 2$ which corresponds to the Gaussian case in which they both coincide.

EXAMPLE 1. Let $a_\lambda \in \mathbb{R}^n$ be such that

$$\sum_{\lambda\in\mathbb{R}^n} |a_\lambda|^{\alpha} < \infty,$$

and $\{\varepsilon_\lambda, \lambda\in\mathbb{R}^n\}$ be a set of independent α-stable symmetric variables. Consider the process

$$X_t = \sum_{\lambda\in\mathbb{R}^n} a_\lambda \varepsilon_\lambda e^{i<t,\lambda>}, \quad t\in\mathbb{R}^n. \tag{4.3}$$

It may be verified that $\{X_t, t\in I = \mathbb{R}^n\}$ is a strongly stationary α-stable process and if $0<\alpha\le 2$ the spectral measure $G(\cdot)$ given by

$$G(\cdot) = \sum_{\lambda\in\mathbb{R}^n} |a_\lambda|\delta_\lambda(\cdot) \tag{4.4}$$

where $\delta_\lambda(\cdot)$ is the Dirac measure at $\lambda\in\mathbb{R}^n$.

An interesting outcome of this example is that if $\alpha = 2$ then ε_λ must be Gaussian, and if $0<\alpha<2$ it is a stable process, so the ε_λ include both classes.

We noted earlier that integrals of the form

$$\int_S f\,d\mu$$

can be defined for random measures μ (with independent values on disjoint sets) on $(S, \mathbb{S}, \nu)$ for $f: S\to\mathbb{R}$ (or $\mathbb{C}$) of bounded measurable class from $L^0(\nu)$.

In particular, if $S=\mathbb{R}$ and $f_\lambda(s) = e^{is\lambda}, \lambda\in\mathbb{R}$, then one has

$$X_t = \int_{\mathbb{R}} e^{it\lambda} d\mu(\lambda), \quad t\in\mathbb{R}. \tag{4.5}$$

Processes with this representation were introduced and studied by Y. Hosoya [**3**], K. Urbanik [**10**] and others. This class is termed *strictly harmonizable*. Although this is similar to strict stationarity, neither includes the other.

DEFINITION 4.1. Let $\mathbb{X}$ be a Banach space and $f: G\to\mathbb{X}$ be a mapping, where G is a locally compact abelian group, so that $G = \mathbb{R}^n, n\ge 1$ is possible. Then f is said to be *V-bounded* (V for variation) provided:

(i) $f(G)$ is bounded, or equivalently contained in a ball of $\mathbb{X}$,

(ii) f is measurable relative to the Borel σ-algebras of $\mathbb{X}$ and G, and that the range of f is separable.

(iii) the set

$$W = \left\{ \int_G f(t)g(t)\,dt : ||\hat{g}||_\infty \le 1, \;\; g \in L^1(G) \right\} \subset \mathbb{X}, \tag{4.6}$$

is such that its closure $\overline{W}$ in the weak topology of $\mathbb{X}$ is compact, where 'dt' is the invariant or *Haar* measure of G, the Lebesgue measure if $G = \mathbb{R}^n$, and $\hat{g}$ is the Fourier transform of g.

In this definition,

$$\hat{g}(s) = \int_{\hat{G}} < g, \gamma > d\gamma,$$

with the point being that f is not required to be positive definite.

THEOREM 4.2. *Let $X : G \to L^\alpha(P), \alpha \ge 1$ be a process. Then*

$$X_t = \int_{\hat{G}} < t, s > dZ(s), t \in G,$$

so it is strictly harmonizable if and only if X is V-bounded and weakly continuous.

A consequence of this representation is that $\{X_t, t \in G\}$ under the stated conditions is an integral of a vector measure $Z : \mathbb{B}(\hat{G}) \to L^\alpha(P), \alpha \ge 1$.

If $\{X_t, t \in G\}$ is strictly stationary, then some special properties of Z can be obtained.

THEOREM 4.3. *If $\{X_t, t \in \mathbb{R}\}$ is a strictly harmonizable α-stable process with representing measure $Z : \mathbb{B}(\mathbb{R}) \to L^\alpha(P)$, which is also isotropic, then $\{X_t, t \in \mathbb{R}\}$ is strongly stationary α-stable. Conversely, if the process is strongly stationary α-stable, $1 < \alpha < 2$, then it is V-bounded and is strictly harmonizable with random measure isotropic.*

5. Second order processes

A process with two moments finite is termed *second order* and we can consider the natural parameters of the process X_t, its mean and covariance functions m and r :

$$m(t) = E(X_t), \qquad r(s,t) = \text{Cov}(X_s, X_t).$$

Second-order stochastic processes play key roles in many areas of the applied and natural sciences. The simplest and best understood is the stationary class. A process is called *weakly stationary* if

$$m(t) = \text{ constant and } r(s,t) = \tilde{r}(s-t)$$

with $\tilde{r}$ assumed as a Borel function. Recall that we defined strictly stationary processes as those whose finite-dimensional distributions are invariant under a shift of the time axis.

Strict sense stationarity implies the weak sense version when the distribution functions have two moments finite.

Now since r is positive definite, if $I = \mathbb{R}$, then by the Bochner-Riesz theorem, for a weakly stationary process $\{X_t, t \in \mathbb{R}\}$ we have

$$r(s-t) = \int_{\mathbb{R}} e^{i(s-t)\lambda} F(d\lambda) \tag{5.1}$$

for almost all $s-t \in \mathbb{R}$, and if r is also continuous, then (5.1) holds for all $s-t \in \mathbb{R}$. F is a bounded nondecreasing nonnegative function, called the *spectral function* of the process, and it is uniquely determined by r. In many applications, the assumption of stationarity is not always valid, this provides the motivation for the following.

A process $\{X_t, t \in \mathbb{R}\} \subset L^2(P)$ with means zero and covariance r is termed *strongly* (or *Loève*) *harmonizable* if

$$r(s,t) = \int_{\mathbb{R}}\int_{\mathbb{R}} e^{is\lambda - it\lambda'} F(d\lambda, d\lambda'), \quad s,t \in \mathbb{R}, \tag{5.2}$$

where $F : \mathbb{R}^2 \to \mathbb{C}$ is a covariance function of bounded *Vitali variation* in the plane, that is

(5.3)

$$|F|(\mathbb{R}^2) = v(F)$$
$$= \sup\{\sum_{i=1}^{m}\sum_{j=1}^{m} |F(A_i, B_j)| : \{A_i\}_1^n, \{B_i\}_1^n \text{ are disjoint intervals of } \mathbb{R}\} < \infty.$$

Here

$$F(A, B) = \int\int_{\mathbb{R}^2} \chi_A(\lambda)\chi_B(\lambda')F(d\lambda, d\lambda')$$

and F is called the *spectral function* (*bimeasure*) of the process.

Now it is easy to see that every weakly stationary process is strongly harmonizable noting that when $F(\cdot,\cdot)$ concentrates on the diagonal $\lambda = \lambda'$. A simple harmonizable process which is not weakly stationary is the following.

EXAMPLE 2. Let $f \in L^1(\mathbb{R})$ and $\hat{f}$ be its Fourier transform:

$$\hat{f}(t) = \int_{\mathbb{R}} e^{it\lambda} f(\lambda)\, d\lambda.$$

If ξ is an r.v. with mean zero and unit variance, and $X_t = \xi\hat{f}(t)$, then $\{X_t, t \in \mathbb{R}\}$ is a strongly harmonizable process.

A process $\{X_t, t \in \mathbb{R}\} \subset L^2(P)$ is *weakly harmonizable* if $E(X_t) = 0$, and its covariance can be represented as (5.2) in which the spectral function F is a covariance function of bounded variation in *Fréchet's sense*:

$$||F|| = \sup\left\{\sum_{i=1}^{n}\sum_{j=1}^{n} a_i\bar{a}_j F(A_i, A_j) : a_i \in \mathbb{C}, |a_i| \le 1, \right.$$
$$\left. \{A_i\}_1^n \text{ are disjoint Borel sets } A_i \subset \mathbb{R}\right\}.$$

Now $||F|| \le v(F) \le \infty$, usually with a strict inequality between the first terms. With this,

$$r(s,t) = \int_{\mathbb{R}}\int_{\mathbb{R}} e^{is\lambda - it\lambda'} F(d\lambda, d\lambda'), \quad s,t \in \mathbb{R}, \tag{5.4}$$

but here the integral is defined in the (weaker) sense of M. Morse and W. Transue and it is not an absolute integral, in contrast to Lebesgue's definition used in the strongly harmonizable case.

It is clear that each strongly harmonizable process is weakly harmonizable, and the above examples show that the converse does not hold. Most of the Loève theory extends to this general class, although different methods and techniques of proof are now necessary. The structure theory of these processes is detailed in the first author's paper [**7**].

Several other extensions of second-order processes are also possible. For some of these we refer to Cramér lectures [**2**], Rao [**7**], Chang and Rao [**1**], with Swift [**9**] containing an extensive treatment of several classes of nonstationary processes.

References

[1] Derek K. Chang and M. M. Rao, *Bimeasures and nonstationary processes*, Real and stochastic analysis, Wiley Ser. Probab. Math. Statist. Probab. Math. Statist., Wiley, New York, 1986, pp. 7–118. MR856580

[2] Harald Cramér, *Structural and statistical problems for a class of stochastic processes*, Princeton University Press, Princeton, N. J., 1971. The first Samuel Stanley Wilks lecture at Princeton University, Princeton, N. J., March 17, 1970; With an introduction by Frederick Mosteller. MR0400370

[3] Yuzo Hosoya, *Harmonizable stable processes*, Z. Wahrsch. Verw. Gebiete **60** (1982), no. 4, 517–533, DOI 10.1007/BF00535714. MR665743

[4] P. Lévy, *Théorie de l'addition des variables*, Gauthier-Villars, Paris, 1937.

[5] Paul Lévy, *Processus stochastiques et mouvement brownien* (French), Suivi d'une note de M. Loève. Deuxième édition revue et augmentée, Gauthier-Villars & Cie, Paris, 1965. MR0190953

[6] M. B. Marcus and G. Pisier, *Characterizations of almost surely continuous p-stable random Fourier series and strongly stationary processes*, Acta Math. **152** (1984), no. 3-4, 245–301, DOI 10.1007/BF02392199. MR741056

[7] M. M. Rao, *Harmonizable processes: structure theory*, Enseign. Math. (2) **28** (1982), no. 3-4, 295–351. MR684239

[8] M. M. Rao and R. J. Swift, *Probability theory with applications*, 2nd ed., Mathematics and Its Applications (Springer), vol. 582, Springer, New York, 2006. MR2205794

[9] Randall J. Swift, *Some aspects of harmonizable processes and fields*, Real and stochastic analysis, Probab. Stochastics Ser., CRC, Boca Raton, FL, 1997, pp. 303–365. MR1464225

[10] K. Urbanik, *Random measures and harmonizable sequences*, Studia Math. **31** (1968), 61–88, DOI 10.4064/sm-31-1-61-88. MR246340

[11] A. Zygmund, *Trigonometric series: Vols. I, II*, Second edition, reprinted with corrections and some additions, Cambridge University Press, London-New York, 1968. MR0236587

Department of Mathematics, University of California, Riverside, California 92521
Email address: `rao@ucr.edu`

Department of Mathematics & Statistics, California State Polytechnic University, Pomona, California 91768
Email address: `rjswift@cpp.edu`

Contemporary Mathematics
Volume **774**, 2021
https://doi.org/10.1090/conm/774/15575

The exponential-dual matrix method: Applications to Markov chain analysis

Gerardo Rubino and Alan Krinik

Dedicated to two beloved colleagues of M. M. Rao at Univ. of California, Riverside: Professor Neil E. Gretsky (1941-2015; also a graduate student of M. M. Rao at Carnegie-Mellon Univ., graduated 1967) and Professor Victor L. Shapiro (1924-2013)

ABSTRACT. Classic performance evaluation using queueing theory is usually done assuming a stable model in equilibrium. However, there are situations where we are interested in the transient phase. In this case, the main metrics are built around the model's state distribution at an arbitrary point in time. In dependability, a significant part of the analysis is done in the transient phase. In previous work, we developed an approach to derive distributions of some continuous time Markovian models, built around uniformization (also called Jensen's method), transforming the problem into a discrete time one, and the concept of stochastic duality. This combination of tools provides significant simplifications in many cases. However, stochastic duality does not always exist. Recently, we discovered that an idea of algebraic duality, formally similar to stochastic duality, can be defined and applied to any linear differential system (or equivalently, to any matrix). In this case, there is no limitation, the transformation is always possible. We call it the exponential-dual matrix method. In the article, we describe the limitations of stochastic duality and how the exponential-dual matrix method operates for any system, stochastic or not. These concepts are illustrated throughout our article with specific examples, including the case of infinite matrices.

1. Introduction

In this article, we first review how the concept of stochastic duality in continuous time Markov chains with discrete state spaces, as defined in [**16**] and used in [**1**], coupled with the uniformization method [**6**], allows us to obtain analytical expressions of transient distributions of fundamental Markovian queueing models. The quantity of literature on the transient analysis of these stochastic models is huge, both in the number of research papers and in books [**12**], [**4**], [**2**], [**15**]. There are many uses of the word "duality" in science, and also inside probability theory (see for instance [**5**] in the Markovian case). When we say "dual" here, or more precisely "stochastic dual", we mean for Markov chains, as defined in [**16**] and developed by Anderson in [**1**].

2020 *Mathematics Subject Classification.* Primary 60J27, 60J35, 60K25, 15A04, 15A16.
Key words and phrases. Markov chains, Markov processes, transient analysis, dual process, exponential-dual matrix, duality, generator, exponential matrix, transition rate matrix, catastrophes, uniformization, closed-forms.

Analyzing transient behavior in stochastic models is extremely important in performance and in dependability analysis, in many engineering branches including computer science and communication networks. The combination of duality with uniformization (also called randomization or Jensen's method) has proved very useful in obtaining closed-form expressions, see our previous articles [**3**, **8**–**10**]. In particular, in [**10**] the $M/M/1$ and $M/M/1/H$ queueing systems are analyzed using these methods, together with a variant of the $M/M/1/H$ where "catastrophes" (transitions from every state to the empty state) are also included.

However, the use of the duality transformation has some limitations. For example, the stochastic dual exists only if a strong monotonicity property of the original chain holds, see [**1**] (a property that makes sense with respect to a total order relation on the state space). The problem is that in some cases, there is no such monotonicity and thus, no dual process. Moreover, the ordering matters, which means that the dual may exist for a specific ordering of the states, and not for a different one. Another possible restriction is that the dual may exist if and only if the transition rates of the original Markov chain satisfy some specific inequalities.

In this article, we first describe the stochastic dual concept and review how the method is used to find analytic expressions of the transient distribution of some fundamental Markov chain models. In particular, we highlight the main structure of the duality/uniformization approach. Examples when the stochastic dual does not exist and the dependency upon ordering of states or other required conditions are explored. Another possibility not usually covered in the literature, occurs when the state space is infinite making the dual transformation a nonconservative matrix. We illustrate this with an example and show how to deal with it and still obtain transient distributions (Subsection 4.1).

In our work with stochastic duality, we realized that many nice properties of duality rely only upon algebraic relations and not upon the stochastic properties of the models such as monotonicity. In fact, a similar transformation can be defined for any matrix, that is, not necessarily just for infinitesimal generators. This generalization of the stochastic dual is what we call "exponential-dual". It turns out that many of the main properties of the dual concept for Markov chains also hold unchanged for the exponential-dual. However, the exponential-dual always exists, independently of the specific ordering of the states and without any condition of the model's parameters. Of course, when we deal with a Markov model and the stochastic dual exists, the exponential-dual coincides with it.

Our article has the following outline. This introductory Section 1 describes the article's context and content. Uniformization is discussed in Section 2 and Section 3 is devoted to stochastic duality. The *uniformization plus duality* approach to finding transient solutions is explained and illustrated by examples in Section 4. The generalization of the stochastic dual to the exponential-dual matrix, along with its main properties and results, is presented in Section 5. In this article, we mainly consider finite matrices. Conclusions and future work comprise Section 6.

2. Uniformization

The uniformization procedure is a way of transforming a problem specified on a continuous time Markov chain $X = \{X(t)\}_{t\geq 0}$ into a similar problem defined on an associated discrete time Markov chain $Y = \{Y_n\}_{n\geq 0}$. The transformation is such that solving the problem defined on Y allows to get immediately an answer to the original problem defined in terms of X. Both chains X and Y share the same discrete state space S and the same initial distribution α (when writing matrix relations using α, it will be considered a row vector). Let A be the infinitesimal generator matrix of X, that is, $A_{i,j} \geq 0$ is the transition rate from i to j for $i \neq j$; $A_{i,i} = -d_i \leq 0$ is the negative rate from state i to i. We assume that the transformation has a scalar real parameter Λ that must satisfy $\Lambda \geq \sup_{i\in S} d_i$. This is why X is said to be *uniformizable* if and only if $\sup_{i\in S} d_i < \infty$. We are only interested in uniformizable processes here. Let U be the transition probability matrix of Markov chain Y. Its 1-step transition matrix is $U = I + A/\Lambda$, where I is the identity matrix indexed on S.

If $N = \{N(t)\}_{t\geq 0}$ is the counting process of a Poisson process having rate Λ and independent of Y, then the process $Z = \{Z(t)\}_{t\geq 0}$ defined on S by $Z(t) = Y_{N(t)}$ is stochastically equivalent to X. In particular, this means that the distribution of $X(t)$, seen as a row vector $p = (p(t))$, satisfies

$$p(t) = \sum_{n\geq 0} \mathrm{e}^{-\Lambda t} \frac{(\Lambda t)^n}{n!} q_n, \tag{2.1}$$

where (q_n) is the distribution of Y_n also considered a row vector indexed on S. To evaluate, for instance, the law of $X(t)$ with an absolute error $< \varepsilon$, we use the previous relation and look first for integer N defined by

$$N = \min\left\{K \geq 0 : \sum_{k=0}^{K} \mathrm{e}^{-\Lambda t} \frac{(\Lambda t)^k}{k!} \geq 1 - \varepsilon\right\}.$$

This can be done in negligible computing time. Using the vector norm $\|\cdot\| = \|\cdot\|_1$, we have

$$\begin{aligned}
\left\|p(t) - \sum_{n=0}^{N} \mathrm{e}^{-\Lambda t} \frac{(\Lambda t)^n}{n!} q_n\right\| &= \left\|\sum_{n>N} \mathrm{e}^{-\Lambda t} \frac{(\Lambda t)^n}{n!} q_n\right\| \\
&\leq \sum_{n>N} \mathrm{e}^{-\Lambda t} \frac{(\Lambda t)^n}{n!} \|q_n\| \\
&\leq \sum_{n>N} \mathrm{e}^{-\Lambda t} \frac{(\Lambda t)^n}{n!} \\
&= 1 - \sum_{n=0}^{N} \mathrm{e}^{-\Lambda t} \frac{(\Lambda t)^n}{n!} \\
&\leq 1 - \big(1 - \varepsilon\big) = \varepsilon.
\end{aligned}$$

Now, let us illustrate what happens if we attack the problem of evaluating the transient distribution of the $M/M/1$ queueing system using uniformization. Consider this chain, illustrated in Figure 1, with parameters $\lambda > 0$ and $\mu > 0$. This chain is uniformizable, and we can use any uniformization rate $\Lambda \geq \lambda + \mu$.

Uniformizing this chain with respect to the uniformizing rate $\Lambda = \lambda + \mu$ leads to the discrete time chain depicted in Figure 2. If we can calculate the transient distribution of Y, which we denoted (q_n), we obtain an expression of that of the original $M/M/1$ queueing system using (2.1). This can be done using counting path techniques as described in [**9, 10**] for basic queueing systems such as the $M/M/1$ or the $M/M/1/H$ having a finite storage capacity, and variations.

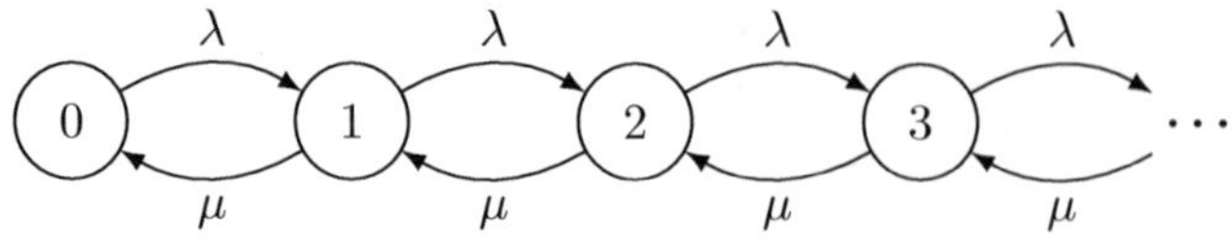

FIGURE 1. The $M/M/1$ model with arrival rate λ and service rate μ.

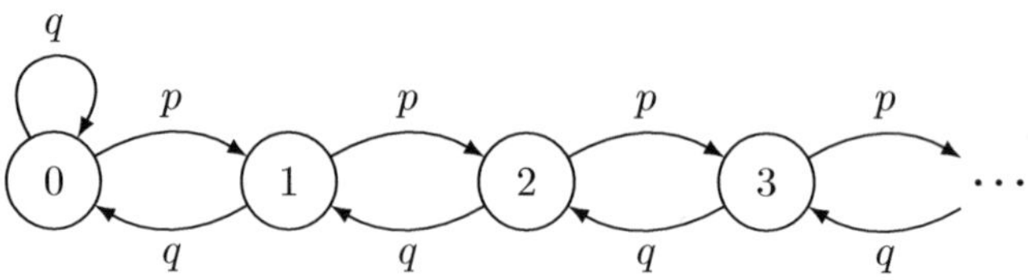

FIGURE 2. The uniformized $M/M/1$ model of Figure 1, with respect to the uniformizing rate $\Lambda = \lambda + \mu$, with the notation $p = \lambda/\Lambda$ and $q = 1 - p = \mu/\Lambda$.

In next section, we explore the concept of duality and its limitations.

3. Stochastic duality

In this section we follow [**1**] using a notation slightly different from Anderson's. We start from a continuous time Markov chain X *where the state space is* $S = \{0, 1, 2, 3, \ldots\}$, the nonnegative integers or $S = \{0, 1, 2, \ldots, n-1\}$, for some integer $n \geq 1$. Define matrix $P = P(t)$ as having entries $P_{i,j}(t) = \mathbb{P}(X(t) = j \mid X(0) = i)$, often also denoted $\mathbb{P}_i(X(t) = j)$. Recall that A denotes X's generator. When $\exp(A)$ exists (for instance, when S is finite), we have $P(t) = \exp(At)$. Seen as a function of t, matrix $P(t)$ is called the matrix of transition functions of X. In our case, where we avoid all possible "pathologies" of X, the matrix of functions $P(t)$ has all the information that we need to work with X.

To fix the ideas and simplify the presentation, consider the case of $S = \{0, 1, 2, 3, \ldots\}$. The transition function P of the continuous time Markov chain X is said to be *stochastically increasing* if and only if for all $t \geq 0$ and for all states $i, j, k \in S$, the ordering $\mathbb{P}_i(X(t) \geq k) \leq \mathbb{P}_{i+1}(X(t) \geq k)$ holds. In [**16**], Siegmund proved that if the transition function P of Markov chain X is stochastically increasing, then there exists another Markov process X^* defined also on S, such that for all $t \geq 0$ and $i, j \in S$, $\mathbb{P}_i(X^*(t) \leq j) = \mathbb{P}_j(X(t) \geq i)$. Process X^* is the dual of X. We will say here stochastic dual, to underline the difference with our generalization in Section 5. Matrix function P^* is also stochastically increasing. Between P and P^*, the following relations hold:

$$P^*_{i,j}(t) = \sum_{k=0}^{i-1} \Big[P_{j-1,k}(t) - P_{j,k}(t) \Big], \tag{3.1}$$

$$P_{i,j}(t) = \sum_{k=0}^{i} \Big[P^*_{j,k}(t) - P^*_{j+1,k}(t) \Big]. \tag{3.2}$$

By convention, if some index takes a value not in S, then the corresponding term is 0, and if the index summation space is empty, then the sum is 0. In the finite case where the state space of X has n states (we will fix them to the set $\{0, 1, \dots, n-1\}$), then the state space of X^* is $\{0, 1, \dots, n\}$. Relation (3.1) holds for $j \leq n-1$, and this makes that $P^*_{n,j}(t) = 0$. When $j = n$, we have $P^*_{i,n}(t) = 1 - \sum_{k=0}^{n-1} P^*_{i,k}(t)$, and $P^*_{n,n}(t) = 1$.

Relations (3.1) and (3.2) can be also written

$$\begin{aligned} \mathbb{P}_i(X(t) = j) &= \mathbb{P}_j(X^*(t) \leq i) - \mathbb{P}_{j+1}(X^*(t) \leq i) \\ &= \mathbb{P}_{j+1}(X^*(t) > i) - \mathbb{P}_j(X^*(t) > i). \end{aligned} \tag{3.3}$$

$$\begin{aligned} \mathbb{P}_i(X^*(t) = j) &= \mathbb{P}_{j-1}(X(t) \leq i-1) - \mathbb{P}_j(X(t) \leq i-1) \\ &= \mathbb{P}_j(X(t) > i) - \mathbb{P}_{j-1}(X(t) > i). \end{aligned} \tag{3.4}$$

where we see monotonicity at work. This is the central result of [**16**], called the Duality Theorem in [**8**].

Between X and X^* we also have other similar relations: if A^* is the generator of X^*,

$$A^*_{i,j} = \sum_{k=0}^{i-1} \Big[A_{j-1,k} - A_{j,k} \Big] = \sum_{k \geq i} \Big[A_{j,k} - A_{j-1,k} \Big]. \tag{3.5}$$

$$A_{i,j} = \sum_{k=0}^{i} \Big[A^*_{j,k} - A^*_{j+1,k} \Big] = \sum_{k \geq i-1} \Big[A^*_{j+1,k} - A^*_{j,k} \Big]. \tag{3.6}$$

If we consider discrete time Markov chains instead, the construction remains valid, changing (3.1) to (3.6) to their immediate discrete versions (in the last two relations (3.5) and (3.6), replacing generators by transition probability matrices).

Suppose that the continuous time Markov chain X has generator A, and that we define A^* using (3.5). If A^* is also a generator and if we build a continuous time Markov chain X^* on $\{0, 1, 2, 3, \dots\}$ having generator A^*, then X^* is the dual of X. The analogous reciprocal property holds if we start from A^* and use (3.6) to build A. These relations also hold if we consider discrete time Markov chains, see [**7**] for some applications of stochastic duality in discrete time.

The stochastic dual matrix A^* has some interesting properties that will be addressed in later sections. For example, it turns out that A and A^* share the same spectrum (see [**11**] for details). In Section 5, some of these properties arise again with regard to exponential-dual matrices.

3.1. On the stochastic dual's existence. The monotonicity condition leading to the dual existence is a strong one. Let us illustrate this with a few very simple examples, using a generic one taken from [**3**]. See first Figure 3, where $\alpha, \beta, \gamma > 0$. The cyclic structure clearly suggests that the monotonicity property can't hold, which can easily be verified, and we then have an example of a Markovian model without a dual, for any value of the transition rates α, β, γ.

In Figure 4, we have a small birth-death process that satisfies the monotonicity condition for any value of its transition rates if the states are numbered as shown in part (a). This is valid for a general birth-death process [**1**]. If instead we number them as in part (b) of the same figure, the new process never has a dual. The verification of these claims is straightforward using the definition of duality and previous relations.

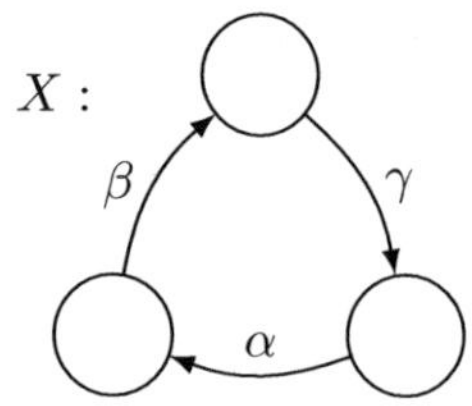

FIGURE 3
In this circular Markov process the transition diagram is not stochastically increasing, whatever the value of the rates *and whatever the numbering of the states.*

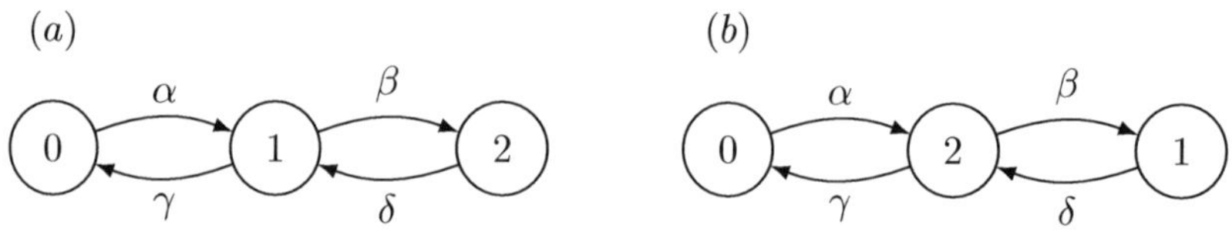

FIGURE 4. Assume that $\alpha, \beta, \gamma, \delta$ are all > 0.
In (a), we have a birth-death process that is always stochastically increasing. Changing the numbering of the states in (b) leads to a process that is never stochastically increasing, no matter which values α, β, γ and δ take. Consider the Markov process in drawn in Figure 5. In previous examples, the dual process either existed or not, regardless of the values of the transition rates. Here, the dual exists if and only if $\beta > \nu$.

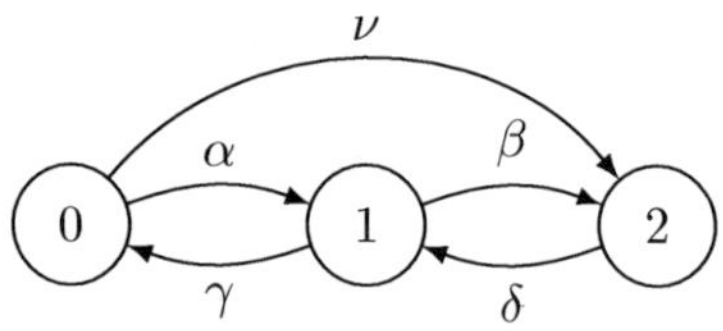

FIGURE 5. Assume that $\alpha, \beta, \gamma, \delta, \nu$ are all > 0. The dual of this process exists if and only if $\beta > \nu$ (see [**3**]).

Before leaving this section, let us point out that the concept of stochastic duality appears in many studies after its definition by Siegmund in [**16**]. For a sample of recent work, see [**7**], [**13**], [**14**], [**18**] and the references therein.

4. Transient analysis using uniformization and duality

Before going to the main topic of this section, observe that the basic uniformization relation (2.1) can be written in matrix form. Using the notation of previous section, we have

$$P(t) = \sum_{n\geq 0} \mathrm{e}^{-\Lambda t} \frac{(\Lambda t)^n}{n!} U^n. \tag{4.1}$$

The basic idea of using the dual combined with uniformization to obtain the transient distribution of a given uniformizable continuous time Markov chain X

defined on $\mathbb{N}$ and having generator A, goes as follows. We first construct A^* and check if it is a generator. If it is, then we form the dual chain X^* of X and uniformize it. Call Y^* the result. We then determine the transient distribution of Y^*, which allows us through (2.1) to obtain that of X^*. Finally, (3.2) is used to obtain the distribution of X.

REMARK 4.1. In the process of determining Y^*, the order of the transformations can be reversed. That is, we can first uniformize X, to obtain Y, and then, in discrete time, construct the dual Y^* of Y. The result will be the same Y^* as before. Figure 6 illustrates that these operations commute. This is immediate from the definitions.

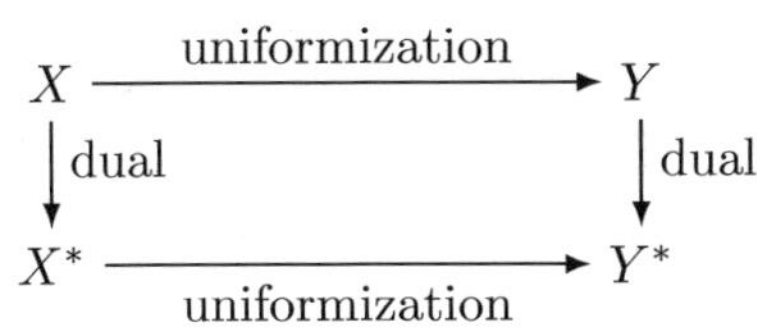

FIGURE 6. Commutativity of the operators "dual" and "uniformization".

Here is a simple example to illustrate the preceding material. Consider the chain depicted in Figure 7. The dual of this chain is X^*, depicted in Figure 8.

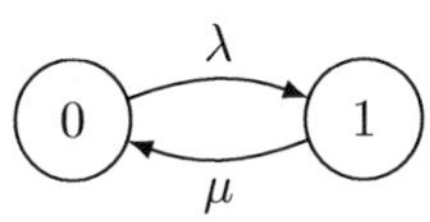

$$A = \begin{pmatrix} -\lambda & \lambda \\ \mu & -\mu \end{pmatrix}$$

FIGURE 7. The two-state irreducible continuous time Markov chain X, where $\lambda > 0$ and $\mu > 0$.

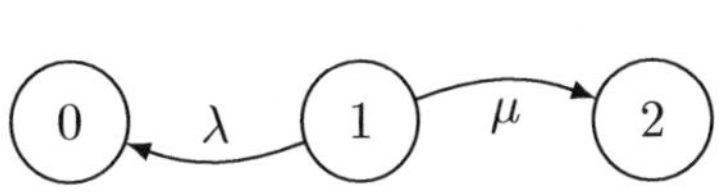

$$A^* = \begin{pmatrix} 0 & 0 & 0 \\ \lambda & -(\lambda+\mu) & \mu \\ 0 & 0 & 0 \end{pmatrix}$$

FIGURE 8. The dual of the chain X given in Figure 7.

Let us check the basic relations (3.3) and (3.4). It is straightforward to compute the transition functions of X and X^*. Using the notation $\Lambda = \lambda + \mu$, $p = \lambda/\Lambda$ and $q = \mu/\Lambda = 1 - p$, we obtain:

$$P(t) = \mathrm{e}^{At} = \begin{pmatrix} p\mathrm{e}^{-\Lambda t} + q & p(1 - \mathrm{e}^{-\Lambda t}) \\ q(1 - \mathrm{e}^{-\Lambda t}) & q\mathrm{e}^{-\Lambda t} + p \end{pmatrix} \tag{4.2}$$

and

$$P^*(t) = \mathrm{e}^{A^* t} = \begin{pmatrix} 1 & 0 & 0 \\ p(1 - \mathrm{e}^{-\Lambda t}) & \mathrm{e}^{-\Lambda t} & q(1 - \mathrm{e}^{-\Lambda t}) \\ 0 & 0 & 1 \end{pmatrix}. \tag{4.3}$$

Then, by (3.2), we have

$$\begin{aligned}
P_{0,0}(t) &= P^*_{0,0}(t) - P^*_{1,0}(t) = 1 - p(1-\mathrm{e}^{-\Lambda t}) = q + p\mathrm{e}^{-\Lambda t},\\
P_{1,0}(t) &= \big[P^*_{0,0}(t) - P^*_{1,0}(t)\big] + \big[P^*_{0,1}(t) - P^*_{1,1}(t)\big]\\
&= \big[q + p\mathrm{e}^{-\Lambda t}\big] + \big[0 - \mathrm{e}^{-\Lambda t}\big] = q\big(1-\mathrm{e}^{-\Lambda t}\big),\\
P_{0,1}(t) &= P^*_{0,0}(t) - P^*_{1,0}(t) = p\big(1-\mathrm{e}^{-\Lambda t}\big) - 0 = p\big(1-\mathrm{e}^{-\Lambda t}\big),\\
P_{1,1}(t) &= \big[P^*_{1,0}(t) - P^*_{2,0}(t)\big] + \big[P^*_{1,1}(t) - P^*_{2,1}(t)\big]\\
&= \big[p\big(1-\mathrm{e}^{-\Lambda t}\big) - 0\big] + \big[\mathrm{e}^{-\Lambda t} - 0\big] = p + q\mathrm{e}^{-\Lambda t}.
\end{aligned}$$

We can also consider the uniformization of X and that of X^* with respect to the uniformization rate $\Lambda = \lambda + \mu$, respectively denoted by Y and Y^*. They are depicted in Figures 9 and 10.

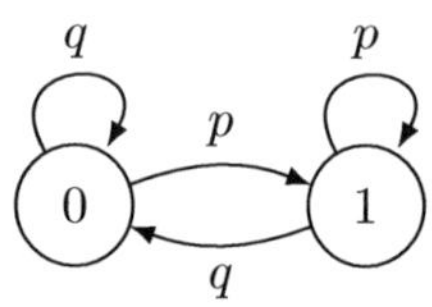

$$U = \begin{pmatrix} q & p \\ q & p \end{pmatrix} = U^n \text{ when } n \geq 1$$

FIGURE 9. The uniformization of the chain X of Figure 7, denoted by Y, with respect to the uniformization rate $\Lambda = \lambda + \mu$, using the notation $\Lambda = \lambda + \mu$, $p = \lambda/\Lambda$ and $q = \mu/\Lambda = 1 - p$.

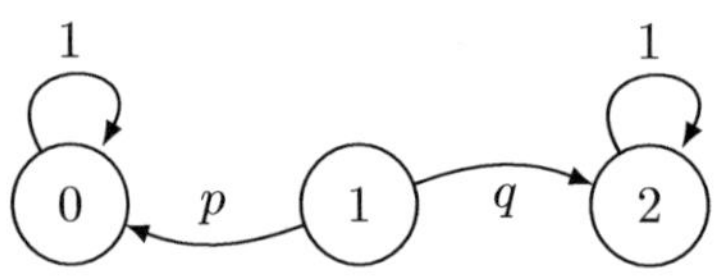

$$U^* = \begin{pmatrix} 1 & 0 & 0 \\ q & 0 & p \\ 0 & 0 & 1 \end{pmatrix} = (U^*)^n \text{ when } n \geq 1$$

FIGURE 10. The uniformization of the chain X^* of Figure 8, denoted by Y^*, with respect to the uniformization rate $\Lambda = \lambda + \mu$, using the notation $\Lambda = \lambda + \mu$, $p = \lambda/\Lambda$ and $q = \mu/\Lambda = 1 - p$.

As a final check, consider (2.1), choosing the case of X. First, observe that (4.2) can be written

$$P(t) = U + \mathrm{e}^{-\Lambda t}\begin{pmatrix} p & -p \\ -q & q \end{pmatrix} = U + \mathrm{e}^{-\Lambda t}(I - U).$$

Now, following (4.1),

$$\begin{aligned}
P(t) &= \sum_{n=0}^{\infty} \mathrm{e}^{-\Lambda t}\frac{(\lambda+\mu)^n t^n}{n!}U^n\\
&= \mathrm{e}^{-\Lambda t}I + \sum_{n=1}^{\infty} \mathrm{e}^{-\Lambda t}\frac{(\lambda+\mu)^n t^n}{n!}U\\
&= \mathrm{e}^{-\Lambda t}I + \big(1-\mathrm{e}^{-\Lambda t}\big)U.
\end{aligned}$$

The second equality uses the fact that $U^n = U$ when $n \geq 1$. The same computation can be checked for the pair X^*, Y^*.

REMARK 4.2. The use of the dual plus uniformization is useful for evaluating the distribution of $X(t)$ when finding the discrete time distribution of Y_n^* is easier than doing the same with $X(t)$, or with Y_n where Y is the uniformization of X (with respect to some appropriate rate Λ). This occurs, for instance, for queueing systems $M/M/1$ and $M/M/1/H$, where formal representations (closed forms) were obtained in this way [**10**].

4.1. Problems with infinite state spaces. Consider a simple immigration process with catastrophes, denoted as usual by X, our target, as illustrated in Figure 11.

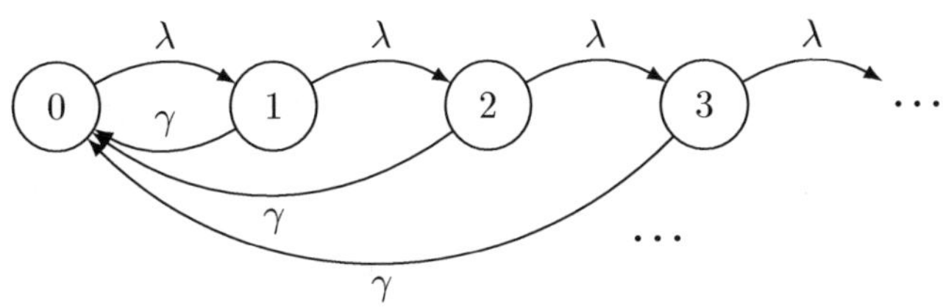

FIGURE 11. Immigration (or birth) process with "catastrophes".

The transient analysis of this process has been the object of many papers, using different approaches (see for instance, [**17**]). It can also be done by means of the procedure described here. First, the generator of X is

$$A = \begin{pmatrix} -\lambda & \lambda & 0 & 0 & 0 & \cdots \\ \gamma & -(\lambda+\gamma) & \lambda & 0 & 0 & \cdots \\ \gamma & 0 & -(\lambda+\gamma) & \lambda & 0 & \cdots \\ \gamma & 0 & 0 & -(\lambda+\gamma) & \lambda & \cdots \\ & & \cdots & & & \end{pmatrix}. \tag{4.4}$$

Applying (3.5), we obtain

$$A^* = \begin{pmatrix} 0 & 0 & 0 & 0 & 0 & \cdots \\ \lambda & -(\lambda+\gamma) & 0 & 0 & 0 & \cdots \\ 0 & \lambda & -(\lambda+\gamma) & 0 & 0 & \cdots \\ 0 & 0 & \lambda & -(\lambda+\gamma) & 0 & \cdots \\ & & \cdots & & & \end{pmatrix}. \tag{4.5}$$

This is not a generator. If we use the trick of adding an artificial auxiliary state Δ (as in [**1**, Proposition 1.1]) and the transitions that convert the new transition rate matrix into a generator, we obtain a Markov process whose graph is depicted in Figure 12.

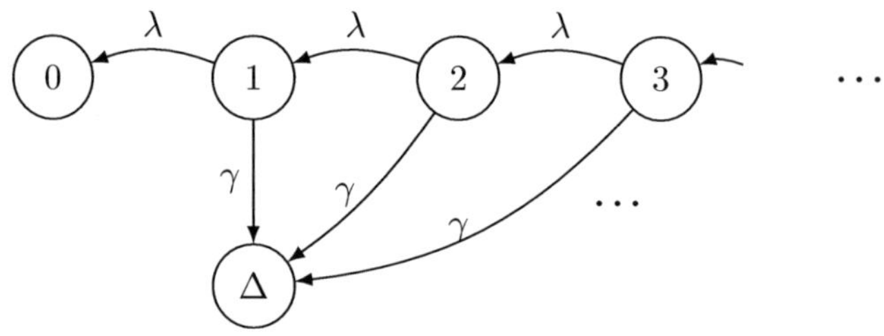

FIGURE 12. Dual of the immigration (or birth) process with "catastrophes" given in Figure 11.

If we decide that, by construction, Δ is greater than any integer i, then, the generator of X^* is symbolically given in (4.6),

$$A^* = \begin{pmatrix} 0 & 0 & 0 & 0 & 0 & \cdots & 0 \\ \lambda & -(\lambda+\gamma) & 0 & 0 & 0 & \cdots & \gamma \\ 0 & \lambda & -(\lambda+\gamma) & 0 & 0 & \cdots & \gamma \\ 0 & 0 & \lambda & -(\lambda+\gamma) & 0 & \cdots & \gamma \\ & & \cdots & & & & \end{pmatrix}. \tag{4.6}$$

where the index runs on $\{0, 1, 2, \ldots\}$ followed by Δ. Now, it is easy to verify that recovering $A_{i,j}$ for $i, j \in \{0, 1, 2, 3, \ldots\}$ from A^* works as before.

Denoting by Y^* the uniformization of the dual chain with respect to the rate $\Lambda = \lambda + \gamma$, with the additional notation $p = \lambda/\Lambda$ and $q = \gamma/\Lambda = 1 - p$, the resulting chain is shown in Figure 13.

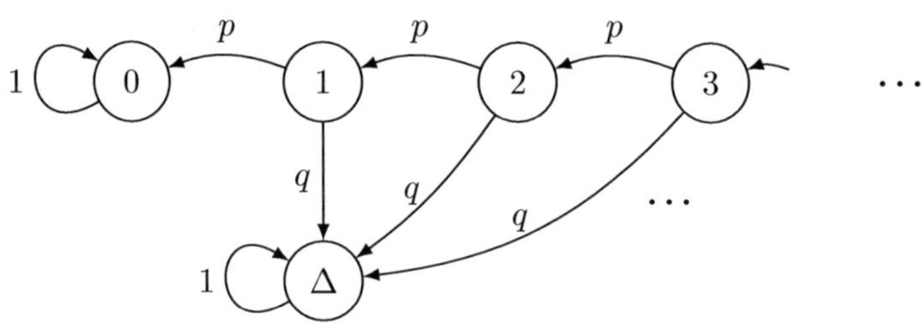

FIGURE 13. uniformization of the chain given in Figure 12, with respect to the uniformization rate $\Lambda = \lambda + \gamma$, with $p = \lambda/\Lambda$ and $q = \gamma/\Lambda = 1 - p$.

It is clear now how to obtain the transition function of this discrete time chain. If U^* is its transition probability matrix, and if we denote

$$\left((U^*)^n\right)_{i,j} = \mathbb{P}(Y^* = j \mid Y_0^* = i),$$

for any $n \geq 0$ and possible states i, j of Y^*, with $i \neq 0$, $i \neq \Delta$ and $j \in \{0, 1, 2, 3, \ldots\} \cup \{\Delta\}$, we have:

- If $n = 0$, $\left((U^*)^0\right)_{i,j} = 1(i = j)$; for any $n \geq 0$, we have $\left((U^*)^n\right)_{0,0} = \left((U^*)^n\right)_{\Delta,\Delta} = 1$.

- From this point, consider $n \geq 1$ and $i \neq 0$, $i \neq \Delta$ (so, $i \geq 1$).
  - Starting from i, we have that $\big((U^*)^n\big)_{i,j} \neq 0$ if and only if $j \neq \Delta$, $i > j$ and $n = i - j$, and in that case, its value is $p^{i-j} = p^n$.
  - For $n = 1, 2, \ldots, i-1$, $\big((U^*)^n\big)_{i,\Delta} = 1 - p^{n-1}$; then, for all $n \geq i$, $\big((U^*)^n\big)_{i,\Delta} = 1 - p^i$.

From these expressions, the distribution of X^* follows using (2.1), and then, that of X from the dual inversion formula (3.2). For instance, let us just check the $P_{0,0}$ case to compare with other papers, e.g. with [**17**]). First, $P_{0,0}(t) = P_{0,0}(t) - P_{1,0}(t)$ using (3.2). We have $P^*_{0,0}(t) = 1$ and $P^*_{1,0}(t) = (1 - \mathrm{e}^{-\Lambda t})p$. So,

$$P_{0,0}(t) = 1 - (1 - \mathrm{e}^{-\Lambda t})p = q + p\mathrm{e}^{-\Lambda t}.$$

5. Generalization of the stochastic-dual: The exponential-dual matrix

The transient distribution of the continuous time Markov process X is the solution to the Chapman-Kolmogorov equation $p'(t) = p(t)A$ where $p(t)$ is the distribution of $X(t)$ seen as a row vector, or in matrix form, using the transition function of X, $P'(t) = P(t)A$. In this section, we consider the finite case only. We know that the transition function $P(t)$ of X, the solution to the previous matrix differential equation, is $P(t) = \mathrm{e}^{At}$ (in the infinite state space case, this exponential may not exist).

Assume now that we are given *an arbitrary square matrix* A (possibly of complex numbers). Our goal is to solve the linear differential system whose matrix is A, that is, to compute e^{At}, that we denote here by $E(t)$, to avoid the notation $P(t)$ reserved for the Markov setting. We then define the exponential-dual of A as a new matrix A^* following the same formal rules as before.

We will also use the notation $E^*(t) = \mathrm{e}^{A^*t}$ instead of $P^*(t)$ for the same reason as before: here, we have left the stochastic setting.

As already said, we will limit ourselves to the case of finite matrices, to focus on the algebraic work and not on the particular applications of the analysis of queueing systems, where the state space may be infinite.

As before, our vectors are row vectors. When we will need a column vector we will use the transpose operator denoted $()^{\mathrm{T}}$. For the sake of clarity, we denote by $\mathbf{0}$ a (row) vector only composed of 0's, and by $\mathbf{1}$ a (row) vector only composed of 1's.

5.1. Exponential-dual definition.

Let A be an arbitrary square matrix of reals (or of complex numbers). We are interested in the computation of e^{At} for $t \geq 0$. Let $n < \infty$ be the dimension of A, whose elements are indexed on $\{0, 1, \ldots, n-1\}$.

DEFINITION 5.1. We define the exponential-dual of A as the matrix of dimension $n+1$, indexed on $\{0, 1, \ldots, n\}$, defined as follows: for any $i, j \in \{0, 1, \ldots, n-1, n\}$,

$$A^*_{i,j} = \sum_{k=i}^{n-1} \Big(A_{j,k} - A_{j-1,k}\Big),$$

where $A_{u,v} = 0$ if (u,v) is out of the A-range (that is, if $(u,v) \notin \{0, 1, \ldots, n-1\}^2$), and where we adopt the usual convention that $\sum_{k=u}^{v} \cdots = 0$ if $u > v$.

We now describe some immediate properties of the exponential-dual matrix.

LEMMA 5.2 (Basic properties of the exponential-dual matrix). *Matrix A^* satisfies the following properties:*

- *the sum of the elements of any row of A^* is 0;*
- *the last row of A^* is only composed by 0's.*

PROOF.

Case 1. Let $0 \le i \le n-1$. Summing the elements of the ith row of A^*,

$$\begin{aligned}
\sum_{j=0}^{n} A^*_{i,j} &= \sum_{j=0}^{n}\sum_{k=i}^{n-1}\Big(A_{j,k} - A_{j-1,k}\Big)\\
&= \sum_{j=0}^{n}\sum_{k=i}^{n-1} A_{j,k} - \sum_{j=0}^{n}\sum_{k=i}^{n-1} A_{j-1,k}\\
&= \sum_{j=0}^{n-1}\sum_{k=i}^{n-1} A_{j,k} - \sum_{\ell=0}^{n-1}\sum_{k=i}^{n-1} A_{\ell,k}\\
&= 0,
\end{aligned}$$

where in the penultimate '=' we use the fact that rows "-1" and "n" of A are out of the index space of the matrix (which is $\{0,\dots n-1\}$), so, the corresponding elements of the matrix are all null.

Case 2. For the last row of A^*, the definition makes that the sum defining element $A_{n,j}$ is empty for any $j \in \{0,1,\dots,n\}$, and then $A^*_{n,j} = 0$. □

Let us illustrate the previous definition.

If $A = \begin{pmatrix} a & b \\ c & d \end{pmatrix}$, then $A^* = \begin{pmatrix} a+b & c+d-(a+b) & -(c+d) \\ b & d-b & -d \\ 0 & 0 & 0 \end{pmatrix}$.

A numerical example: if $A = \begin{pmatrix} 1 & -2 \\ 3 & -4 \end{pmatrix}$, then $A^* = \begin{pmatrix} -1 & 0 & 1 \\ -2 & -2 & 4 \\ 0 & 0 & 0 \end{pmatrix}$.

Another one: $A = \begin{pmatrix} -1 & 1 \\ 2 & -2 \end{pmatrix}$ (a generator), leads to $A^* = \begin{pmatrix} 0 & 0 & 0 \\ 1 & -3 & 2 \\ 0 & 0 & 0 \end{pmatrix}$ (which is also a generator).

5.2. The exponential of the exponential-dual. Remember that we denote $E^*(t) = \mathrm{e}^{A^*t}$. The given properties of A^* imply some general properties for $E^*(t)$.

LEMMA 5.3 (Initial properties of $E^*(t)$). *Matrix $E^*(t)$ satisfies the following properties:*

- *the sum of the elements of any row of $E^*(t)$ is 1;*
- *the last row of $E^*(t)$ is composed of 0's except for its last element, $E^*_{n,n}(t)$, which is equal to 1.*

PROOF.

Case 1. Let $0 \le i \le n$. Recall that $\mathbf{1}^{\mathrm{T}}$ denotes a column vector only composed of 1's, whose dimension is defined by the context. We must prove that $E^*(t)\mathbf{1}^{\mathrm{T}} = \mathbf{1}^{\mathrm{T}}$. We know that $A^*\mathbf{1}^{\mathrm{T}} = \mathbf{0}^{\mathrm{T}}$, where $\mathbf{0}^{\mathrm{T}}$ is a column vector only composed of 0's, whose dimension is defined by the context.

By definition,

$$E^*(t) = \mathrm{e}^{A^*t} = I + \sum_{\ell \geq 1} \frac{t^\ell}{\ell!}(A^*)^\ell.$$

After right-multiplying by $\mathbf{1}^{\mathrm{T}}$, we have by Lemma 5.2

$$E^*(t)\mathbf{1}^{\mathrm{T}} = \mathbf{1}^{\mathrm{T}} + \sum_{\ell \geq 1} \frac{t^\ell}{\ell!}(A^*)^\ell \mathbf{1}^{\mathrm{T}} = \mathbf{1}^{\mathrm{T}} + \mathbf{0}^{\mathrm{T}} = \mathbf{1}^{\mathrm{T}}.$$

Case 2. For the last row of E^*, consider the decomposition of A^* in blocks as follows:

$$A^* = \left(\begin{array}{c|c} \widetilde{A^*} & -\widetilde{A^*}\mathbf{1}^{\mathrm{T}} \\ \hline \mathbf{0} & 0 \end{array}\right).$$

This decomposition of A^* in blocks corresponds to the partition of $\{0, 1, \dots, n\}$ in two sets, $\{0, 1, \dots, n-1\}$ and $\{n\}$. If we index the block-decomposition on $\{0, 1\}$, block $(0, 0)$ is $\widetilde{A^*}$, with denotes the restriction of A^* to its first $n-1$ elements (square sub-matrix with dimension n); block $(0, 1)$ is the column vector $-\widetilde{A^*}\mathbf{1}^{\mathrm{T}}$ (this follows from the fact that $A^*\mathbf{1}^{\mathrm{T}} = \mathbf{0}^{\mathrm{T}}$); block $(1, 0)$ is $\mathbf{0}$, a row vector of 0's (size $n-1$) and block $(1, 1)$ is the number 0.

A basic property of matrix exponentials then says that

$$\mathrm{e}^{A^*} = \left(\begin{array}{c|c} \mathrm{e}^{\widetilde{A^*}} & \mathbf{1}^{\mathrm{T}} - \mathrm{e}^{\widetilde{A^*}}\mathbf{1}^{\mathrm{T}} \\ \hline \mathbf{0} & \mathrm{e}^0 = 1 \end{array}\right),$$

so, in the same way,

$$\mathrm{e}^{A^*t} = \left(\begin{array}{c|c} \mathrm{e}^{\widetilde{A^*t}} & \mathbf{1}^{\mathrm{T}} - \mathrm{e}^{\widetilde{A^*t}}\mathbf{1}^{\mathrm{T}} \\ \hline \mathbf{0} & \mathrm{e}^0 = 1 \end{array}\right).$$

$\square$

Consider the previous numerical example $A = \begin{pmatrix} 1 & -2 \\ 3 & -4 \end{pmatrix}$, leading to $A^* = \begin{pmatrix} -1 & 0 & 1 \\ -2 & -2 & 4 \\ 0 & 0 & 0 \end{pmatrix}$. We have:

$$\mathrm{e}^{A^*t} = \begin{pmatrix} \mathrm{e}^{-t} & 0 & 1 - \mathrm{e}^{-t} \\ -2\mathrm{e}^{-t} + 2\mathrm{e}^{-2t} & \mathrm{e}^{-2t} & 1 + 2\mathrm{e}^{-t} - 3\mathrm{e}^{-2t} \\ 0 & 0 & 1 \end{pmatrix}.$$

5.3. Inversion lemma for the exponential-dual. Knowing the exponential-dual A^* of A, we can recover A using the following result:

LEMMA 5.4 (Inversion lemma for the exponential-dual matrix). *For $0 \le i, j \le n-1$, we have*

$$A_{i,j} = \sum_{k=0}^{i} \Big(A^*_{j,k} - A^*_{j+1,k} \Big).$$

PROOF. Re-write the definition of A^* using the following notation: for $0 \le j, k \le n$,

$$A^*_{j,k} = \sum_{\ell=j}^{n-1} \Big(A_{k,\ell} - A_{k-1,\ell} \Big).$$

Summing the first $i+1$ elements of row j of A^* gives

$$\begin{aligned}
\sum_{k=0}^{i} A^*_{j,k} &= \sum_{k=0}^{i} \sum_{\ell=j}^{n-1} \Big(A_{k,\ell} - A_{k-1,\ell} \Big) \\
&= \sum_{\ell=j}^{n-1} \sum_{k=0}^{i} \Big(A_{k,\ell} - A_{k-1,\ell} \Big) \\
&= \sum_{\ell=j}^{n-1} \Big[A_{0,\ell} + A_{1,\ell} + \cdots + A_{i,\ell} - \big(A_{-1,\ell} + A_{0,\ell} + \cdots + A_{i-1,\ell} \big) \Big] \\
&= \sum_{\ell=j}^{n-1} A_{i,\ell}.
\end{aligned}$$

Writing the obtained equality $\sum_{k=0}^{i} A^*_{j,k} = \sum_{\ell=j}^{n-1} A_{i,\ell}$ again but replacing j by $j+1$ produces $\sum_{k=0}^{i} A^*_{j+1,k} = \sum_{\ell=j+1}^{n-1} A_{i,\ell}$. Subtracting now both equalities gives

$$\sum_{k=0}^{i} A^*_{j,k} - \sum_{k=0}^{i} A^*_{j+1,k} = \sum_{\ell=j}^{n-1} A_{i,\ell} - \sum_{\ell=j+1}^{n-1} A_{i,\ell},$$

that is,

$$\sum_{k=0}^{i} \Big(A^*_{j,k} - A^*_{j+1,k} \Big) = A_{i,j}. \qquad \square$$

5.4. Main result for the exponential-dual. Our main result is the following one:

THEOREM 5.5. *Define matrix function F using the following relation: for any i and j belonging to the index set $\{0, 1, \dots, n-1\}$,*

$$F_{i,j}(t) = \sum_{k=0}^{i} \Big(E^*_{j,k}(t) - E^*_{j+1,k}(t) \Big).$$

Then, $F(t) = \mathrm{e}^{At}$.

Before proving this main theorem, we need the following lemma.

LEMMA 5.6 (Inversion lemma for matrix function E^*). *Knowing F, we can recover matrix E^* using the following relations: for $0 \leq i \leq n$ and $1 \leq j \leq n-1$, we have (we are omitting 't' here, for more clarity in the text)*

$$E^*_{i,j} = \sum_{k=i}^{n-1} \Big(F_{j,k} - F_{j-1,k} \Big)$$

for the last column of E^*, we have, for any $i \in \{0, 1, \ldots, n\}$,

$$E^*_{i,n} = 1 - \sum_{k=i}^{n-1} F_{n-1,k}.$$

PROOF. Let us re-write the definition of F with the following notation:

$$F_{j,k} = \sum_{\ell=0}^{j} \Big(E^*_{k,\ell} - E^*_{k+1,\ell} \Big),$$

Summing the last components of row j (which is $\leq n-1$) of F, starting at column i, gives

$$\begin{aligned}
\sum_{k=i}^{n-1} F_{j,k} &= \sum_{k=i}^{n-1} \sum_{\ell=0}^{j} \Big(E^*_{k,\ell} - E^*_{k+1,\ell} \Big) \\
&= \sum_{\ell=0}^{j} \sum_{k=i}^{n-1} \Big(E^*_{k,\ell} - E^*_{k+1,\ell} \Big) \\
&= \sum_{\ell=0}^{j} \Big[E^*_{i,\ell} + E^*_{i+1,\ell} + \cdots + E^*_{n-1,\ell} - \big(E^*_{i+1,\ell} + E^*_{i+2,\ell} + \cdots + E^*_{n,\ell} \big) \Big] \\
&= \sum_{\ell=0}^{j} E^*_{i,\ell}.
\end{aligned}$$

We used the fact that, since in the sums on ℓ, we always have $\ell < n$, then $E^*_{n,\ell} = 0$.

Writing now the obtained equality $\sum_{k=i}^{n-1} F_{j,k} = \sum_{\ell=0}^{j} E^*_{i,\ell}$ and the equality obtained by changing j by $j-1$, that is, $\sum_{k=i}^{n-1} F_{j-1,k} = \sum_{\ell=0}^{j-1} E^*_{i,\ell}$ and subtracting them, we get

$$\sum_{k=i}^{n-1} \Big(F_{j,k} - F_{j-1,k} \Big) = E^*_{i,j}.$$

For the case of $j = n-1$, we start from the same expression (replacing j by $n-1$):

$$F_{n-1,k} = \sum_{\ell=0}^{n-1} \Big(E^*_{k,\ell} - E^*_{k+1,\ell} \Big),$$

We sum on k from i to $n-1$:

$$\begin{aligned}
\sum_{k=i}^{n-1} F_{n-1,k} &= \sum_{k=i}^{n-1}\sum_{\ell=0}^{n-1}\Big(E^*_{k,\ell} - E^*_{k+1,\ell}\Big)\\
&= \sum_{\ell=0}^{n-1}\sum_{k=i}^{n-1}\Big(E^*_{k,\ell} - E^*_{k+1,\ell}\Big)\\
&= \sum_{\ell=0}^{n-1}\Big[E^*_{i,\ell} + E^*_{i+1,\ell} + \cdots + E^*_{n-1,\ell} - \big(E^*_{i+1,\ell} + E^*_{i+2,\ell} + \cdots + E^*_{n,\ell}\big)\Big]\\
&= \sum_{\ell=0}^{n-1} E^*_{i,\ell}\\
&= 1 - E^*_{i,n},
\end{aligned}$$

leading to

$$E^*_{i,n} = 1 - \sum_{k=i}^{n-1} F_{n-1,k}. \qquad \square$$

Proof of the Main Theorem 5.5. Starting from some given matrix A, we construct A^*, compute $E^*(t) = \mathrm{e}^{A^*t}$, and construct F. We must prove that $F(t) = \mathrm{e}^{At}$, or equivalently, that $F' = FA$, or that $F' = AF$.

Let us use the notation $G(t) = E^*(t)$ to avoid all these *, and omit 't' as before for the sake of clarity.

We will prove that

$$F' = AF. \tag{5.1}$$

Fix i, j with $0 \le i, j \le n-1$, and focus on the right hand side of (5.1).

$$\big(AF\big)_{i,j} = \sum_{u=0}^{n-1} A_{i,u}F_{u,j} = \sum_{u=0}^{n-1} A_{i,u}\sum_{v=0}^{u}\big(G_{j,v} - G_{j+1,v}\big). \tag{5.2}$$

Now, we look at the left hand side of (5.1) From the definition of F,

$$F'_{i,j} = \sum_{k=0}^{i}\big(G'_{j,k} - G'_{j+1,k}\big). \tag{5.3}$$

We know that $G' = GA^* = A^*G$. Let us use here the second equality.

$$\begin{aligned}
G'_{j,k} &= \sum_{v=0}^{n} G_{j,v}A^*_{v,k}\\
&= \sum_{v=0}^{n-1} G_{j,v}A^*_{v,k}\\
&\quad \nwarrow \text{ because the last row of } A^* \text{ is only composed of 0's}\\
&= \sum_{v=0}^{n-1} G_{j,v}\sum_{u=v}^{n-1}\big(A_{k,u} - A_{k-1,u}\big).
\end{aligned}$$

Coming back to $F'_{i,j}$, we have

$$\begin{aligned}
F'_{i,j} &= \sum_{k=0}^{i}\big(G'_{j,k} - G'_{j+1,k}\big) = \sum_{k=0}^{i}\sum_{v=0}^{n-1}\big(G_{j,v} - G_{j+1,v}\big)\sum_{u=v}^{n-1}\big(A_{k,u} - A_{k-1,u}\big) \\
&= \sum_{v=0}^{n-1} G_{j,v}\sum_{u=v}^{n-1}\sum_{k=0}^{i}\big(A_{k,u} - A_{k-1,u}\big) - \sum_{v=0}^{n-1} G_{j+1,v}\sum_{u=v}^{n-1}\sum_{k=0}^{i}\big(A_{k,u} - A_{k-1,u}\big) \\
&= \sum_{v=0}^{n-1}\big(G_{j,v} - G_{j+1,v}\big)\sum_{u=v}^{n-1}\sum_{k=0}^{i}\big(A_{k,u} - A_{k-1,u}\big) \\
&\quad \nwarrow \text{ after moving the first sum on } k \text{ to the end} \\
&= \sum_{v=0}^{n-1}\big(G_{j,v} - G_{j+1,v}\big)\sum_{u=v}^{n-1} A_{i,u} \\
&\quad \nwarrow \text{ after observing that } \sum_{k=0}^{i}\big(A_{k,u} - A_{k-1,u}\big) \text{ is a telescopic series} \\
&= \sum_{u=0}^{n-1} A_{i,u}\sum_{v=0}^{u}\big(G_{j,v} - G_{j+1,v}\big) \\
&\quad \nwarrow \text{ after interchanging the summation order,}
\end{aligned}$$

which is exactly (5.2). □

Observe now that we have proved the equivalent of relations (3.1) and (3.2) but in our generalization, where A is an arbitrary square matrix: using the notation E and E^*, we have

$$E^*_{i,\,j}(t) = \sum_{k=0}^{i-1}\Big[E_{j-1,\,k}(t) - E_{j,\,k}(t)\Big],$$

$$E_{i,\,j}(t) = \sum_{k=0}^{i}\Big[E^*_{j,\,k}(t) - E^*_{j+1,\,k}(t)\Big].$$

This is equivalent to the Duality Theorem mentioned in [**8**], here a direct consequence of Theorem 5.5.

EXAMPLE. Let us look at the generic 2-state case. If

$$E^* = \begin{pmatrix} E^*_{0,0} & E^*_{0,1} & 1 - E^*_{0,0} - E^*_{0,1} \\ E^*_{1,0} & E^*_{1,1} & 1 - E^*_{1,0} - E^*_{1,1} \\ 0 & 0 & 1 \end{pmatrix},$$

then we have:

$$A = \begin{pmatrix} E^*_{0,0} - E^*_{1,0} & E^*_{1,0} \\ E^*_{0,0} + E^*_{0,1} - \big(E^*_{1,0} + E^*_{1,1}\big) & E^*_{1,0} + E^*_{1,1} \end{pmatrix}.$$

Reconsider example $A = \begin{pmatrix} 1 & -2 \\ 3 & -4 \end{pmatrix}$, whose exponential-dual was

$$A^* = \begin{pmatrix} -1 & 0 & 1 \\ -2 & -2 & 4 \\ 0 & 0 & 0 \end{pmatrix}.$$

We had

$$E^*(t) = \begin{pmatrix} e^{-t} & 0 & 1-e^{-t} \\ -2e^{-t}+2e^{-2t} & e^{-2t} & 1+2e^{-t}-3e^{-2t} \\ 0 & 0 & 1 \end{pmatrix}.$$

Using the inversion formulas and Theorem 5.5, we obtain

$$F(t) = \begin{pmatrix} 3e^{-t}-2e^{-2t} & -2e^{-t}+2e^{-2t} \\ 3e^{-t}-3e^{-2t} & -2e^{-t}+3e^{-2t} \end{pmatrix} = e^{At}.$$

6. Conclusions

In Section 3, we first describe stochastic duality (or simply duality, and meaning in the sense of [**16**] or [**1**]), and use it as a tool to find transient distributions of basic Markovian queuing models when combined with uniformization. This approach makes sense when the analysis of the dual is simpler than that of the initial model, and this has been the case in several articles [**8**], [**9**], [**10**] dealing with fundamental queuing systems. However, there is a limitation to this approach, which is the fact that some Markovian models have no dual, or that they come with restrictions.

After discussing these drawbacks of dual usage, we define an algebraically similar concept but without any reference to stochastic processes. Because of its similarity to the dual and its role in computing matrix exponentials, we call it the exponential-dual, and we show that it coincides with the dual when we are in a Markov setting and the dual exists. The advantage of the exponential-dual is that it exists for any given matrix.

Future work will explore this concept in more depth by separating the discrete and continuous "time" cases, and the connections between them. Another direction that deserves attention is the exploration of the relations with other basic matrix transformations, and more generally, with spectral analysis, see [**7**].

References

[1] William J. Anderson, *Continuous-time Markov chains*, Springer Series in Statistics: Probability and its Applications, Springer-Verlag, New York, 1991. An applications-oriented approach, DOI 10.1007/978-1-4612-3038-0. MR1118840

[2] Rabi N. Bhattacharya and Edward C. Waymire, *Stochastic processes with applications*, Classics in Applied Mathematics, vol. 61, Society for Industrial and Applied Mathematics (SIAM), Philadelphia, PA, 2009. Reprint of the 1990 original [MR1054645], DOI 10.1137/1.9780898718997.ch1. MR3396216

[3] Michael L. Green, Alan Krinik, Carrie Mortensen, Gerardo Rubino, and Randall J. Swift, *Transient probability functions: a sample path approach*, Discrete random walks (Paris, 2003), Discrete Math. Theor. Comput. Sci. Proc., AC, Assoc. Discrete Math. Theor. Comput. Sci., Nancy, 2003, pp. 127–136. MR2042380

[4] Joti Lal Jain, Sri Gopal Mohanty, and Walter Böhm, *A course on queueing models*, Statistics: Textbooks and Monographs, Chapman & Hall/CRC, Boca Raton, FL, 2007. MR2257773

[5] Sabine Jansen and Noemi Kurt, *On the notion(s) of duality for Markov processes*, Probab. Surv. **11** (2014), 59–120, DOI 10.1214/12-PS206. MR3201861

[6] Arne Jensen, *Markoff chains as an aid in the study of Markoff processes*, Skand. Aktuarietidskr. **36** (1953), 87–91, DOI 10.1080/03461238.1953.10419459. MR57488

[7] Alan Krinik, Hubertus von Bremen, Ivan Ventura, Uyen Vietthanh Nguyen, Jeremy J. Lin, Thuy Vu Dieu Lu, Chon In (Dave) Luk, Jeffrey Yeh, Luis A. Cervantes, Samuel R. Lyche, Brittney A. Marian, Saif A. Aljashamy, Mark Dela, Ali Oudich, Pedram Ostadhassanpanjehali, Lyheng Phey, David Perez, John Joseph Kath, Malachi C. Demmin, Yoseph Dawit, Christine Carmen Marie Hoogendyk, Aaron Kim, Matthew McDonough, Adam Trevor Castillo, David Beecher, Weizhong Wong, and Heba Ayeda, *Explicit transient probabilities of various Markov models*, in Stochastic Processes and Functional Analysis, New Perspectives, AMS Contemporary Mathematics Series, Volume 774, edited by Randall J. Swift, Alan Krinik, Jennifer Switkes and Jason Park, November 2021, pp. 97–151.

[8] Alan Krinik and Sri Gopal Mohanty, *On batch queueing systems: a combinatorial approach*, J. Statist. Plann. Inference **140** (2010), no. 8, 2271–2284, DOI 10.1016/j.jspi.2010.01.023. MR2609486

[9] Alan Krinik, Carrie Mortensen, and Gerardo Rubino, *Connections between birth-death processes*, Stochastic processes and functional analysis, Lecture Notes in Pure and Appl. Math., vol. 238, Dekker, New York, 2004, pp. 219–240. MR2059909

[10] Alan Krinik, Gerardo Rubino, Daniel Marcus, Randall J. Swift, Hassan Kasfy, and Holly Lam, *Dual processes to solve single server systems*, J. Statist. Plann. Inference **135** (2005), no. 1, 121–147, DOI 10.1016/j.jspi.2005.02.010. MR2202343

[11] S. Lyche, *On Deep Learning and Neural Networks*, Master's thesis, California State Polytechnic University, Pomona, California, US, 2018.

[12] J. Medhi, *Stochastic models in queueing theory*, 2nd ed., Academic Press, Amsterdam, 2003. MR1991930

[13] Anthony G. Pakes, *Convergence rates and limit theorems for the dual Markov branching process*, J. Probab. Stat., posted on 2017, Art. ID 1410507, 13, DOI 10.1155/2017/1410507. MR3628142

[14] Paweł Lorek, *Generalized gambler's ruin problem: explicit formulas via Siegmund duality*, Methodol. Comput. Appl. Probab. **19** (2017), no. 2, 603–613, DOI 10.1007/s11009-016-9507-6. MR3649560

[15] Gerardo Rubino and Bruno Sericola, *Markov chains and dependability theory*, Cambridge University Press, Cambridge, 2014, DOI 10.1017/CBO9781139051705. MR3469975

[16] D. Siegmund, *The equivalence of absorbing and reflecting barrier problems for stochastically monotone Markov processes*, Ann. Probability **4** (1976), no. 6, 914–924, DOI 10.1214/aop/1176995936. MR431386

[17] Randall J. Swift, *A simple immigration-catastrophe process*, Math. Sci. **25** (2000), no. 1, 32–36. MR1771175

[18] Pan Zhao, *Siegmund duality for continuous time Markov chains on $\mathbb{Z}_+^d$*, Acta Math. Sin. (Engl. Ser.) **34** (2018), no. 9, 1460–1472, DOI 10.1007/s10114-018-7064-3. MR3836232

Gerardo Rubino, INRIA, Campus de Beaulieu, 35042 Rennes, France
Email address: gerardo.rubino@inria.fr

Alan Krinik, Cal Poly Pomona, 3801 West Temple Ave., Pomona, California 91768
Email address: ackrinik@cpp.edu

Contemporary Mathematics
Volume **774**, 2021
https://doi.org/10.1090/conm/774/15576

Two moment closure techniques for an interacting species model

Jennifer Switkes

ABSTRACT. We explore a stochastic analogue for a generalized deterministic interacting species model for two species. By applying two moment closure techniques, we approximate the expected values, variances, and covariance for the two populations and compare the results. First, we assume a multivariate normal distribution with standard additive variances and covariance. Next, we assume a multivariate lognormal distribution with multiplicative variances and covariance. There is good agreement between the two moment closure techniques. For the stable equilibria, the expected values of the populations converge to values that are similar to, but not equal to, the values of the deterministic equilibria. The variances and covariance also converge over time.

1. Introduction

Deterministic interacting species models have been studied widely for many years. In recent years, stochastic interacting species models have been developed. In 2002, Swift developed a stochastic version of a classical deterministic predator-prey model, using a birth-death formulation [**Sw**]. He noted that the stochastic model resulted in a system of differential equations for the expected population sizes of the two species, with the same structure as the corresponding deterministic model. Whereas the deterministic model contains a classical mass action term, the stochastic model contains an expected value of the product of the population sizes. This stochastic system is not closed, since upon accounting for the expected value of the product, higher order moments are introduced. In 2004, Lloyd discussed moment closure techniques in the context of epidemic models [**L**], models that have the same mathematical structure as interacting species models. In appendices to his paper, Lloyd presented the theory and details for use of two moment closure techniques: (1) multivariate normal assumptions using standard additive variances and covariance; (2) multivariate lognormal assumptions using multiplicative variances and covariance. These techniques allow approximation of the stochastic system through a closed, truncated system involving higher order moments. In 2017 and 2018, Curtis, Trakoolthai, and Switkes [**C, T**] employed the moment closure techniques discussed by Lloyd [**L**] to the predator-prey model explored by Swift [**Sw**], constructing a closed system in terms of the expected populations sizes, variances,

2020 *Mathematics Subject Classification.* 60J80, 92B05.
Key words and phrases. Interacting species, birth-death process, moment closure, differential equations, expected value.

and covariance, and finding strong agreement between the deterministic model and closed stochastic model. In particular, the classical cyclical predator-prey behavior observed in the deterministic model is visible in the stochastic model, but with spread in the results. Here, we employ the techniques used by Curtis, Trakoolthai, and Switkes [**C**, **T**] to develop a stochastic version of an interacting species model with bifurcation explored by Switkes and Szypowski [**Sz**], a model that depending on parameter values can describe either competitive exclusion or stable competition.

In Section 2 we introduce the deterministic generalized interacting species model from [**Sz**]. In Section 3, we develop the corresponding stochastic model, obtaining a system of differential equations for the expected population sizes of the two species. We note that the system is not closed, since higher-order moments are involved implicitly. In Section 4, we assume that the two populations are dstributed according to a multivariate normal distribution. Using standard additive variances and covariance, we apply the moment closure technique described in [**L**], obtaining an approximate closed system in terms of the expected values, variances, and covariance. In Section 5, we assume a multivariate lognormal distribution, use multiplicative variances and covariance, and again obtain an approximate closted system. In both Section 4 and 5, we compare our results with the dynamics of the original deterministic model. In Section 6, we make some concluding remarks and suggest areas for further work.

2. A generalized interacting species model

Consider the following symmetric generalized deterministic interacting species model [**Sz**] for populations $x(t)$ and $y(t)$ of species X and species Y, respectively:

$$\frac{dx}{dt} = [a-(k+1)x-(2-k)y]\,x, \tag{2.1}$$

$$\frac{dy}{dt} = [a-(2-k)x-(k+1)y)]\,y \tag{2.2}$$

For $-\infty < k < 1/2$, the model exhibits competitive exclusion. For $1/2 < k < \infty$, the model exhibits stable competition. A bifurcation takes place at $k = 1/2$. The equilibria for system (2.1)–(2.2) are $(0,0)$, $(a/3, a/3)$, $(0, a/(k+1)$, $(a/(k+1), 0)$ for $k \neq -1$ and $k \neq 1/2$.

We will focus here on $k = 0$ and $k = 1$. For $k = 0$ (competitive exclusion), the equilibria $(0,a)$ and $(a,0)$ are stable. For $k = 1$ (stable competition), the equilibrium $(a/3, a/3)$ is stable. The equilibrium at $(0,0)$ is stable always, but uninteresting.

3. Stochastic interacting species model

We now develop a corresponding stochastic model [**C**, **Sw**, **T**]. Let $X(t)$ and $Y(t)$ be random variables governing the number of individuals in species X and species Y, respectively, at time t. Let the probability that the populations of species X and species Y are x and y, respectively, at time t be denoted by

$$P_{x,y}(t) = P[X(t) = x, Y(t) = y], \quad \text{for} \quad x = 0,1,2,3,\ldots \text{ and } y = 0,1,2,3,\ldots$$

The probability that the population of species X increases from x to $x+1$ during a time interval of length Δt is $ax\Delta t + o(\Delta t)$, and the probability that the population of species X decreases from x to $x-1$ during a time interval of length Δt is $[(k+1)x + (2-k)y]x\Delta t + o(\Delta t)$. The probability that the population of species Y increases from y to $y+1$ during a time interval of length Δt is $ay\Delta t + o(\Delta t)$, and the probability that the population of species Y decreases from y to $y-1$ during a time interval of length Δt is $[(2-k)x+(k+1)y]y\Delta t + o(\Delta t)$. We assume that at most one change occurs in Δt. Thus,

$$\begin{aligned} P_{x,y}(t+\Delta t) &= [1-(ax+[(k+1)x+(2-k)y]x+ay \\ &\quad +[(2-k)x+(k+1)y]y+o(\Delta t))\Delta t]P_{x,y}(t) \\ &\quad +(a(x-1)+o(\Delta t))\,\Delta t P_{x-1,y}(t) \\ &\quad +(a(y-1)+o(\Delta t))\,\Delta t P_{x,y-1}(t) \\ &\quad +([(k+1)(x+1)+(2-k)y](x+1)+o(\Delta t))\,\Delta t P_{x+1,y}(t) \\ &\quad +([(2-k)x+(k+1)(y+1)](y+1)+o(\Delta t))\,\Delta t P_{x,y+1}(t). \end{aligned}$$

Rearranging terms, taking the limit as Δt goes to 0, and noting that $\lim_{\Delta t\longrightarrow 0}\frac{o(\Delta t)}{\Delta t} = 0$, we obtain

$$\begin{aligned} P'_{x,y}(t) &= -(ax+[(k+1)x+(2-k)y]x+ay+[(2-k)x+(k+1)y]y)\,P_{x,y}(t) \\ &\quad +a(x-1)P_{x-1,y}(t)+a(y-1)P_{x,y-1}(t) \\ &\quad +[(k+1)(x+1)+(2-k)y](x+1)P_{x+1,y}(t) \\ &\quad +[(2-k)x+(k+1)(y+1)](y+1)P_{x,y+1}(t). \end{aligned}$$

We define a probability generating function (p.g.f.) by

$$\phi(z_1,z_2,t)=\sum_{x=0}^{\infty}\sum_{y=0}^{\infty}P_{x,y}(t)z_1^x z_2^y.$$

Note that

$$\begin{aligned} \frac{\partial\phi(z_1,z_2,t)}{\partial t} &= \sum_{x=0}^{\infty}\sum_{y=0}^{\infty}P'_{x,y}(t)z_1^x z_2^y \\ &= -\sum_{x=0}^{\infty}\sum_{y=0}^{\infty}(ax+[(k+1)x+(2-k)y]x \\ &\qquad\qquad +ay+[(2-k)x+(k+1)y]y)P_{x,y}(t)z_1^x z_2^y \\ &\quad +\sum_{x=0}^{\infty}\sum_{y=0}^{\infty}a(x-1)P_{x-1,y}(t)z_1^x z_2^y+\sum_{x=0}^{\infty}\sum_{y=0}^{\infty}a(y-1)P_{x,y-1}(t)z_1^x z_2^y \\ &\quad +\sum_{x=0}^{\infty}\sum_{y=0}^{\infty}[(k+1)(x+1)+(2-k)y](x+1)P_{x+1,y}(t)z_1^x z_2^y \\ &\quad +\sum_{x=0}^{\infty}\sum_{y=0}^{\infty}[(2-k)x+(k+1)(y+1)](y+1)P_{x,y+1}(t)z_1^x z_2^y. \end{aligned}$$

Note also that

$$\begin{aligned}
\frac{\partial \phi(1,1,t)}{\partial z_1} &= \sum_{x=0}^{\infty}\sum_{y=0}^{\infty} x P_{x,y}(t) = E\left[X(t)\right], \\
\frac{\partial \phi(1,1,t)}{\partial z_2} &= \sum_{x=0}^{\infty}\sum_{y=0}^{\infty} x P_{x,y}(t) = E\left[Y(t)\right], \\
\frac{\partial^2 \phi(1,1,t)}{\partial z_2 \partial z_1} &= \sum_{x=0}^{\infty}\sum_{y=0}^{\infty} xy P_{x,y}(t) = E\left[X(t)Y(t)\right], \\
\frac{\partial^2 \phi(1,1,t)}{\partial z_1^2} &= \sum_{x=0}^{\infty}\sum_{y=0}^{\infty} x(x-1) P_{x,y}(t) = E\left[(X(t))^2 - X(t)\right], \\
\frac{\partial^2 \phi(1,1,t)}{\partial z_2^2} &= \sum_{x=0}^{\infty}\sum_{y=0}^{\infty} y(y-1) P_{x,y}(t) = E\left[(Y(t))^2 - Y(t)\right].
\end{aligned}$$

Carrying out the needed calculus, we obtain

$$\begin{aligned}
\frac{\partial \phi(z_1,z_2,t)}{\partial t} &= a(z_1^2 - z_1)\frac{\partial \phi(z_1,z_2,t)}{\partial z_1} + a(z_2^2 - z_2)\frac{\partial \phi(z_1,z_2,t)}{\partial z_2} \\
&\quad +(2-k)[z_1 + z_2 - 2z_1 z_2]\frac{\partial^2 \phi(z_1,z_2,t)}{\partial z_2 \partial z_1} \\
&\quad +(k+1)(z_1 - z_1^2)\frac{\partial^2 \phi(z_1,z_2,t)}{\partial z_1^2} + (k+1)(1-z_1)\frac{\partial \phi(z_1,z_2,t)}{\partial z_1} \\
&\quad +(k+1)(z_2 - z_2^2)\frac{\partial^2 \phi(z_1,z_2,t)}{\partial z_2^2} + (k+1)(1-z_2)\frac{\partial \phi(z_1,z_2,t)}{\partial z_2}.
\end{aligned} \tag{3.1}$$

Taking the partial derivative of (3.1) with respect to z_1 and letting $z_1 = z_2 = 1$, we obtain

$$\frac{dE\left[X(t)\right]}{dt} = aE\left[X(t)\right] - (2-k)E\left[X(t)Y(t)\right] - (k+1)E\left[(X(t))^2\right]. \tag{3.2}$$

Similarly, taking the partial derivative of (3.1) with respect to z_2 and letting $z_1 = z_2 = 1$, we obtain

$$\frac{dE\left[Y(t)\right]}{dt} = aE\left[Y(t)\right] - (2-k)E\left[X(t)Y(t)\right] - (k+1)E\left[(Y(t))^2\right]. \tag{3.3}$$

The partial derivative notation is no longer needed since z_1 and z_2 are no longer involved and the expected value is deterministic. Note the similarity in structure to system (2.1)-(2.2). However, in general $E[X(t)Y(t)] \neq E[X(t)]E[Y(t)]$, $E[(X(t))^2] \neq (E[X(t)])^2$, and $E[(Y(t))^2] \neq (E[Y(t)])^2$.

In the moment closure procedures that we will pursue below, we will need the derivatives of $E[X^2 - X]$, $E[Y^2 - Y]$, and $E[XY]$.

Taking the second partial derivative of (3.1) with respect to z_1 twice and letting $z_1 = z_2 = 1$, we obtain

$$\begin{aligned}
\frac{d\left[E\left[X^2 - X\right]\right]}{dt} &= (2a + 2(k+1))E[X^2] - 2(k+1)E[X^3] \\
&\quad -2(2-k)E[X^2 Y] + 2(2-k)E[XY].
\end{aligned}$$

Similarly, taking the second partial derivative of (3.1) with respect to z_2 twice and letting $z_1 = z_2 = 1$, we obtain

$$\begin{aligned}\frac{d\left[E\left[Y^2 - Y\right]\right]}{dt} &= (2a + 2(k+1))E[Y^2] - 2(k+1)E[Y^3] \\ &\quad -2(2-k)E[Y^2X] + 2(2-k)E[XY].\end{aligned}$$

Finally, taking the mixed second partial derivative of (3.1) with respect to z_1 and z_2 and letting $z_1 = z_2 = 1$, we obtain

$$\frac{d\left[E\left[XY\right]\right]}{dt} = 2aE[XY] - 3E[X^2Y] - 3E[Y^2X]$$

In what follows, we use two different moment closure techniques to obtain approximate closed systems of differential equations for the expected values, variances, and covariance of the populations of species X and species Y. We first pursue a moment closure technique based on assuming the populations to be distributed according to a multivariate normal distribution, using standard variances and covariance. We then pursue a moment closure technique based on assuming the populations to be distributed according to a multivariate lognormal distribution, using multiplicative variances and covariance.

4. Moment closure using normal distribution

Suppose that the populations of species X and species Y are distributed according to a multivariate normal distribution. The multivariate normal distribution is known [**L**, **C**] to have vanishing third order central moments $E[(X - E[X])^j (Y - E[Y])^k]$, where $j + k = 3$. Thus, we assume that

$$\begin{aligned}E\left[(X - E\left[X\right])^3\right] &= 0, \qquad E\left[(Y - E\left[Y\right])^3\right] = 0, \\ E\left[(X - E\left[X\right])^2 (Y - E\left[Y\right])\right] &= 0, \qquad E\left[(X - E\left[X\right])(Y - E\left[Y\right])^2\right] = 0.\end{aligned}$$

Using the standard additive definitions of variance and covariance we have that

$$E\left[X^2\right] = Var[X] + (E\left[X\right])^2, \qquad E\left[Y^2\right] = Var[Y] + (E\left[Y\right])^2, \tag{4.1}$$

$$E\left[XY\right] = Cov\left[X,Y\right] + E\left[X\right]E\left[Y\right]. \tag{4.2}$$

Using the assumption that third order central moments vanish, and doing some algebra, we obtain

$$\begin{aligned}E\left[X^2Y\right] &= Var[X]E\left[Y\right] + (E\left[X\right])^2 E\left[Y\right] + 2Cov\left[X,Y\right]E\left[X\right], \\ E\left[Y^2X\right] &= Var[Y]E\left[X\right] + (E\left[Y\right])^2 E\left[X\right] + 2Cov\left[X,Y\right]E\left[Y\right], \\ E\left[X^3\right] &= 3Var[X]E\left[X\right] + (E\left[X\right])^3, \\ E\left[Y^3\right] &= 3Var[Y]E\left[Y\right] + (E\left[Y\right])^3.\end{aligned}$$

Using (4.1)-(4.2) in system (3.2)-(3.3), we obtain

$$\begin{aligned}\frac{dE\left[X\right]}{dt} &= aE\left[X\right] - (2-k)\left(Cov\left[X,Y\right] + E\left[X\right]E\left[Y\right]\right) \\ &\quad -(k+1)\left(Var[X] + (E\left[X\right])^2\right)\end{aligned} \tag{4.3}$$

$$\begin{aligned}\frac{dE\left[Y\right]}{dt} &= aE\left[Y\right] - (2-k)\left(Cov\left[X,Y\right] + E\left[X\right]E\left[Y\right]\right) \\ &\quad -(k+1)\left(Var[Y] + (E\left[Y\right])^2\right).\end{aligned} \tag{4.4}$$

We now find $dVar[X]/dt$ and $dVar[Y]/dt$. Since

$$Var[X] = E[X^2] - (E[X])^2 = E[X^2 - X] + E[X] - (E[X])^2,$$

we have that

$$\frac{dVar[X]}{dt} = \frac{d\left[E\left[X^2 - X\right]\right]}{dt} + \frac{dE[X]}{dt} - 2E[X]\frac{dE[X]}{dt}.$$

Substituting into the righthand side and simplifying, we obtain equation (4.5):

$$\begin{aligned}\frac{dVar[X]}{dt} &= (2a + 2(k+1))(Var[X] + (E[X])^2) \\ &\quad -2(k+1)(3Var[X]E[X] + (E[X])^3) \\ &\quad -2(2-k)(Var[X]E[Y] + (E[X])^2E[Y] + 2Cov[X,Y]E[X]) \\ &\quad +2(2-k)(Cov[X,Y] + E[X]E[Y]) \\ &\quad +(1 - 2E[X])(aE[X] - (2-k)(Cov[X,Y] + E[X]E[Y]) \\ &\quad -(k+1)(Var[X] + (E[X])^2)).\end{aligned}$$

A similar process yields equation (4.6):

$$\begin{aligned}\frac{dVar[Y]}{dt} &= (2a + 2(k+1))(Var[Y] + (E[Y])^2) \\ &\quad -2(k+1)(3Var[Y]E[Y] + (E[Y])^3) \\ &\quad -2(2-k)(Var[Y]E[X] + (E[Y])^2E[X] + 2Cov[X,Y]E[Y]) \\ &\quad +2(2-k)(Cov[X,Y] + E[X]E[Y]) \\ &\quad +(1 - 2E[Y])(aE[Y] - (2-k)(Cov[X,Y] + E[X]E[Y]) \\ &\quad -(k+1)(Var[Y] + (E[Y])^2)).\end{aligned}$$

Finally we find $dCov[X,Y]/dt$. Since

$$Cov[X,Y] = E[XY] - E[X]E[Y],$$

we have that

$$\frac{dCov[X,Y]}{dt} = \frac{d\left[E\left[XY\right]\right]}{dt} - E[X]\frac{dE[Y]}{dt} - E[Y]\frac{dE[X]}{dt}.$$

Substituting into the expressions on the righthand side and simplifying, we obtain equation (4.7):

$$\begin{aligned}\frac{dCov[X,Y]}{dt} &= 2a(Cov[X,Y] + E[X]E[Y]) \\ &\quad -3(Var[X]E[Y] + (E[X])^2E[Y] + 2Cov[X,Y]E[X]) \\ &\quad -3(Var[Y]E[X] + (E[Y])^2E[X] + 2Cov[X,Y]E[Y]) \\ &\quad -E[X](aE[Y] - (2-k)(Cov[X,Y] + E[X]E[Y])) \\ &\quad -E[X](k+1)(Var[Y] + (E[Y])^2) \\ &\quad -E[Y](aE[X] - (2-k)(Cov[X,Y] + E[X]E[Y])) \\ &\quad -E[Y](k+1)(Var[X] + (E[X])^2).\end{aligned}$$

System (4.3)–(4.7) represents a closed system of differential equations for $E[X]$, $E[Y]$, $Var[X]$, $Var[Y]$, and $Cov[X,Y]$.

In Figure 1, we show simulations for several sets of initial conditions with $k = 1$, and $a = 3000$. The deterministic system (2.1)-(2.2) has a stable equilibrium

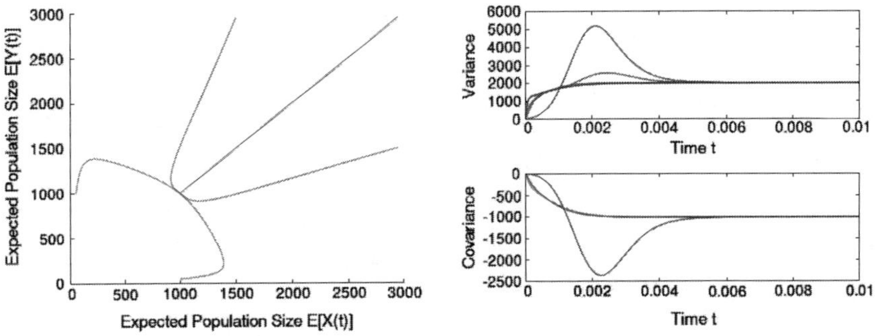

FIGURE 1. Multivariate normal expected population values (left), variances (upper right), and covariance (lower right): stable competition ($k = 1$, $a = 3000$).

at $(1000, 1000)$. In contrast, the stochastic system (4.3)–(4.7) shows a stable population equilibrium at approximately $(999, 999)$. Interestingly, the variances and covariance appear to reach stable equilibria as well, at values around 2005 (standard deviation of about 45) for the variances, and around -1004 for the covariance.

We will explore further the stable equilibrium located at $(a/3, a/3)$ in the deterministic model with $k = 1$. Due to the symmetry of the deterministic model, in the stochastic formulation as well this equlibrium has $E[X] = E[Y]$. Assuming that $Var[X]$ and $Var[Y]$ converge to a common value V as $t \to \infty$, and that $Cov[X, Y]$ converges to a value C, and setting $E[X] = E[Y]$, we obtain in the limit that the expected values for the equilibrium with both species present are given by

$$E[X] = E[Y] = \frac{a + \sqrt{a^2 - 12C - 24V}}{6}$$

With $V \approx 2005$ and $C \approx -1004$, we obtain $E[X] = E[Y] \approx 999$, as indeed is shown in our plots.

In Figure 2, we show simulations for several sets of initial conditions with $k = 0$, and $a = 3000$. The deterministic system (2.1)-(2.2) has stable equilibria at $(0, 3000)$ and $(3000, 0)$. In contrast, the stochastic system (4.3)–(4.7) shows stable population equilibria at approximately $(0, 2999)$ and $(2999, 0)$. Interestingly, here as well, the variances and covariance appear to reach stable equilibria again, at values around 3001 for the variance of the population that survives, and 0 for the variance of the population that goes extinct and for the covariance.

We will explore further the stable equilibria located at $(0, a)$ and $(a, 0)$ in the deterministic model with $k = 0$. Due to the symmetry of the deterministic model, in the stochastic formulation as well these equilibria will be located symmetrically. Assuming that the variance of the surviving species converges to a value V as $t \to \infty$, and that $Cov[X, Y]$ converges to 0, we obtain in the limit that the expected value for the population of the species that survives is given by

$$E[\cdot] = \frac{a + \sqrt{a^2 - 4V}}{2}$$

for the surviving species X or Y, respectively. With $V \approx 3001$, we obtain $E[\cdot] \approx 2999$, as is shown in our plots.

We now pursue a lognormal moment closure approach.

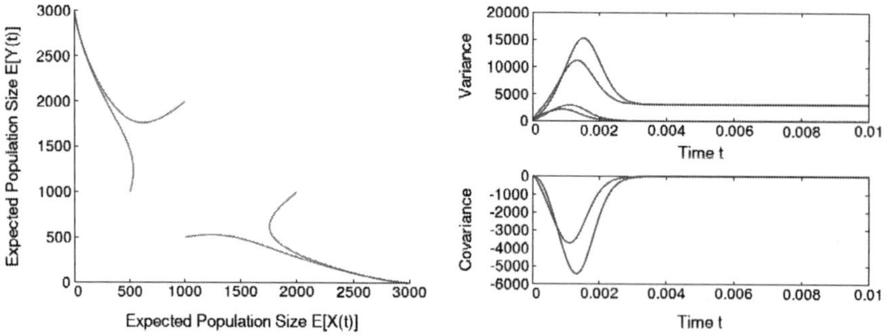

FIGURE 2. Multivariate normal expected population values (left), variances (upper right), and covariance (lower right): competitive exclusion ($k = 0$, $a = 3000$).

5. Moment closure using lognormal distribution

We now will assume that the populations are distributed according to a multivariate lognormal distribution. Also, we will use a concept of multiplicative variances and covariance. We will denote these by V_X, V_Y, and C_{XY}, respectively. We write

$$E\left[X^2\right] = V_X\left(E\left[X\right]\right)^2, \qquad E\left[Y^2\right] = V_Y\left(E\left[Y\right]\right)^2, \tag{5.1}$$

$$E\left[XY\right] = C_{XY}E\left[X\right]E\left[Y\right]. \tag{5.2}$$

The assumption of a multivariate lognormal distribution is known [**L**, **T**] to lead to the following approximations, which we will take with equality:

$$\begin{aligned} E\left[X^2Y\right] &= V_X(C_{XY})^2\left(E\left[X\right]\right)^2E\left[Y\right], \\ E\left[Y^2X\right] &= V_Y(C_{XY})^2\left(E\left[Y\right]\right)^2E\left[X\right], \\ E\left[X^3\right] &= (V_X)^3\left(E\left[X\right]\right)^3, \\ E\left[Y^3\right] &= (V_Y)^3\left(E\left[Y\right]\right)^3. \end{aligned}$$

Using (5.1)-(5.2) in system (3.2)-(3.3), we obtain

$$\frac{dE\left[X\right]}{dt} = aE\left[X\right] - (2-k)C_{XY}E\left[X\right]E\left[Y\right] - (k+1)V_X(E\left[X\right])^2, \tag{5.3}$$

$$\frac{dE\left[Y\right]}{dt} = aE\left[Y\right] - (2-k)C_{XY}E\left[X\right]E\left[Y\right] - (k+1)V_Y(E\left[Y\right])^2. \tag{5.4}$$

We now find dV_X/dt and dV_Y/dt. Since

$$V_X = \frac{E[X^2]}{(E[X])^2} = \frac{E[X^2-X]+E[X]}{(E[X])^2} = \frac{E[X^2-X]}{(E[X])^2} + \frac{1}{E[X]},$$

we have that

$$\frac{dV_X}{dt} = \frac{1}{(E[X])^2}\left[\frac{dE[X^2-X]}{dt} - \frac{dE[X]}{dt}\right] - \frac{2}{(E[X])^3}\frac{dE[X]}{dt}E[X^2-X].$$

Substituting into the righthand side and simplifying, we obtain

$$(5.5)\quad \frac{dV_X}{dt} = \left[(k+1)+2(2-k)E[Y](C_{XY}-(C_{XY})^2)\right]V_x + 2(k+1)E[X]\left((V_X)^2-(V_X)^3\right)+\frac{a+(2-k)C_{XY}E[Y]}{E[X]}$$

A similar process yields

$$(5.6)\quad \frac{dV_Y}{dt} = \left[(k+1)+2(2-k)E[X](C_{XY}-(C_{XY})^2)\right]V_Y + 2(k+1)E[Y]\left((V_Y)^2-(V_Y)^3\right)+\frac{a+(2-k)C_{XY}E[X]}{E[Y]}$$

Finally we find dC_{XY}/dt. Since

$$C_{XY} = \frac{E[XY]}{E[X]E[Y]},$$

we have that

$$\frac{dC_{XY}}{dt} = \frac{1}{E[X]E[Y]}\frac{dE[XY]}{dt} - \frac{C_{XY}}{E[X]}\frac{dE[X]}{dt} - \frac{C_{XY}}{E[Y]}\frac{dE[Y]}{dt}.$$

Substituting into the righthand side and simplifying, we obtain

$$(5.7)\quad \frac{dC_{XY}}{dt} = (k+1)(V_xE[X]+V_YE[Y])C_{XY} + ((2-k-3V_x)E[X]+(2-k-3V_Y)E[Y])(C_{XY})^2$$

System (5.3)–(5.7) represents a closed system of differential equations for $E[X]$, $E[Y]$, V_X, V_Y, and C_{XY}.

In Figure 3, we show simulations for several sets of initial conditions with $k=1$, and $a=3000$. The deterministic system (2.1)-(2.2) has a stable equilibrium at $(1000,1000)$. In contrast, the stochastic system (5.3)–(5.7) shows a stable population equilibrium at approximately $(999,999)$. The multiplicative variances and multiplicative covariance appear to reach stable equilibria as well, at values around 1.002 for the variances, and around 0.999 for the covariance.

We note that these multiplicative variance and multiplicative covariance values, when converted to standard additive variance and additive covariance values, agree closely with the multivariate normal model results. Working with the formulas for $Var[X]$, $Var[Y]$, $Cov[X,Y]$, V_X, V_Y, and C_{XY}, we find that the following should approximately hold:

$$Var[X] \approx (V_X-1)(E[X])^2, \qquad Var[Y] \approx (V_Y-1)(E[Y])^2$$

and

$$Cov[X,Y] \approx E[XY]\left(1-\frac{1}{C_{XY}}\right).$$

Using values for the $k=1$ models at equilibrium, $E[X]\approx 999$, $E[Y]\approx 999$, $E[XY]\approx(999)^2$, $V_X=V_Y\approx 1.002$, and $C_{XY}\approx 0.999$, we obtain

$$(V_X-1)(E[X])^2 = (V_Y-1)(E[Y])^2 \approx 1996$$

which is similar to $Var[X]=Var[Y]\approx 2005$. Also,

$$E[XY]\left(1-\frac{1}{C_{XY}}\right)\approx -999$$

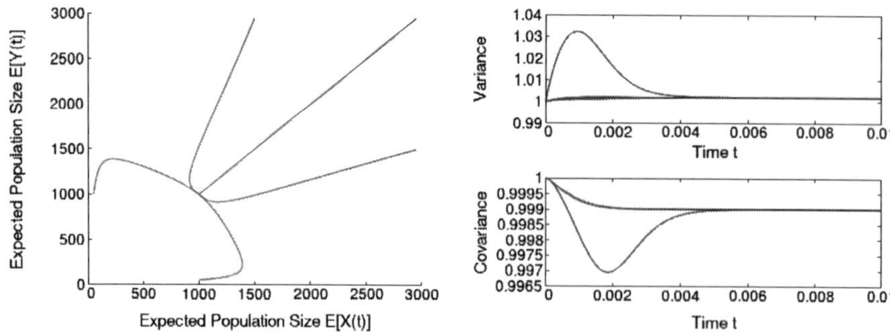

FIGURE 3. Multivariate lognormal expected population values (left), multiplicative variances (upper right), and multiplicative covariance (lower right): stable competition ($k = 1$, $a = 3000$).

which is similar to $Cov[X, Y] \approx -1004$.

Returning to the multivariate lognormal model, we will explore further the stable equilibrium located at $(a/3, a/3)$ in the deterministic model with $k = 1$. Due to the symmetry of the deterministic model, in this stochastic multivariate lognormal formulation as well this equlibrium has $E[X] = E[Y]$. Assuming that V_X and V_Y converge to a common value v as $t \to \infty$, and that C_{XY} converges to a value c, and setting $E[X] = E[Y]$, we obtain in the limit that the expected values for the equilibrium with both species present are given by

$$E[X] = E[Y] = \frac{a}{2v + c}$$

With $v \approx 1.002$ and $c \approx 0.999$, we obtain $E[X] = E[Y] \approx 999$, as indeed is shown in our plots. We note that the multiplicative structure used here for the variances and covariance results in simpler structure for examining the equilibrium, as compared to the multivariate normal moment closure results obtained using an additive structure.

In Figure 4, we show simulations for several sets of initial conditions with $k = 0$, and $a = 3000$. The deterministic system (2.1)-(2.2) has stable equilibria at $(0, 3000)$ and $(3000, 0)$. In contrast, the stochastic system (5.3)–(5.7) shows stable population equilibria at approximately $(0, 2999)$ and $(2999, 0)$. Interestingly, here as well, the multiplicative variance for the population that survives and the multiplicative covariance appear to reach stable equilibria again, at values around 1.003 and 0.999, respectively. The multiplicative variance for the population that goes extinct heads towards ∞.

Here, analysis in the limit as $t \to \infty$ appears to be complicated by the multiplicative structure that pushes towards infinity the multiplicative variance of the population of the species that goes extinct. Although algebraic exploration such as we did earlier does not seem to work here, we point out that our numerical results, again, predict an expected value for the species that survives that is slightly lower than the population level in the deterministic model.

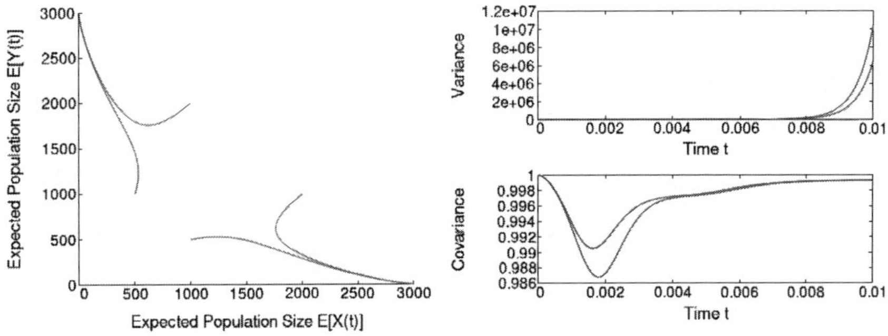

FIGURE 4. Multivariate lognormal expected population values (left), multiplicative variances (upper right), and multiplicative covariance (lower right): competitive exclusion ($k = 0$, $a = 3000$).

6. Conclusions

In summary, we note that both the multivariate normal moment closure technique of Section 4 and the multivariate lognormal moment closure technique of Section 5 recapture the behavior of the deterministic model in terms of expected values for the populations. Each stochastic moment closure model does so with slightly lower equilibrium population values as compared to the deterministic model. Results obtained appear to be good for both the stable competition model ($k = 1$) and the competitive exclusion model ($k = 0$).

The power of the moment closure stochastic models is that they provide estimates of the variances and covariance for the populations of the two species. In comparing the variance values provided by the two stochastic models, note that positive variances in the multivariate normal model with a standard additive structure correspond to multiplicative variances in the multivariate lognormal model that are larger than 1. The negative covariance values in the multivariate normal model with a standard additive structure correspond to multiplicative covariance values in the multivariate lognormal model that are smaller than 1. These measures of spread provide insight into the interacting species models explored here.

Previous work in [**C**, **T**] focused on a model in which trajectories in the phase plane are closed curves about an equilibrium. In our models with $k = 0$ and $k = 1$, we are exploring the dynamics of a stochastic model in a scenario in which population trajectories converge to an equilibrium, either through competitive exclusion or through stable competition, allowing study of the spread of surviving population values. Further work could include extending this analysis to explore the bifurcation at $k = 1/2$, or exploring other ranges of k values as described in [**Sz**], in order to study spread in various types of species interactions.

References

[C] Diana Curtis and Jennifer Switkes, *A moment closure technique for a stochastic predator-prey model*, Math. Sci. **42** (2017), no. 3, 157–168. MR3727550

[L] A. Lloyd, *Estimating variability in models for recurrent epidemics: assessing the use of moment closure techniques*, Theor. Pop. Bio. 65 (2004), 49–65.

[Sw] Randall J. Swift, *A stochastic predator-prey model*, Irish Math. Soc. Bull. **48** (2002), 57–63. MR1930526

[Sz] Jennifer Switkes and Ryan Szypowski, *Bifurcation in an interacting species model*, Math. Sci. **42** (2017), no. 2, 104–110. MR3586104

[T] Tanawat Trakoolthai, Diana Curtis, and Jennifer Switkes, *A multivariate log-normal moment closure technique for the stochastic predator-prey model*, Math. Sci. **43** (2018), no. 2, 71–81. MR3888116

Department of Mathematics and Statistics, California State Polytechnic University, Pomona, California

Email address: `jmswitkes@cpp.edu`

Selected Published Titles in This Series

774 **Randall J. Swift, Alan Krinik, Jennifer M. Switkes, and Jason H. Park, Editors,** Stochastic Processes and Functional Analysis, 2021

773 **Nicholas R. Baeth, Thiago H. Freitas, Graham J. Leuschke, and Victor H. Jorge Pérez, Editors,** Commutative Algebra, 2021

772 **Anatoly M. Vershik, Victor M. Buchstaber, and Andrey V. Malyutin, Editors,** Topology, Geometry, and Dynamics, 2021

771 **Nicolás Andruskiewitsch, Gongxiang Liu, Susan Montgomery, and Yinhuo Zhang, Editors,** Hopf Algebras, Tensor Categories and Related Topics, 2021

770 **Stéphane Ballet, Gaetan Bisson, and Irene Bouw, Editors,** Arithmetic, Geometry, Cryptography and Coding Theory, 2021

769 **Kiyoshi Igusa, Alex Martsinkovsky, and Gordana Todorov, Editors,** Representations of Algebras, Geometry and Physics, 2021

768 **Dražen Adamović, Andrej Dujella, Antun Milas, and Pavle Pandžić, Editors,** Lie Groups, Number Theory, and Vertex Algebras, 2021

767 **Moshe Jarden and Tony Shaska, Editors,** Abelian Varieties and Number Theory, 2021

766 **Paola Comparin, Eduardo Esteves, Herbert Lange, Sebastián Reyes-Carocca, and Rubí E. Rodríguez, Editors,** Geometry at the Frontier, 2021

765 **Michael Aschbacher,** Quaternion Fusion Packets, 2021

764 **Gabriel Cunningham, Mark Mixer, and Egon Schulte, Editors,** Polytopes and Discrete Geometry, 2021

763 **Tyler J. Jarvis and Nathan Priddis, Editors,** Singularities, Mirror Symmetry, and the Gauged Linear Sigma Model, 2021

762 **Atsushi Ichino and Kartik Prasanna,** Periods of Quaternionic Shimura Varieties. I., 2021

761 **Ibrahim Assem, Christof Geiß, and Sonia Trepode, Editors,** Advances in Representation Theory of Algebras, 2021

760 **Olivier Collin, Stefan Friedl, Cameron Gordon, Stephan Tillmann, and Liam Watson, Editors,** Characters in Low-Dimensional Topology, 2020

759 **Omayra Ortega, Emille Davie Lawrence, and Edray Herber Goins, Editors,** The Golden Anniversary Celebration of the National Association of Mathematicians, 2020

758 **Jan Šťovíček and Jan Trlifaj, Editors,** Representation Theory and Beyond, 2020

757 **Kaïs Ammari and Stéphane Gerbi, Editors,** Identification and Control: Some New Challenges, 2020

756 **Joeri Van der Veken, Alfonso Carriazo, Ivko Dimitrić, Yun Myung Oh, Bogdan D. Suceavă, and Luc Vrancken, Editors,** Geometry of Submanifolds, 2020

755 **Marion Scheepers and Ondřej Zindulka, Editors,** Centenary of the Borel Conjecture, 2020

754 **Susanne C. Brenner, Igor Shparlinski, Chi-Wang Shu, and Daniel B. Szyld, Editors,** 75 Years of Mathematics of Computation, 2020

753 **Matthew Krauel, Michael Tuite, and Gaywalee Yamskulna, Editors,** Vertex Operator Algebras, Number Theory and Related Topics, 2020

752 **Samuel Coskey and Grigor Sargsyan, Editors,** Trends in Set Theory, 2020

751 **Ashish K. Srivastava, André Leroy, Ivo Herzog, and Pedro A. Guil Asensio, Editors,** Categorical, Homological and Combinatorial Methods in Algebra, 2020

750 **A. Bourhim, J. Mashreghi, L. Oubbi, and Z. Abdelali, Editors,** Linear and Multilinear Algebra and Function Spaces, 2020